DICK WINKLER

Practical Calculus
for the Social and Managerial Sciences

Practical Calculus
for the Social and Managerial Sciences

Laurence D. Hoffmann
Claremont Men's College

McGraw-Hill Book Company
New York St. Louis San Francisco Auckland Düsseldorf
Johannesburg Kuala Lumpur London Mexico Montreal New Delhi
Panama Paris São Paulo Singapore Sydney Tokyo Toronto

Practical Calculus
for the Social and Managerial Sciences

Copyright © 1975 by McGraw-Hill, Inc. All rights reserved.
Printed in the United States of America. No part of this publication may be reproduced, stored in a retrieval system, or transmitted, in any form or by any means, electronic, mechanical, photocopying, recording, or otherwise, without the prior written permission of the publisher.

4 5 6 7 8 9 0 M A M M 7 9 8 7 6

Library of Congress Cataloging in Publication Data

Hoffmann, Laurence D   date
    Practical calculus for the social and managerial sciences.

    1. Calculus. I. Title.
QA303.H569   515   74-10786
ISBN 0-07-029315-5

This book was set in Times New Roman.
The editors were A. Anthony Arthur and Shelly Levine Langman;
the designer was Jo Jones;
the production supervisor was Leroy A. Young.
The drawings were done by Vantage Art, Inc.
The Maple Press Company was printer and binder.

Contents

Contents

Preface
viii

Functions and Graphs

1 Functions
1

2 Graphs
12

3 Linear Functions
22

4 Limits and Continuity
35

5 Polynomials
49

6 Rational Functions
60

7 Intersections of Graphs
73

8 Review Problems
82

Differentiation

1 Introduction
87

2 The Derivative
89

3 Techniques of Differentiation
101

4 The Chain Rule
112

5 Implicit Differentiation
119

6 Relative Maxima and Minima
123

7 Absolute Maxima and Minima
137

8 Practical Optimization Problems
141

9 The Second Derivative
154

10 Rate of Change
168

11 Related Rates
181

12 Review Problems
186

Exponential and Logarithmic Functions

1 The Number e
191

2 Exponential Functions
195

3 The Natural Logarithm
205

4 Differentiation of Logarithmic and Exponential Functions
213

5 Review Problems
223

Integration

1 Antiderivatives
225

2 Techniques of Integration
231

3 Integration by Substitution
240

4 Integration by Parts
247

5 Tables of Integrals
252

6 Elementary Differential Equations
256

7 Separable Differential Equations
267

8 The Definite Integral
279

9 Area and Integration
285

10 Computation of Areas
291

11 The Definite Integral as a Limit of a Sum
297

12 Further Applications of the Definite Integral
307

13 Review Problems
317

Functions of Several Variables

1 Functions and Graphs
321

2 Partial Derivatives
329

3 The Chain Rule
338

4 Level Curves
343

5 Relative Maxima and Minima
351

6 Lagrange Multipliers
360

7 Review Problems
371

Appendix
375

1 Exponential Function
376

2 Natural Logarithms (Base e)
377

3 Squares, Cubes, Square Roots, and Cube Roots
378

Answers to Selected Problems
379

Index
403

Preface

If you are a student of economics, business, biology, or one of the social sciences, and if your mathematical background includes little more than high school algebra, then this book was written for you. The primary goal of the book is to give you, in one semester, a firm working knowledge of elementary calculus.

To achieve this goal, the material in the text has been chosen with utility as the main criterion, and some traditional topics in calculus that are of theoretical interest have been omitted. Special emphasis has been placed on the techniques and strategies you will need in solving practical problems in the social and managerial sciences.

Since you are primarily interested in the applications of calculus, you may be impatient with mathematical explanations of why new results work, and with precise, detailed statements of these results. However, without a clear understanding of why a rule is valid and of the circumstances under which it can legitimately be applied, your ability to use the rule will be limited. Therefore, the main results in this text are stated carefully and completely. Most of them are proved or justified, but only *after* they have been illustrated and used. Technical jargon and notational formalism have been kept to a minimum and, whenever possible, explanations are informal, intuitive, and geometric.

Each section in the text is accompanied by a set of problems. A collection of review problems appears at the end of each chapter. Many of the problems involve routine computation and are designed to help you master new techniques. Others ask you to apply the new techniques to practical situations. Problems involving the translation of practical situations into their mathematical models occur throughout the text. And a few of the problems are purely theoretical. An asterisk (*) is used to identify the most challenging problems.

You should not expect to be able to complete all five chapters of the text during a one-semester course. The length of the book gives it the flexibility necessary for use in two essentially different courses: a specialized course for business and economics students; and a more traditional course for social science students in general. If yours is the traditional course, you should cover most of the material in the first four chapters and omit Chapter 5 entirely. Chapter 5 deals with differential calculus for functions of several variables, a subject that plays a central role in mathematical economics. If you are in a course designed especially for students of business and economics, you should cover the first three chapters, parts of Chapter 4, and everything in Chapter 5.

Professor Jerry St. Dennis is a mathematical economist at Claremont Men's College. His generous advice and constant encouragement during the preparation of this book have been invaluable. I am also indebted to my colleagues, Professor Gerald L. Bradley and Professor John A. Ferling of Claremont Men's College who used a preliminary version of the text in their classes and offered many helpful comments and suggestions. Mr. Kenneth R. Drew checked the entire manuscript for accuracy and worked all the problems. His solutions form the basis for the answer section at the back of the book. Finally, the students at the Claremont Colleges who used preliminary versions of the text have been most helpful. For their enthusiasm, keen perception, and healthy irreverence I am especially grateful.

Laurence D. Hoffmann

Functions and Graphs

Complicated relationships arising in practical situations often become tractable when expressed mathematically. In this chapter, we investigate procedures for translating practical relationships into mathematical functions and develop efficient techniques for representing functions geometrically as graphs.

1. FUNCTIONS

A *function* is a rule that associates with each object in a set A, one and only one object in a set B. Throughout the first four chapters of this book, we will work exclusively with functions for which the sets A and B are collections of numbers. For our purposes, then, the following informal definition of function will be sufficient.

FUNCTION

A *function* is a rule according to which "new" numbers are associated with "old" numbers. To be called a function, the rule must have the property that it associates one and only one "new" number with each "old" number.

Consider the following elementary example.

Example 1.1

A certain function is defined by the rule that the "new" number is obtained by adding 4 to the square of the "old" number. What number does this function associate with 3?

Solution

The number associated with 3 is $3^2 + 4$, or 13. □

Often a function can be conveniently abbreviated by a mathematical formula. Traditionally, we let the letter x denote the "old" number and y the "new" number, and we write an equation relating x and y. For instance, the function in Example 1.1 can be expressed by the equation $y = x^2 + 4$. The letters x and y that appear in such an equation are called *variables*. The numerical value of the variable y is determined by that of the variable x. Because of this, y is sometimes referred to as the *dependent variable* and x as the *independent variable*.

There is an alternative notation for functions that is widely used and somewhat more versatile. A letter such as f is chosen to stand for the function itself, and the value that the function associates with x is denoted by $f(x)$ instead of y. The symbol $f(x)$ is read "f of x." Let us rewrite Example 1.1 using this functional notation.

Example 1.2

If $f(x) = x^2 + 4$, find $f(3)$.

Solution

$$f(3) = 3^2 + 4 = 13$$

Observe the convenience and simplicity of this notation. In Example 1.2, the compact formula $f(x) = x^2 + 4$ completely defines our function, and the simple equation $f(3) = 13$ suffices to indicate that 13 is the number that the function associates with 3.

In the next example, a letter other than f is used to denote the function, and a letter other than x is used for the independent variable.

Example 1.3

If $g(t) = \sqrt{t - 2}$, find $g(5)$, $g(27)$, $g(2)$, and $g(1)$.

Solution

$$g(5) = \sqrt{5 - 2} = \sqrt{3}$$
$$g(27) = \sqrt{27 - 2} = 5$$

and

$$g(2) = \sqrt{2 - 2} = 0$$

We cannot compute $g(1)$, because if $t = 1$, then $t - 2 = -1$ and there is no (real) number whose square is -1. In fact, our function assigns no value to any t for which $t - 2$ is negative. That is, $g(t)$ is defined only for $t \geq 2$.

1. FUNCTIONS   3

The set of numbers for which a function is defined is called the *domain* of the function. In Example 1.3, for instance, the domain of g consists of all numbers that are greater than or equal to 2.

The domain of the function in Example 1.3 was restricted because negative numbers do not have (real) square roots. The functions in the next example are quotients. Since division by zero is impossible, these functions will be undefined whenever their denominators are zero.

Example 1.4

Find the domains of the following functions.

(a) $f(x) = 1/(x - 3)$

(b) $g(x) = 1/(1 - \sqrt{x - 3})$

Solution

(a) The only value of x for which $f(x)$ cannot be computed is $x = 3$, the value that would make the denominator zero. Hence the domain of f consists of all numbers other than 3.

(b) The square root that appears in the formula for $g(x)$ is defined only for $x \geq 3$. Moreover, if $x = 4$, the denominator of g is $1 - \sqrt{4 - 3} = 1 - 1 = 0$. Hence, $g(x)$ is not defined when $x = 4$, and so the domain of g consists of all numbers greater than or equal to 3 except $x = 4$.   □

Throughout this text, we will be developing techniques for analyzing functions. Before we can apply these techniques to the solution of practical problems, we must be able to translate practical relationships into mathematical functions. This translation process is illustrated in the following examples.

Example 1.5

A manufacturer can produce cassette tape recorders at a cost of $20 each. He estimates that if he sells them for x dollars apiece he will be able to sell approximately $120 - x$ tape recorders a month. The manufacturer's monthly profit is a function of the price x at which he sells the tape recorders.

(a) Express this function mathematically.

(b) Compute the profit if the tape recorders are sold for $70 apiece.

Solution

(a) We begin by stating the desired relationship in words.

Profit = revenue − cost

where

Revenue = (number of recorders)(price per recorder)

and

Cost = (number of recorders)(cost per recorder)

We now express these quantities in terms of the variable x.

Number of recorders = $120 - x$

Price per recorder = x

and

Cost per recorder = 20

Letting f denote the profit function, we conclude that

$$f(x) = (120 - x)(x) - (120 - x)(20) = (120 - x)(x - 20)$$

(b) To calculate the profit if the selling price is $70, we simply evaluate our profit function when $x = 70$. We get

$$f(70) = (120 - 70)(70 - 20) = 2{,}500$$

That is, the manufacturer's profit will be $2,500 when the tape recorders are sold for $70 apiece. □

In the next example, we will need three formulas to define the desired function.

Example 1.6

A bus company has adopted the following pricing policy for groups wishing to charter its buses. Groups containing no more than 40 people will be charged a fixed amount of $2,400 (40 times $60). In groups containing between 40 and 80 people, everyone will pay $60 minus 50 cents for each person in excess of 40. The company's lowest fare of $40 per person will be offered to groups that have 80 members or more. Express the bus company's revenue as a function of the size of the group.

Solution

Let x denote the number of people in the group and $f(x)$ the corresponding revenue. If $0 \leq x \leq 40$, the revenue is simply $f(x) = 2{,}400$. If $x \geq 80$, each person pays $40 and so the corresponding revenue is $f(x) = 40x$. The expression for $f(x)$ when $40 < x < 80$ is slightly more complicated. Let us begin our analysis of this situation with the basic relationship

Revenue = (number of people)(fare per person)

Since x denotes the total number of people in the group, $x - 40$ is the number of people in excess of 40. The fare per person is the original \$60 reduced by $\frac{1}{2}$ dollar for each of the $x - 40$ extra people. Thus,

Fare per person $= 60 - \frac{1}{2}(x - 40) = 80 - \frac{1}{2}x$

To get the revenue, we simply multiply this expression by x, the number of people in the group. Thus, for $40 < x < 80$, $f(x) = 80x - \frac{1}{2}x^2$.

We can summarize all three cases compactly as follows:

$$f(x) = \begin{cases} 2,400 & \text{if } 0 \leq x \leq 40 \\ 80x - \frac{1}{2}x^2 & \text{if } 40 < x < 80 \\ 40x & \text{if } x \geq 80 \end{cases}$$

Although this function $f(x)$ is defined for all non-negative values of x, it represents the bus company's revenue only when x is a non-negative *integer*. (Why?) □

In the next example, the quantity we are seeking is expressed most naturally in terms of two variables. One of these must be eliminated before we can write the quantity as a function of a single variable.

Example 1.7

A rectangular lot is to be fenced off on three sides. If the area of the lot is to be 3,200 square yards, express the length of the fencing as a function of the length of the unfenced side.

Solution

It is natural to start by introducing two variables, say x and y, to denote the lengths of the sides of the lot (Figure 1). Then,

Length of fencing $= x + 2y$

Since we want the length of the fencing expressed as a function of x only, we must find a way of expressing y in terms of x; that is, we must find an equation relating x and y. The fact that the area is to be 3,200 square yards provides us with this equation. Specifically,

$xy = 3,200$

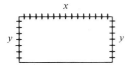

Figure 1

FUNCTIONS AND GRAPHS

Solving this for y we get

$$y = \frac{3{,}200}{x}$$

which we then substitute into the formula for the length of the fencing. This gives

$$f(x) = x + \frac{6{,}400}{x}$$

where f denotes the length of the fencing.

The function $f(x)$ is defined for all values of x except $x = 0$, and represents the length of the fencing if x is positive. □

Geometry may facilitate the translation of a practical relationship into a mathematical function. Consider the following example.

Example 1.8

Two cars leave an intersection at the same time. One travels east at a constant speed of 50 miles per hour, while the other goes north at 30 miles per hour. Express the distance between the cars as a function of time.

Solution

Let t denote the number of hours since the cars were together at the intersection. In t hours, the first car travels $50t$ miles to the east, while the second travels $30t$ miles to the north. The relative positions of the two cars after t hours are shown in Figure 2. The distance $D(t)$ between the cars at this time is the length of the hypotenuse of the right triangle. By the pythagorean theorem,

$$[D(t)]^2 = (30t)^2 + (50t)^2$$

so that

$$D(t) = (10\sqrt{34})t$$

The function $D(t)$ is defined for all t and gives the distance between the cars when $t \geq 0$.

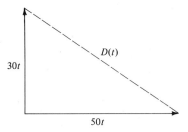

Figure 2

In the next example, a quantity is given as a function of one variable which, in turn, can be written as a function of a second variable. By combining the functions in an appropriate way, we can express the original quantity as a function of the second variable. This process is known as the *composition of functions*.

Example 1.9

At a certain factory, the total cost of manufacturing u items during the daily production run is $3u^2 + u + 9$ dollars. Suppose that by noon 40 items have already been produced, and that during the afternoon additional items are manufactured at a rate of 10 per hour. Let x denote the number of hours past noon.

(a) Express the total manufacturing cost as a function of x.

(b) How much has been spent on production by 1:00 P.M.?

Solution

(a) We know that the total cost is related to the variable u by the equation

Total cost $= 3u^2 + u + 9$

Since our goal is to express this cost as a function of x, we begin by writing u in terms of x. In particular,

$u = 40 + 10x$

If we now substitute the expression $40 + 10x$ for u in the cost equation, and let $f(x)$ denote the cost at time x, we get the desired function of x:

$f(x) = 3(40 + 10x)^2 + (40 + 10x) + 9$
$= 300x^2 + 2{,}410x + 4{,}849$

(b) At 1:00 P.M., $x = 1$, and (assuming production continues throughout the lunch hour) the corresponding total cost is $f(1) = 300 + 2{,}410 + 4{,}849 = \$7{,}559$. □

The computation of the function $f(x)$ in part (a) of Example 1.9 is a special case of the following general procedure.

COMPOSITION OF FUNCTIONS

A function $g(u)$ may be combined with another function $h(x)$ to produce a *composite function*

$f(x) = g(h(x))$

In particular, $f(x)$ is the function obtained by substituting $h(x)$ for u in the expression for $g(u)$.

In Example 1.9, total cost $f(x)$ was the composite function $g(h(x))$, where $g(u) = 3u^2 + u + 9$ and $h(x) = 40 + 10x$. Some of the other functions we have already encountered can be thought of as composites. For instance, the function $f(x) = 1/(x - 3)$ in Example 1.4 is the composite function $g(h(x))$, where $g(u) = 1/u$ and $h(x) = x - 3$. There are other ways to write this function as a composite. Can you think of two other functions, $g(u)$ and $h(x)$, for which $1/(x-3) = g(h(x))$?

In many cases, the domain of $g(h(x))$ is restricted. In fact, for $g(h(x))$ to be defined, x must be in the domain of h, while $h(x)$ must be in the domain of g.

Example 1.10

Let $g(u) = 1/u$ and $h(x) = \sqrt{x - 1}$. Find the composite function $f(x) = g(h(x))$ and specify its domain.

Solution

Replacing u by $\sqrt{x - 1}$ in the formula for g, we get

$$f(x) = g(h(x)) = \frac{1}{\sqrt{x - 1}}$$

The domain of g consists of all u except $u = 0$. Hence the domain of the composite function f will consist of all x in the domain of h for which $h(x) \neq 0$; in particular, all x greater than 1. □

Problems

1. For each of the following functions, compute the values indicated.
   (a) $f(x) = 3x^2 + 5x - 2$; $f(1), f(0), f(-2)$
   (b) $g(x) = x/(x^2 + 1)$; $g(2), g(0), g(-1)$
   (c) $h(t) = \sqrt{t^2 + 2t + 4}$; $h(2), h(0), h(-4)$
   (d) $g(u) = (u + 1)^{3/2}$; $g(0), g(-1), g(8)$
   (e) $f(t) = (2t - 1)^{-3/2}$; $f(1), f(2), f(5)$
   (f) $h(x) = \begin{cases} -2x + 4 & \text{if } x \leq 1 \\ x^2 + 1 & \text{if } x > 1 \end{cases}$ $[h(3), h(1), h(0), h(-3)]$
   (g) $f(n) = \begin{cases} 3 & \text{if } n < -5 \\ n + 1 & \text{if } -5 \leq n \leq 5 \\ \sqrt{n} & \text{if } n > 5 \end{cases}$ $[f(-6), f(-5), f(16)]$

2. Specify the domain of each of the following functions.
   (a) $f(x) = x^3 - 3x^2 + 2x + 5$ (b) $g(x) = (x^2 + 5)/(x + 2)$
   (c) $f(t) = (t + 1)/(t^2 + 2t - 3)$ (d) $y = \sqrt{x^2 + 4}$
   (e) $h(u) = \sqrt{u^2 - 4}$ (f) $g(x) = \sqrt{2x - 4}$
   (g) $f(t) = (2t - 4)^{3/2}$ (h) $y = (x - 1)/\sqrt{x^2 + 4}$
   (i) $f(x) = (x^2 - 4)^{-1/2}$ (j) $h(t) = \sqrt{t^2 - 4}/\sqrt{t - 4}$

3. For each of the following functions $f(x)$, find functions $h(x)$ and $g(u)$ such that $f(x) = g(h(x))$.
   (a) $f(x) = (x^5 - 3x^2 + 12)^3$
   (b) $f(x) = \sqrt{3x - 5}$
   (c) $f(x) = (x - 1)^2 + 2(x - 1) + 3$
   (d) $f(x) = 1/(x^2 + 1)$
   (e) $f(x) = 1/\sqrt{x^2 + 1}$
   (f) $f(x) = \sqrt{3 - 2x^2} + 1/(3 - 2x^2)$

4. Find the composite function $g(h(x))$ and specify its domain.
   (a) $g(u) = 1/u$, $h(x) = x^2 + x - 2$
   (b) $g(u) = 1/u^2$, $h(x) = x - 1$
   (c) $g(u) = u^2$, $h(x) = 1/(x - 1)$
   (d) $g(u) = u^2 + u - 2$, $h(x) = 1/x$
   (e) $g(u) = u^2 + u - 2$, $h(x) = x + 1$
   (f) $g(u) = \sqrt{u}$, $h(x) = x - 1$
   (g) $g(u) = 1/\sqrt{u}$, $h(x) = x^2 - 9$
   *(h) $g(u) = \sqrt{u}$, $h(x) = x^2$
   *(i) $g(u) = u^2$, $h(x) = \sqrt{x}$
   *(j) $g(u) = 1/u$, $h(x) = 1/x$

5. A function g is said to be an *inverse* of a function f if $g(f(x)) = x$ for all x in the domain of f. (For example, $g(x) = x^2$ is an inverse of the function $f(x) = \sqrt{x}$.) Find an inverse of each of the following functions.
   (a) $f(x) = x^{1/3}$
   (b) $f(x) = x^3$
   (c) $f(x) = 1/x$
   (d) $f(x) = 1/\sqrt{x}$
   (e) $f(x) = x + 5$

6. A manufacturer can produce waterbed frames at a cost of $10 each. He estimates that if he sells them for x dollars each he will be able to sell approximately $50 - x$ frames a month.
   (a) Express his monthly profit as a function of the selling price x.
   (b) What is his profit if the frames are sold for $20 each? For $40 each?
   (c) For what values of x does this function have an economic interpretation?

7. A college bookstore can obtain the book *Social Groupings of the American Dragonfly* from the publisher at a cost of $3 per book. The bookstore estimates that it can sell 200 copies at a price of $15, and that it will be able to sell 10 more copies for each 50-cent reduction in the price. (For example, 220 copies will be sold if the price is $14.)
   (a) Compute the bookstore's total profit on this book if the selling price is $10.
   (b) Express the store's profit as a function of the selling price.

8. A manufacturer is currently selling 2,000 lamps a month to retailers at a price of $2 per lamp. He estimates that for each 1-cent increase in the price he will sell 10 fewer lamps each month. The manufacturer's costs consist of a fixed overhead of $500 per month plus 40 cents a lamp for labor and materials.
   (a) Express the manufacturer's monthly profit as a function of his selling price.
   (b) Compute his profit if the lamps are sold for $2.50 apiece.

9. The cost of automobile rental is $15 plus 15 cents per mile.
   (a) What is the cost of renting a car for a 50-mile trip?

(b) Express the rental cost as a function of the number of miles driven. (Assume, for simplicity, that when fractions of miles are involved the cost is prorated. For example, a trip of $\frac{2}{3}$ mile would cost $15.10.)

10. Suppose that the car in Problem 9 averages 12 miles per gallon and that gasoline costs 72 cents per gallon.
   (a) What is the total cost (rental plus gasoline) of renting the car for a 50-mile trip?
   (b) Express the total cost of renting the car as a function of the number of miles driven.

11. A record club offers the following special sale. If five records are bought at the full price of $6 each, additional records can then be bought at half price. There is a limit of nine records per customer.
   (a) What is the cost of buying seven records?
   (b) Express the cost of the records as a function of the number bought.

12. A truck is 300 miles due east of a car and is traveling west at a constant speed of 30 miles per hour. Meanwhile, the car is going north at 60 miles per hour. Express the distance between the car and the truck as a function of time.

13. A cable is to be run from a power plant on one side of a river 900 feet wide to a factory on the other side, 3,000 feet downstream. The cable will be run in a straight line from the power plant to some point P on the opposite bank, and then along the bank to the factory. The cost of running the cable across the water is $5 per foot, while the cost over land is $4 per foot. Let x be the distance from P to the point directly across the river from the power plant, and express the cost of installing the cable as a function of x.

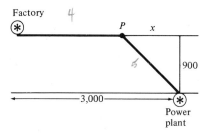

14. A city recreation department plans to build a rectangular playground 3,600 square yards in area. The playground is to be surrounded by a fence.
   (a) Express the length of the fencing as a function of the length of one of the sides.
   (b) Use the function obtained in part (a) to calculate the amount of fencing needed if the dimensions of the playground are 100 by 36 yards.

15. An open box is to be made from an 18- by18-inch square piece of cardboard by removing a small square from each corner and folding up the flaps to form the sides. Express the volume of the resulting box as a function of the length x of a side of the removed squares.

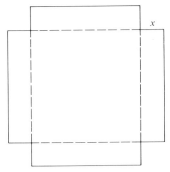

16. A closed box with a square base is to have a volume of 250 cubic feet. The material for the top and bottom of the box costs $2 per square foot, and the material for the sides costs $1 per square foot. Express the construction cost of the box as a function of the length of its base.

*17. A plastics firm has received an order from the city recreation department to manufacture 8,000 special Styrofoam kickboards for its summer swimming program. The firm owns several machines, each of which can produce 30 kickboards an hour. The cost of setting up the machines to produce these particular kickboards is $20 per machine. Once the machines have been set up, the operation is fully automated and can be supervised by a single foreman earning $4.80 per hour. Express the cost of producing the 8,000 kickboards as a function of the number of machines used.

18. At a certain factory, the total cost of manufacturing q units during the daily production run is $0.2q^3 - 0.1q^2 + 0.5q + 600$ dollars. After t hours on a typical workday, $100\sqrt{t^2 + 1}$ units have been produced. Express the total manufacturing cost as a function of t.

19. An environmental study of a certain suburban community suggests that the average daily level of carbon monoxide in the air will be $0.5p + 1$ parts per million when the population is p thousand people. It is estimated that t years from now the population of the community will be $10 + 0.1t^2$ thousand people. Express the carbon monoxide level as a function of t.

20. A manufacturer has found that the cost of producing his first n units is $C(n) = 3n^2 + n + 9$ dollars.
    (a) What is the cost of manufacturing the first 20 units?
    (b) What is the cost of manufacturing the twentieth unit?

2. GRAPHS

Functions can be represented geometrically as graphs drawn on a rectangular coordinate system. Traditionally, the independent variable is represented on the horizontal axis and the dependent variable on the vertical axis.

THE GRAPH OF A FUNCTION

The *graph* of a function f consists of all points whose coordinates (x, y) satisfy the equation $y = f(x)$.

In subsequent sections we will develop sophisticated techniques for sketching graphs of functions. For many functions, however, we get a fairly good sketch by the elementary method of plotting points. This familiar method of curve sketching is illustrated in the following example.

Example 2.1

Graph the function $f(x) = x^3$.

Solution

We begin by computing $f(x)$ for several convenient values of x. The results of these computations are summarized in the following table:

x	0	$\frac{1}{2}$	$-\frac{1}{2}$	1	-1	2	-2
$f(x)$	0	$\frac{1}{8}$	$-\frac{1}{8}$	1	-1	8	-8

We then plot the corresponding points $(x, f(x))$ and connect them by a smooth curve to obtain a sketch of the graph of the function (Figure 3).

□

This primitive method of curve sketching has obvious shortcomings. The computation of a table of values may be tedious and time-consuming. Even worse, no matter how many points we plot, we cannot be sure of the behavior of the graph between these points. These difficulties can be partially overcome by the use of techniques to be introduced later in this chapter. However, not until we develop differential calculus in Chapter 2 will we have a really effective procedure for graphing functions.

In the next example, we graph the revenue function that arose in Example 1.6. The fact that this function cannot be represented by a single formula poses no particular difficulties for graphing.

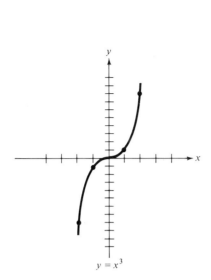

$y = x^3$

Figure 3

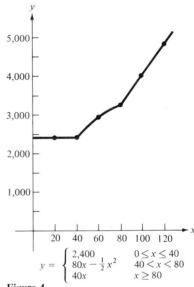

$$y = \begin{cases} 2{,}400 & 0 \le x \le 40 \\ 80x - \tfrac{1}{2}x^2 & 40 < x < 80 \\ 40x & x \ge 80 \end{cases}$$

Figure 4

Example 2.2

Graph the revenue function from Example 1.6.

$$f(x) = \begin{cases} 2{,}400 & \text{if } 0 \le x \le 40 \\ 80x - \tfrac{1}{2}x^2 & \text{if } 40 < x < 80 \\ 40x & \text{if } x \ge 80 \end{cases}$$

Solution

When computing a table of values for this function, we must remember to use the formula that is appropriate for the particular value of x. Using the formula $f(x) = 2{,}400$ when $0 \le x \le 40$, the formula $f(x) = 80x - \tfrac{1}{2}x^2$ when $40 < x < 80$, and the formula $f(x) = 40x$ when $x \ge 80$, we can compile the following table:

x	0	20	40	60	80	100	120
$f(x)$	2,400	2,400	2,400	3,000	3,200	4,000	4,800

The corresponding graph is sketched in Figure 4. □

Certain algebraic alterations of a function produce simple geometric transformations of its graph. By exploiting these transformations, we can sometimes obtain the graph of a complicated function fairly easily from that of a simpler function. As our first transformation of this type, let us consider *vertical translation*.

Example 2.3

Graph the function $g(x) = x^3 - 2$.

Solution

This function is closely related to the function $f(x) = x^3$ in Example 2.1. Indeed, for any choice of x, the corresponding function value $g(x)$ is 2 less than the function value $f(x)$.

x	0	$\frac{1}{2}$	$-\frac{1}{2}$	1	-1	2	-2
$g(x)$	-2	$-\frac{15}{8}$	$-\frac{17}{8}$	-1	-3	6	-10

Hence the graph of g (Figure 5b) can be obtained by simply shifting the graph of f (Figure 5a) *down* by 2 units.

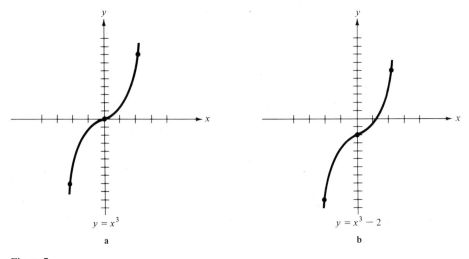

Figure 5

Example 2.3 illustrates the following general situation.

VERTICAL TRANSLATION

If $g(x) = f(x) \pm c$ for some positive constant c, then the graph of g is obtained by shifting the graph of f vertically c units. If the sign preceding c is plus, the shift is upward; if the sign is minus, the shift is downward.

The related concept of *horizontal translation* is illustrated in the next example.

Example 2.4

Graph the function $g(x) = (x - 2)^3$.

Solution

Using the following table of values, we obtain the graph in Figure 6b.

x	2	$\frac{5}{2}$	$\frac{3}{2}$	3	1	4	0
$g(x)$	0	$\frac{1}{8}$	$-\frac{1}{8}$	1	-1	8	-8

We notice that this graph resembles the graph of the function $f(x) = x^3$, shown again in Figure 6a. Indeed, it looks as if the graph of g can be obtained by shifting the graph of f to the right by 2 units. A little thought reveals why this is so. For both of the functions $f(x) = x^3$ and $g(x) = (x - 2)^3$, the function values are obtained by cubing. In the case of f, the variable x itself is cubed, while for g, this variable is reduced by 2 and *then* cubed. In other words, g takes on the same values as f, but it does so 2 units "later." Hence, the graphs of f and g are identical, except that the graph of g is located 2 units to the right of the graph of f.

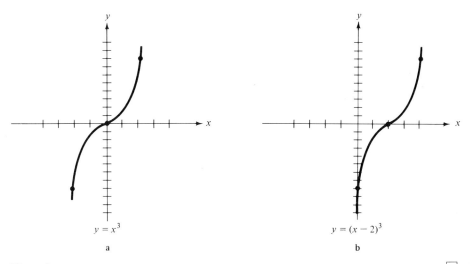

Figure 6

The function $g(x) = (x - 2)^3$ in the preceding example is related to the function $f(x) = x^3$ by the equation $g(x) = f(x - 2)$. [For example, if $x = 4$, then $g(4) = 8 = f(2)$.] This type of relationship characterizes functions whose graphs are horizontal translations of one another.

HORIZONTAL TRANSLATION

If $g(x) = f(x \pm c)$ for some positive constant c, then the graph of g is obtained by shifting the graph of f horizontally c units. If the sign preceding c is plus, the shift is to the left; if the sign is minus, the shift is to the right.

The result of changing the sign of a function is to reflect its graph across the x axis. This is because the point $(x, -y)$ is the reflection across the x axis of the point (x, y), as illustrated in Figure 7.

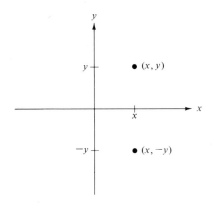

Figure 7

REFLECTION ACROSS THE x AXIS

If $g(x) = -f(x)$, then the graph of g is obtained by reflecting the graph of f across the x axis.

Example 2.5

Graph the function $g(x) = -x^3$.

Solution

Since $g(x) = -f(x)$ where $f(x) = x^3$, the graph of g (Figure 8b) can be obtained by reflecting the graph of f (Figure 8a) across the x axis. □

As illustrated in Figure 9, the point $(-x, y)$ is the reflection across the y axis of the point (x, y). This observation leads to the following algebraic characterization of reflection of graphs across the y axis. (Do you see why?)

2. GRAPHS   17

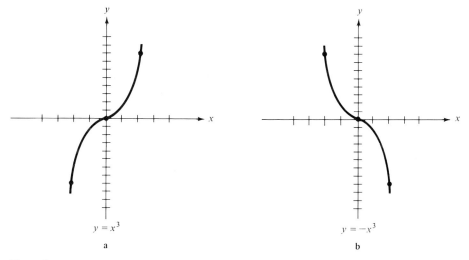

Figure 8

REFLECTION ACROSS THE y AXIS

If $g(x) = f(-x)$, then the graph of g is obtained by reflecting the graph of f across the y axis.

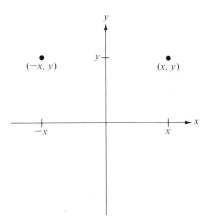

Figure 9

Example 2.6

Graph the function $g(x) = (-x)^3$.

Solution

We notice that $g(x) = f(-x)$, where $f(x) = x^3$. Hence the graph of g (Figure 10b) is obtained by reflecting the graph of f (Figure 10a) across the y axis.

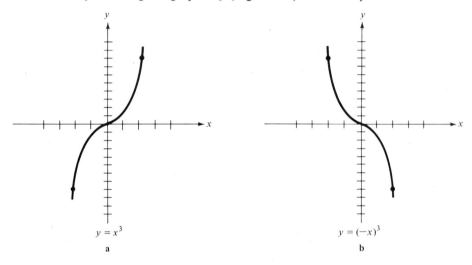

Figure 10

This is the same graph we obtained in Example 2.5 when we reflected the graph of f across the x axis. Of course, this is to be expected, since the function $g(x) = -x^3$ in Example 2.5 happens to be exactly the same as our present function $g(x) = (-x)^3$. □

Examples 2.5 and 2.6 show that the reflection across the x axis of the graph of the function $f(x) = x^3$ is the same as its reflection across the y axis. Functions with this special property are said to be *symmetric with respect to the origin*. Can you think of some other functions that are symmetric with respect to the origin?

In the next example, we combine several of our new techniques to obtain the graph of a relatively complicated function.

Example 2.7

Graph the function $g(x) = 3 - (x + 2)^3$.

Solution

We begin with the graph of $y = x^3$ (Figure 11). The graph of $y = (x + 2)^3$ is obtained by translating this graph horizontally by -2 units; that is, 2 units to the left (Figure 12). Next, we graph the function $y = -(x + 2)^3$ by reflecting the graph in Figure 12 across the x axis (Figure 13). And finally, we obtain the graph of $y = 3 - (x + 2)^3$ by translating the graph in Figure 13 vertically by 3 units (Figure 14).

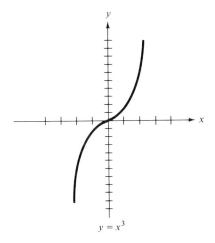

Figure 11 $y = x^3$

Figure 12 $y = (x + 2)^3$

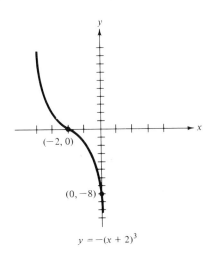

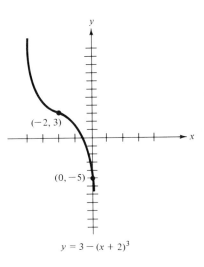

Figure 13 $y = -(x + 2)^3$

Figure 14 $y = 3 - (x + 2)^3$

Although the next function is related algebraically to the function f in Example 2.1, its graph cannot be obtained by a simple geometric transformation of the graph of f.

Example 2.8

Graph the function $g(x) = 1/x^3$.

Solution

We begin with the following table of values:

x	1	-1	2	-2	3	-3
$g(x)$	1	-1	$\frac{1}{8}$	$-\frac{1}{8}$	$\frac{1}{27}$	$-\frac{1}{27}$

Since division by zero is impossible, we cannot calculate $g(x)$ when $x = 0$. Hence there is no point on the graph whose x coordinate is zero, and consequently, there is a break in the graph when $x = 0$. This suggests that we look more closely at the behavior of g for values of x near zero. Continuing our table of values, we get

x	$\frac{1}{2}$	$-\frac{1}{2}$	$\frac{1}{3}$	$-\frac{1}{3}$
$g(x)$	8	-8	27	-27

When the corresponding points are plotted, we obtain the graph in Figure 15.

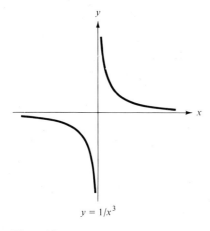

$y = 1/x^3$

Figure 15

Observe that as positive values of x approach zero, the resulting function values $g(x) = 1/x^3$ get larger, increasing without bound. As negative values of x approach zero, $g(x)$ decreases without bound. Moreover, as x itself increases or decreases without bound, $g(x)$ approaches zero. Behavior of this type will be explored in greater detail in Sections 4 and 6, when we discuss the concepts of continuity and limits. □

Problems

1. Graph the following functions.
   (a) $f(x) = x$
   (b) $f(x) = x^2$
   (c) $f(x) = x^3$
   (d) $f(x) = \sqrt{x}$

(e) $f(x) = x^{-1/3}$
(f) $f(x) = 2x - 1$
(g) $f(x) = 2x^2 - 5x - 2$
(h) $f(x) = x(x-1)(x+2)$
(i) $f(x) = 1/x$
(j) $f(x) = 1/x^2$

(k) $f(x) = x/(x+1)$
(l) $f(x) = \begin{cases} 1 & \text{if } 0 \le x < 2 \\ 3 & \text{if } 2 \le x < 4 \\ 5 & \text{if } 4 \le x < 6 \end{cases}$

(m) $f(x) = \begin{cases} x+1 & \text{if } x > 0 \\ x-1 & \text{if } x \le 0 \end{cases}$

(n) $f(x) = \begin{cases} 3 & \text{if } x > 2 \\ x^2 - 1 & \text{if } x \le 2 \end{cases}$

2. A record club offers the following special sale. If five records are bought at the full price of $6 apiece, additional records can then be bought at half price. There is a limit of nine records per customer. (See Problem 11 on page 10.) Express the cost of buying records as a function of the number bought and graph this function.

3. A local amusement park charges admission according to the following policy: Infants under 2 years are free; children between the ages of 2 and 12 are admitted for $1.50; and everyone over 12 years of age pays $3. Express the admission price as a function of a person's age and graph the function.

4. Graph the function $f(x) = x^2$. Then, without additional computation, graph
   (a) $f(x) = x^2 + 3$
   (b) $f(x) = (x+1)^2$
   (c) $f(x) = -x^2$
   (d) $f(x) = 4 - (x+1)^2$

5. Graph the function $f(x) = 1/x^2$. Then, without additional computation, graph
   (a) $f(x) = 2 + 1/x^2$
   (b) $f(x) = 1/(x+2)^2$
   (c) $f(x) = -1/x^2$
   (d) $f(x) = 2 + 1/(x+2)^2$

6. Graph the function $f(x) = 1/x$. Then, without additional computation, graph
   (a) $f(x) = 1/x - 1$
   (b) $f(x) = 1/(x-1)$
   (c) $f(x) = -1/x$
   (d) $f(x) = 3 - 1/(x+2)$

7. Graph the function $f(x) = \sqrt{x}$. Then, without additional computation, graph
   (a) $f(x) = 3 + \sqrt{x}$
   (b) $f(x) = \sqrt{x-2}$
   (c) $f(x) = -\sqrt{x}$
   (d) $f(x) = 2 - \sqrt{x+2}$

*8. Graph the function $f(x) = 1/(x-1)$. Then, without additional computation, graph $g(x) = x/(x-1)$.

9. Let $f(x) = x^5$. Find the function g whose graph is obtained by translating the graph of f to the right by 3 units and then up by 5 units.

10. Let $f(x) = x^4$. Find the function g whose graph is obtained by reflecting the graph of f across the x axis and then translating this reflected graph 2 units to the left.

11. Let $f(x) = 1/x$. Find the function g whose graph is obtained by reflecting the graph of f across the y axis and then translating this reflected graph down by 3 units.

12. A function is said to be symmetric about the y axis if its graph and the reflection of its graph across the y axis are identical.
    (a) Find three functions that are symmetric about the y axis.
    (b) If f is symmetric about the y axis, how are $f(x)$ and $f(-x)$ related? Explain your answer.

13. (a) Find three functions that are symmetric with respect to the origin.
    (b) If f is symmetric with respect to the origin, how are $f(x)$ and $f(-x)$ related? Explain your answer.

14. Find a number c for which the curve $y = x^2 + 2x + c$ passes through the point $(1, 2)$. Is there more than one such number c?

15. Find numbers m and b for which the graph of the function $y = mx + b$ passes through the point $(2, 3)$. What other choices of m and b would work?

16. Prove that the distance d between the two points (x_1, y_1) and (x_2, y_2) is given by the formula

    $$d = \sqrt{(x_2 - x_1)^2 + (y_2 - y_1)^2}$$

    (*Hint:* Apply the pythagorean theorem to a right triangle whose hypotenuse is the line segment joining the two points.)

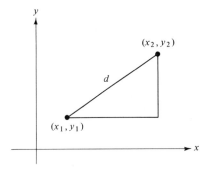

17. In each of the following, compute the distance between the two given points using the formula in Problem 16.
    (a) $(1, 0)$ and $(0, 1)$ (b) $(5, -1)$ and $(2, 3)$ (c) $(2, 6)$ and $(2, -1)$

3. LINEAR FUNCTIONS

In many practical situations, the rate at which one quantity changes with respect to another is constant. Consider the following simple example.

Example 3.1

A truck leaves Cleveland en route to Chicago, 345 miles away. If the truck is driven at a constant speed of 60 miles per hour, express the distance of the truck from its destination as a function of time.

Solution

Let x denote the number of hours the truck has been on the road and $f(x)$ the corresponding distance between the truck and Chicago. Traveling at 60 miles per hour, the truck will go $60x$ miles in x hours, and so $f(x) = 345 - 60x$. ☐

The distance function $f(x) = 345 - 60x$ is an example of a *linear function*. As we shall see, linear functions are characterized by the fact that the rate at which the dependent variable changes with respect to the independent variable is constant. In Example 3.1, the rate at which the distance $f(x)$ decreases with respect to time x is the *constant* speed, 60 miles per hour.

LINEAR FUNCTION

A *linear function* is a function of the form

$$f(x) = a_0 + a_1 x$$

where a_0 and a_1 are constants.

The functions $f(x) = \frac{3}{2} + 2x$, $f(x) = -5x$, and $f(x) = 12$ are all linear. Linear functions are traditionally written in the equivalent form

$$y = mx + b$$

where m and b are constants. We shall adopt this standard notation.

The graphs of linear functions are particularly easy to sketch, since they are always straight lines. Consider, for example, the simple linear function $y = mx$. Since $y = 0$ when $x = 0$, the graph of this function passes through the origin. Moreover, each value of y is m times the corresponding value of x, and so the graph is a straight line that changes m times as fast vertically as it does horizontally. The sign of m tells us whether the line is increasing or decreasing. If m is positive, y increases as x increases; if m is negative, y decreases as x increases. These two possibilities are illustrated in Figure 16.

The situation changes only slightly when we consider the general linear function $y = mx + b$. Indeed, the graph of $y = mx + b$ is obtained by simply translating the graph of $y = mx$ vertically by b units. Thus, instead of passing through the origin, the line $y = mx + b$ crosses the y axis at $(0, b)$, a point known as the *y intercept* of the line (Figure 17).

The constant m in the equation $y = mx + b$ is called the *slope* of the corresponding line. It is the amount by which the y coordinate of a point on the line changes

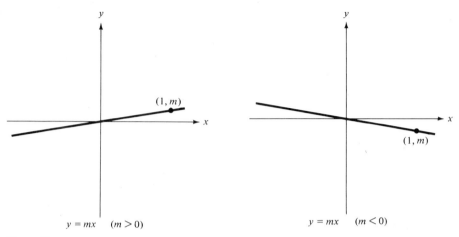

Figure 16

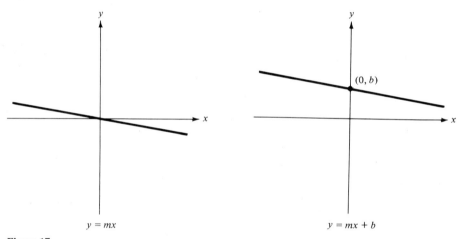

Figure 17

when the x coordinate is increased by 1. In other words, the slope is the ratio of a change in y to the corresponding change in x.

The slope of a nonvertical line can be computed if two points on that line are known. Suppose (x_1, y_1) and (x_2, y_2) lie on a line as in Figure 18. Between these points, x changes by an amount $x_2 - x_1$, and y by an amount $y_2 - y_1$. Hence,

$$\text{Slope} = \frac{\text{change in } y}{\text{change in } x} = \frac{y_2 - y_1}{x_2 - x_1}$$

It is sometimes convenient to use the symbol Δy instead of $y_2 - y_1$ to denote the change in y. The symbol Δy is read "delta y." Similarly, Δx is used to denote $x_2 - x_1$.

3. LINEAR FUNCTIONS

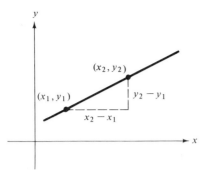

Figure 18

SLOPE

The slope m of the nonvertical line passing through the points (x_1, y_1) and (x_2, y_2) is given by the formula

$$m = \frac{\Delta y}{\Delta x} = \frac{y_2 - y_1}{x_2 - x_1}$$

Example 3.2

Find the slope of the line joining the points $(3, -1)$ and $(-2, 5)$.

Solution

$$m = \frac{\Delta y}{\Delta x} = \frac{5 - (-1)}{-2 - 3} = -\frac{6}{5}$$

SLOPE-INTERCEPT FORM OF THE EQUATION OF A LINE

The graph of the linear function

$$y = mx + b$$

is the straight line with slope m and y intercept $(0, b)$.

Example 3.3

Find the slope and y intercept of the line $y = 3x - 2$ and draw the graph.

Solution

This linear function is of the form $y = mx + b$, with $m = 3$ and $b = -2$. Hence its graph is the straight line with slope 3 and y intercept $(0, -2)$.

A line is determined by two points. Hence, to graph our line, we need to plot only two of its points, say (1, 1) and the y intercept (0, −2). The resulting line is drawn in Figure 19.

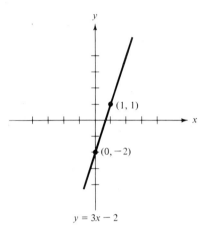

Figure 19

Example 3.4

Find the slope and y intercept of the line $3y + 2x = 6$ and draw the graph.

Solution

The first step is to put our equation in the standard form $y = mx + b$. Solving for y we get

$$y = -\tfrac{2}{3}x + 2$$

from which we conclude that the slope is $-\tfrac{2}{3}$ and the y intercept is (0, 2). The line is graphed in Figure 20.

Horizontal and vertical lines have particularly simple equations. The y coordinates of all the points on a horizontal line are the same. Hence, a horizontal line is the graph of a linear function of the form $y = b$ for some constant b (Figure 21). The slope of a horizontal line is zero, since changes in x produce no change in y whatsoever.

The x coordinates of all the points on a vertical line are equal. Hence vertical lines are characterized by equations of the form $x = c$, where c is constant (Figure 22). (Note that the equation of a vertical line is not a linear function of x; that is, it cannot be put in the form $y = mx + b$.) The slope of a vertical line is undefined. This is because only the y coordinates of points on a vertical line can change, and so the denominator of the quotient (change in y)/(change in x) is zero.

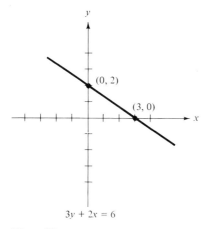

Figure 20

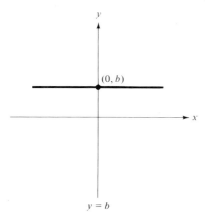

Figure 21

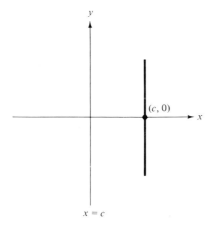

Figure 22

We have seen that certain geometric information about a line can be readily obtained from the equation $y = mx + b$. We now consider the converse problem of finding the equation of a line when geometric properties of that line are known.

Example 3.5

Find the equation of the line that passes through the point (5, 1) and has slope $\frac{1}{2}$.

Solution

We are seeking a linear equation of the form $y = mx + b$. Since the slope of our line is $\frac{1}{2}$, we may substitute $\frac{1}{2}$ for m. We then find the value of b for which the line $y = \frac{1}{2}x + b$ passes through the point (5, 1) by substituting $x = 5$ and $y = 1$ into this equation. We get

$$1 = \tfrac{1}{2}(5) + b$$

or

$$b = -\tfrac{3}{2}$$

and so the desired equation is

$$y = \tfrac{1}{2}x - \tfrac{3}{2} \qquad \square$$

There is a slightly more efficient way to solve the problem in Example 3.5. A point (x, y) will be on the line that passes through (5, 1) and that has slope $\frac{1}{2}$ if the quotient $(y - 1)/(x - 5)$ is equal to the slope $\frac{1}{2}$; that is, if

$$\frac{y - 1}{x - 5} = \frac{1}{2}$$

or, equivalently,

$$y = \tfrac{1}{2}x - \tfrac{3}{2}$$

In general, the coordinates of a point (x, y) on the line that passes through a given point (x_0, y_0) and that has slope m must satisfy the equation

$$\frac{y - y_0}{x - x_0} = m$$

This leads immediately to the following form of the equation of a line.

POINT-SLOPE FORM OF THE EQUATION OF A LINE

$$y - y_0 = m(x - x_0)$$

is an equation of the line that passes through the point (x_0, y_0) and that has slope m.

The point-slope form of the equation of a line is particularly useful when a linear equation is to be determined from geometric information. On the other hand, when the equation is already known, geometric information such as the slope or y intercept is most easily obtained by writing the equation in the slope-intercept form, $y = mx + b$.

Example 3.6

Find an equation of the line that passes through the points $(3, -2)$ and $(1, 5)$.

Solution

To use the formula $y - y_0 = m(x - x_0)$, we need to know a point through which the line passes and the slope of the line. In this case, we are given two points which we use to compute the slope:

$$m = \frac{5 - (-2)}{1 - 3} = -\frac{7}{2}$$

Using $(1, 5)$ as the given point (x_0, y_0), we then obtain the equation

$$y - 5 = -\tfrac{7}{2}(x - 1)$$

or

$$y = -\tfrac{7}{2}x + \tfrac{17}{2}$$

[Convince yourself that the resulting equation would have been the same had we chosen $(3, -2)$ as the given point (x_0, y_0).]   □

Techniques for translating geometric information into linear equations often prove useful in practical problems that ostensibly are not geometric in nature. Consider the following typical examples.

Example 3.7

Students at a state college may preregister for their fall classes by mail during the summer. Those who do not preregister must register in person in September. The registrar can process 35 students per hour during this September registration period. After 4 hours in September, 360 students have been registered.

(a) Express the number of students registered as a function of time.

(b) How many students were registered after 3 hours?

(c) How many students preregistered during the summer?

FUNCTIONS AND GRAPHS

Solution

(a) Let x denote the number of hours during which September registration has been open and y the corresponding total number of registered students. The value of y increases by 35 each time x increases by 1. Thus y is a linear function of x with slope 35. Moreover, when $x = 4$, $y = 360$. Hence the equation defining y as a function of x is simply the equation of the line that passes through $(4, 360)$ and has slope 35. That is,

$$y - 360 = 35(x - 4)$$

or

$$y = 35x + 220$$

The corresponding line is shown in Figure 23.

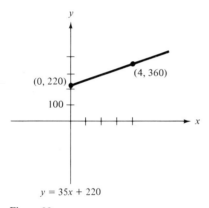

Figure 23

(b) The number of students registered after 3 hours is the value of y corresponding to $x = 3$; namely, $35(3) + 220 = 325$.

(c) The number of students who preregistered is the value of y corresponding to $x = 0$; namely, 220, the y coordinate of the y intercept. □

In the following financial example, the equation of a straight line is determined from knowledge of two points through which it passes.

Example 3.8

A doctor owns $1,500 worth of medical books which, for tax purposes, are assumed to depreciate linearly over a 10-year period. That is, the value of his medical library decreases at a constant rate, so that at the end of 10 years the value will be zero.

(a) Express the value of the books as a function of time.

(b) By how much does the value decrease each year?

Solution

(a) Let y denote the value of the books at the end of x years. Then, $y = 1,500$ when $x = 0$, and $y = 0$ when $x = 10$. The graph depicting the value of the library is therefore a straight line passing through the points (0, 1,500) and (10, 0) (Figure 24). The slope of this line is $(1,500 - 0)/(0 - 10) = -150$, and the y intercept is (0, 1,500). Hence the corresponding linear equation is $y = -150x + 1,500$.

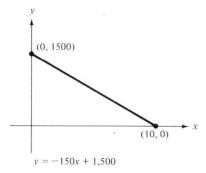

$y = -150x + 1,500$

Figure 24

(b) The amount by which the value of the library changes each year is simply the slope, -150. That is, the value *decreases* by $150 each year. □

In our final example, we consider a familiar practical situation that leads to a function whose graph consists of a collection of disjoint line segments.

Example 3.9

There was a time when the postal rate for airmail letters weighing no more than 7 ounces was 11 cents per ounce or fraction thereof. Express the cost of sending such an airmail letter as a function of its weight and graph this function.

Solution

Let x be the weight of the letter and $f(x)$ the corresponding postage. Then,

$$f(x) = \begin{cases} 11 & \text{if } 0 < x \leq 1 \\ 22 & \text{if } 1 < x \leq 2 \\ \vdots & \quad \vdots \\ 77 & \text{if } 6 < x \leq 7 \end{cases}$$

32 FUNCTIONS AND GRAPHS

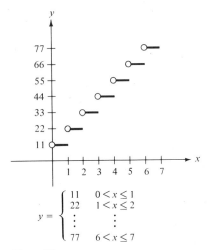

Figure 25

The graph of f, consisting of seven horizontal line segments, is shown in Figure 25. [The circles around the points (0, 11), (1, 22), (2, 33), (3, 44), (4, 55), (5, 66), and (6, 77) indicate that these points are not included as part of the graph of f.] □

Problems

1. Find the slope and y intercept of each of the following lines and draw a graph.
   (a) $y = 3x$
   (b) $y = 3x - 2$
   (c) $y = x$
   (d) $x + y = 0$
   (e) $x + y = 3$
   (f) $2x + 4y = 12$
   (g) $5y - 3x = 4$
   (h) $4x = 2y + 6$
   (i) $y = 2$
   (j) $x = -3$

2. Write equations for the lines with the following properties.
   (a) Through (2, 0) with slope 1
   (b) Through (−1, 2) with slope $\frac{2}{3}$
   (c) Through (5, −2) with slope $-\frac{1}{2}$
   (d) Through (0, 0) with slope 5
   (e) Through (2, 5) and parallel to the x axis
   (f) Through (2, 5) and parallel to the y axis
   (g) Through (1, 0) and (0, 1)
   (h) Through (2, 5) and (1, −2)
   (i) Through (−2, 3) and (0, 5)
   (j) Through (1, 5) and (3, 5)
   (k) Through (1, 5) and (1, −4)

3. A certain oil well yields 400 barrels of crude oil a month that can be sold for $15 per barrel.
   (a) Derive a function that expresses the total revenue from the well over the next x months.
   (b) How much revenue will the oil well generate over the next 6 months?

4. A mountain climber gains altitude at the rate of 120 feet per hour. After 4 hours, he has reached a height of 3,680 feet.
   (a) Express his height as a function of the time spent climbing and graph this function.
   (b) What was his altitude when he began the climb?
   (c) How high will he be at the end of 6 hours?

5. During the summer, a group of students builds kayaks in a converted garage. The rental cost of the garage is $600 for the summer. The materials needed to build a kayak cost $25.
   (a) Express the group's total cost as a function of the number of boats built.
   (b) Graph this function. What is its slope and y intercept?
   (c) Suppose the students sell the kayaks for $175 each. How many must they sell to break even? How many must they sell to make a profit of $450?

6. Membership in a swimming club costs $150 for the 12-week summer season. If a member joins after the start of the season, his fee is prorated; that is, the fee is reduced linearly.
   (a) Express the membership fee as a function of the number of weeks that have elapsed by the time the membership is purchased.
   (b) Graph this function.
   (c) What is the cost of a membership that is purchased 5 weeks after the start of the season?

7. A taxi fleet contains 30 cabs, each of which is driven approximately 200 miles a day and averages 15 miles to the gallon. If the price of gasoline is 70 cents per gallon, derive a function expressing the amount of money the taxi company can expect to spend on gasoline over the next x days.

8. A manufacturer buys $20,000 worth of machinery which depreciates linearly so that its trade-in value after 10 years will be $1,000.
   (a) Express the value of the machinery as a function of its age.
   (b) What is the value of the machinery after 4 years?

9. The value of a certain rare book doubles every 10 years. The book was originally worth $3.
   (a) How much is the book worth when it is 30 years old? When it is 40 years old?
   (b) Can the value of the book be expressed as a linear function of its age? Explain your answer.

10. Two jets bound for Los Angeles leave New York 30 minutes apart. The first travels 550 miles per hour, while the second goes 650 miles per hour. At what time will the second plane pass the first?

11. A local natural history museum charges admission to groups according to the following policy. Groups of fewer than 50 people are charged a rate of $1.50 per person, while groups of 50 people or more are charged a reduced rate of $1 per person.
    (a) Express the amount a group will be charged for admission to the museum as a function of the size of the group.
    (b) Graph this function.
    (c) For what values of the independent variable does this function have a practical interpretation?
    (d) How much money will a group of 49 people save in admission costs if it can recruit one additional member?

12. What is the relationship between the slopes of parallel lines? Explain your answer. Then, write equations of the lines with the following properties.
    (a) Through (1, 3) and parallel to the line $4x + 2y = 7$
    (b) Through (0, 2) and parallel to the line $2y - 3x = 5$
    (c) Through (−2, 5) and parallel to the line through the points (1, 2) and (6, −1)

*13. Show that if a line L_1 with slope m_1 is perpendicular to a line L_2 with slope m_2, then $m_1 = -1/m_2$. (*Hint:* Find expressions for the slopes of the perpendicular lines L_1 and L_2 in the accompanying figure. Then apply the pythagorean theorem, together with the distance formula from Problem 16 on page 22 to the right triangle OAB to obtain the desired relationship between these slopes.)

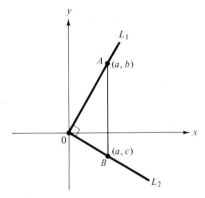

14. Write equations of the lines with the following properties.
    (a) Through (−1, 3) and perpendicular to the line $4x + 2y = 7$
    (b) Through (0, 0) and perpendicular to the line $2y - 3x = 5$
    (c) Through (2, 1) and perpendicular to the line joining (0, 3) and (2, −1)

4. LIMITS AND CONTINUITY

Most of the graphs we have encountered have been unbroken curves. A function whose graph is unbroken is said to be *continuous*. For example, the function $f(x) = x^3$ is continuous (Figure 26a). So is the revenue function

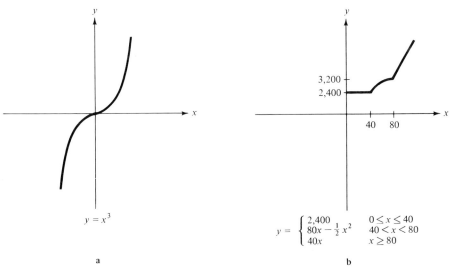

Figure 26

$$f(x) = \begin{cases} 2{,}400 & \text{if } 0 \le x \le 40 \\ 80x - \tfrac{1}{2}x^2 & \text{if } 40 < x < 80 \\ 40x & \text{if } x \ge 80 \end{cases}$$

that we obtained in Example 1.6 (Figure 26b). On the other hand, the function $f(x) = 1/x^3$ from Example 2.8 is not continuous because its graph breaks into two pieces at $x = 0$ (Figure 27).

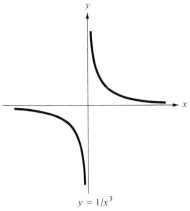

Figure 27

36 FUNCTIONS AND GRAPHS

Gaps or breaks in the graph of a function are called *discontinuities*. The function $f(x) = 1/x^3$ has a discontinuity when $x = 0$. The postal rate function,

$$f(x) = \begin{cases} 11 & \text{if } 0 < x \leq 1 \\ 22 & \text{if } 1 < x \leq 2 \\ \vdots & \vdots \\ 77 & \text{if } 6 < x \leq 7 \end{cases}$$

from Example 3.9 has six discontinuities: when $x = 1, 2, 3, 4, 5,$ and 6 (Figure 28).

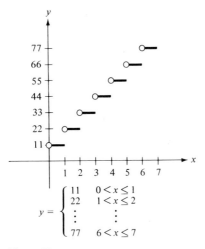

Figure 28

The function $f(x) = 1/x^3$ and the postal rate function illustrate the two most common types of discontinuities. A function that is defined as a quotient will have a discontinuity whenever its denominator is zero. A function that is defined in several pieces will have discontinuities if the graphs of the individual pieces are disjoint from one another. Before discussing these two types of discontinuities in more detail, let us consider some additional examples.

Example 4.1

Locate the discontinuities of the function $f(x) = 1/(x - 3)^3$ and sketch a graph.

Solution

This function has a discontinuity when $x = 3$, since its denominator is zero for this value of x.

The graph of f is obtained by translating the graph of $y = 1/x^3$ to the right 3 units (Figure 29).

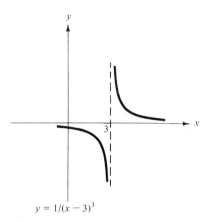

$y = 1/(x-3)^3$

Figure 29

Example 4.2

Locate the discontinuities of the function

$$f(x) = \begin{cases} 1 & \text{if } 0 \leq x \leq 2 \\ x + 1 & \text{if } 2 < x \leq 5 \\ 6 & \text{if } 5 < x \leq 7 \end{cases}$$

and draw a graph.

Solution

The graph of f consists of three line segments: the line $y = 1$ for $0 \leq x \leq 2$; the line $y = x + 1$ for $2 < x \leq 5$; and the line $y = 6$ for $5 < x \leq 7$ (Figure 30).

The only discontinuity occurs at $x = 2$, where the first segment fails to connect with the second.

One way to specify the behavior of a function near a discontinuity is to describe what happens to its graph as the independent variable approaches the discontinuity from either side. Consider, for example, the function

$$f(x) = \begin{cases} 1 & \text{if } 0 \leq x \leq 2 \\ x + 1 & \text{if } 2 < x \leq 5 \\ 6 & \text{if } 5 < x \leq 7 \end{cases}$$

FUNCTIONS AND GRAPHS

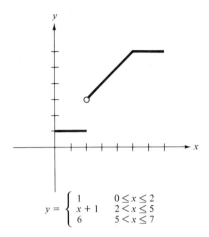

$$y = \begin{cases} 1 & 0 \leq x \leq 2 \\ x+1 & 2 < x \leq 5 \\ 6 & 5 < x \leq 7 \end{cases}$$

Figure 30

that we sketched in Figure 30. The only discontinuity occurs when $x = 2$. As values of x that are greater than 2 get closer and closer to 2, the corresponding function values $f(x)$ get closer and closer to 3. This behavior is sometimes described by saying

"the limit of $f(x)$ as x approaches 2 from the right is 3"

and may be abbreviated as

$$\lim_{x \to 2^+} f(x) = 3$$

The plus sign following the 2 indicates that the values of x under consideration are slightly *larger* than 2; that is, 2 *plus* a small amount.

As values of x that are less than 2 get closer and closer to 2, the corresponding function values $f(x)$ get closer and closer to 1. (In fact, they are *equal* to 1 for all x between 0 and 2.) This can be described by saying

"the limit of $f(x)$ as x approaches 2 from the left is 1"

and abbreviated by writing

$$\lim_{x \to 2^-} f(x) = 1$$

Here, the minus sign following the 2 indicates that the values of x under consideration are slightly *smaller* than 2; that is, 2 *minus* a small amount.

This descriptive limit notation has obvious extensions to more general situations.

ONE-SIDED LIMITS

If $f(x)$ gets closer and closer to some number b as values of x that are greater than a get closer and closer to a, we say that b is the *limit of $f(x)$ as x approaches a from the right*, and we write

$$\lim_{x \to a^+} f(x) = b$$

If $f(x)$ gets closer and closer to some number c as values of x that are less than a get closer and closer to a, we say that c is the *limit of $f(x)$ as x approaches a from the left*, and we write

$$\lim_{x \to a^-} f(x) = c$$

Figure 31 shows three situations that can be described by the limit statement $\lim_{x \to a^+} f(x) = b$. As you examine these illustrations, bear in mind that this limit

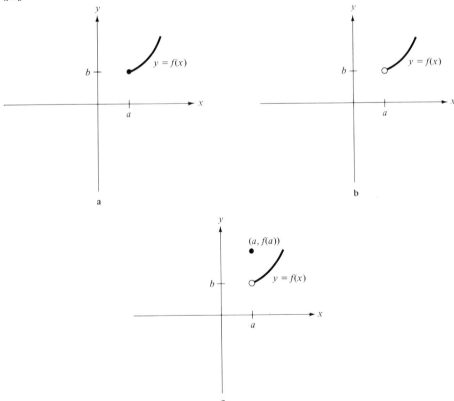

Figure 31

notation describes the behavior of the graph *near* $x = a$, and not necessarily *at* $x = a$. Observe that in Figure 31a, the function $f(x)$ is defined when $x = a$ and, in fact, $f(a) = b$. In Figure 31b, on the other hand, $f(a)$ is not defined. And, in Figure 31c, $f(a)$ is defined but it is not equal to b.

Example 4.3

Locate the discontinuities of the function

$$f(x) = \begin{cases} x^2 & \text{if } 0 \leq x < 2 \\ 5 & \text{if } 2 \leq x < 4 \\ x - 1 & \text{if } 4 \leq x < 6 \end{cases}$$

and use limit notation to describe the behavior of f near each of these discontinuities.

Solution

From the graph of f (Figure 32) we see that there are two discontinuities: when $x = 2$ and $x = 4$. Moreover,

$$\lim_{x \to 2^+} f(x) = 5 \qquad \lim_{x \to 2^-} f(x) = 4$$

and

$$\lim_{x \to 4^+} f(x) = 3 \qquad \lim_{x \to 4^-} f(x) = 5$$

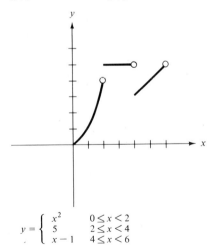

$$y = \begin{cases} x^2 & 0 \leq x < 2 \\ 5 & 2 \leq x < 4 \\ x - 1 & 4 \leq x < 6 \end{cases}$$

Figure 32

□

Our limit notation may be extended to describe discontinuities that arise when the denominator of a quotient becomes zero. Near such discontinuities, the graph of the function is generally unbounded. A typical example is the function $f(x) = 1/(x-3)^3$ which is sketched again in Figure 33. The only discontinuity of f occurs when $x = 3$. As values of x that are greater than 3 get closer to 3, the function values $f(x)$ increase without bound. This behavior may be described by saying

"the limit of $f(x)$ as x approaches 3 from the right is plus infinity"

and abbreviated as

$$\lim_{x \to 3^+} f(x) = +\infty$$

The plus sign preceding ∞ tells us that the function values $f(x)$ are *increasing* (rather than decreasing) without bound.

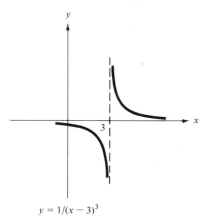

$y = 1/(x-3)^3$

Figure 33

As values of x that are less than 3 get closer to 3, the function in Figure 33 decreases without bound. This can be described by saying

"the limit of $f(x)$ as x approaches 3 from the left is minus infinity"

and abbreviated by writing

$$\lim_{x \to 3^-} f(x) = -\infty$$

Here, the minus sign preceding ∞ tells us that $f(x)$ is *decreasing* without bound.

INFINITE LIMITS

If $f(x)$ increases (or decreases) without bound as values of x that are greater than a get closer and closer to a, we say that the *limit* of $f(x)$ as x approaches a *from the right* is plus (or minus) infinity, and we write

$$\lim_{x \to a^+} f(x) = +\infty \qquad \text{if } f(x) \text{ increases without bound}$$

and

$$\lim_{x \to a^+} f(x) = -\infty \qquad \text{if } f(x) \text{ decreases without bound}$$

Analogous limit notation exists to describe the behavior of $f(x)$ as x approaches a from the left.

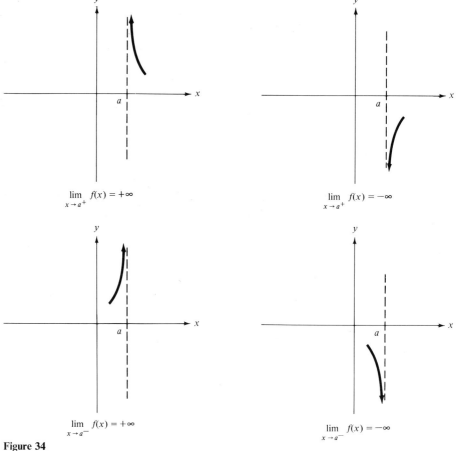

Figure 34

4. LIMITS AND CONTINUITY   43

Illustrated in Figure 34, with the corresponding limit notation, are the four possibilities that can arise if the graph of a function is unbounded near a discontinuity at $x = a$.

We need not compute a long table of values to determine the behavior of a quotient near one of its discontinuities. In analyzing the function $f(x) = 1/(x-3)^3$, for example, we could reason as follows.

The absolute value of a quotient with a fixed numerator increases as the denominator approaches zero. Hence, as x approaches 3 from either side, the absolute value of $1/(x-3)^3$ increases without bound. Suppose that x approaches 3 from the right. Then x is greater than 3 and $x - 3$ is positive, so the corresponding values of the function $f(x) = 1/(x-3)^3$ are also positive. Hence $\lim_{x \to 3^+} f(x) = +\infty$. On the other hand, if x approaches 3 from the left, x is less than 3 and $x - 3$ is negative, so the corresponding values of $f(x) = 1/(x-3)^3$ are negative. Therefore, $\lim_{x \to 3^-} f(x) = -\infty$.

Here is another example to illustrate this approach to the analysis of discontinuities.

Example 4.4

Find all values of x for which the function $f(x) = (x-5)(x-1)/(x+3)$ has a discontinuity, and use limit notation to describe the behavior of f near each of these discontinuities.

Solution

The only discontinuity occurs when $x = -3$. As x approaches -3 from either side, the denominator $x + 3$ approaches zero, while the numerator $(x-5)(x-1)$ approaches $(-3-5)(-3-1) = 32$. It follows that the absolute value of the quotient $f(x) = (x-5)(x-1)/(x+3)$ increases without bound. Thus,

$$\lim_{x \to -3^+} f(x) = +\infty \quad \text{or} \quad -\infty$$

and

$$\lim_{x \to -3^-} f(x) = +\infty \quad \text{or} \quad -\infty$$

To settle the matter, we need to check the sign of $f(x)$ as x approaches -3 from each side.

In computing the limit from the right, we consider values of x that are close to but greater than -3. Since x is *close to* -3,

$x - 5$ is negative and $x - 1$ is negative.

Since x is *greater than* -3,

$x + 3$ is positive.

Hence the quotient

$f(x) = (x - 5)(x - 1)/(x + 3)$ is positive

and so

$$\lim_{x \to -3^+} f(x) = +\infty$$

On the other hand, in computing the limit from the left, we consider values of x that are close to but less than -3. Because x is *close to* -3, it again follows that

$x - 5$ is negative and $x - 1$ is negative.

But now, since x is *less than* -3,

$x + 3$ is negative.

Hence, this time, the quotient

$f(x) = (x - 5)(x - 1)/(x + 3)$ is negative,

and so

$$\lim_{x \to -3^-} f(x) = -\infty$$

□

Some quotients remain bounded near their discontinuities. These functions tend to be somewhat unnatural and rarely arise in practical work. However, quotients of this type will play a brief but important role in our initial development of differential calculus in Chapter 2. Let us consider a typical example of such a quotient.

Example 4.5

Find all values of x for which the function $f(x) = (x - 1)(x + 2)/(x - 1)$ has a discontinuity, and use limit notation to describe the behavior of f near each of these discontinuities.

Solution

The only discontinuity occurs when $x = 1$. If we try to analyze the behavior of the graph near this discontinuity using our previous methods, we quickly run into trouble. For, as x approaches 1, *both* the numerator and the denominator approach zero, giving us no clue to the size of the quotient. Hence we must adopt a different strategy.

4. LIMITS AND CONTINUITY   45

Our function is not defined when $x = 1$. For all other values of x, however, we can simplify the function by dividing numerator and denominator by $x - 1$. (Since $x \neq 1$, we will not be dividing by zero.) Thus, $f(x) = x + 2$ for $x \neq 1$ and is undefined when $x = 1$. Hence its graph consists of the straight line $y = x + 2$, with a hole when $x = 1$ (Figure 35).

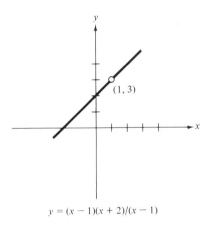

$y = (x - 1)(x + 2)/(x - 1)$

Figure 35

It is now clear from the graph that $\lim_{x \to 1^+} f(x) = 3$ and $\lim_{x \to 1^-} f(x) = 3$. □

The crucial difference between the function in Example 4.5 and the quotients we have previously encountered is that *both* its numerator and denominator vanish at the discontinuity. In general, when this happens, our strategy will be to simplify the function (as we did by canceling $x - 1$ in Example 4.5). In most cases, the simplified form of the function will be valid for all values of x except the particular value corresponding to the discontinuity. Since we are interested in the behavior of the function *near* its discontinuity (not *at* its discontinuity where it is undefined), we may use the simplified form of the function in evaluating the corresponding limits.

We use this technique once more in the next example. This time, the process of algebraic simplification is slightly more sophisticated than in Example 4.5.

Example 4.6

Find all values of x for which the function $f(x) = (\sqrt{x} - 1)/(x - 1)$ has a discontinuity, and use limit notation to describe the behavior of f near each of these discontinuities.

Solution

The only discontinuity occurs at $x = 1$, when both the numerator and denominator vanish. To simplify f, we employ the algebraic trick of multiplying numerator and denominator by $\sqrt{x} + 1$ to get

$$f(x) = \frac{(\sqrt{x} - 1)(\sqrt{x} + 1)}{(x - 1)(\sqrt{x} + 1)}$$

$$= \frac{x - 1}{(x - 1)(\sqrt{x} + 1)}$$

$$= \frac{1}{\sqrt{x} + 1}$$

provided $x \neq 1$.

Now, as x approaches 1 from either side, $\sqrt{x}$ also approaches 1 and so the corresponding values of the function $f(x) = 1/(\sqrt{x} + 1)$ approach $\frac{1}{2}$. That is, $\lim_{x \to 1^+} f(x) = \frac{1}{2}$, and $\lim_{x \to 1^-} f(x) = \frac{1}{2}$. □

When the one-sided limits $\lim_{x \to a^+} f(x)$ and $\lim_{x \to a^-} f(x)$ are equal (and finite), we sometimes omit the superfluous reference to the direction from which x approaches a and simply speak of *the limit* of $f(x)$ as x approaches a.

LIMIT

If $f(x)$ gets closer and closer to some number b as x approaches a both from the left and from the right, we say that b is *the limit* of $f(x)$ as x approaches a and we write

$$\lim_{x \to a} f(x) = b$$

For instance, in Example 4.5, $\lim_{x \to 1} f(x) = 3$; and in Example 4.6, $\lim_{x \to 1} f(x) = \frac{1}{2}$.

Problems

1. For each of the functions f sketched below, determine the limits $\lim_{x \to a^+} f(x)$ and $\lim_{x \to a^-} f(x)$ and decide if the function has a discontinuity at $x = a$.

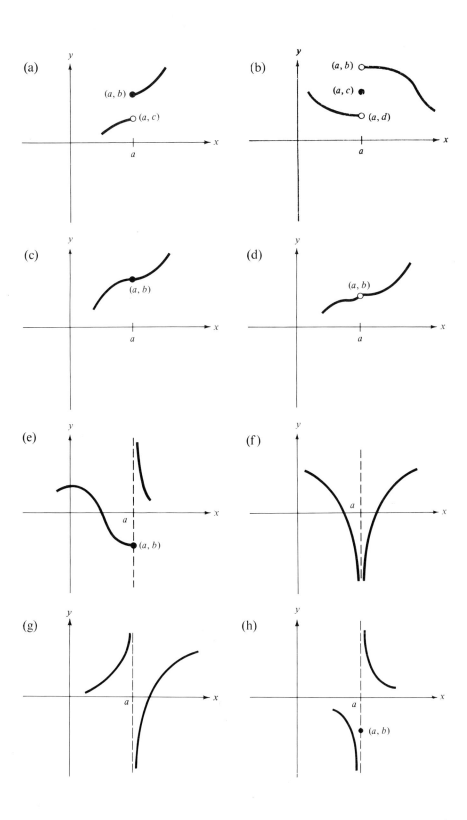

2. Graph the following functions and locate their discontinuities. Use limit notation to describe the behavior of the functions near each of their discontinuities.

(a) $f(x) = \begin{cases} 0 & \text{if } -1 \leq x \leq 1 \\ 2 & \text{if } 1 < x < 3 \\ 5 & \text{if } 3 \leq x < 5 \end{cases}$

(b) $f(x) = \begin{cases} x & \text{if } x < 2 \\ -\frac{3}{2}x + 6 & \text{if } x \geq 2 \end{cases}$

(c) $f(x) = \begin{cases} 2x + 1 & \text{if } x \leq 2 \\ x - 1 & \text{if } x > 2 \end{cases}$

(d) $f(x) = \begin{cases} x^2 & \text{if } 0 \leq x < 3 \\ 6 & \text{if } x = 3 \\ 9 & \text{if } x > 3 \end{cases}$

(e) $f(x) = \begin{cases} -2x + 4 & \text{if } x < 1 \\ x^2 + 1 & \text{if } x \geq 1 \end{cases}$

(f) $f(x) = \begin{cases} 1/x & \text{if } x \neq 0 \\ 0 & \text{if } x = 0 \end{cases}$

(g) $f(x) = \begin{cases} 2 & \text{if } 0 \leq x \leq 3 \\ 1/(x - 3) & \text{if } 3 < x < 5 \\ \frac{1}{2} & \text{if } x \geq 5 \end{cases}$

3. Graph the following functions and locate their discontinuities. Use limit notation to describe the behavior of the functions near each of their discontinuities.
(a) $f(x) = 1/(x + 2)$
(b) $f(x) = 1/(x + 2)^2$
(c) $f(x) = 1 + 1/x$
(d) $f(x) = 1/(x - 4)^3$
(e) $f(x) = (x^2 - 1)/(x + 1)$
(f) $f(x) = (x + 1)/(x^2 - 1)$
(g) $f(x) = 1/(x^2 + 1)$

4. Locate the discontinuities of the following functions and use limit notation to describe the behavior of the functions near each of their discontinuities.

(a) $f(x) = 1/(2 - x)$
(b) $f(x) = -3/(x - 5)^2$
(c) $f(x) = (x + 2)/(x - 1)$
(d) $f(x) = (2 - x)/(x - 1)$
(e) $f(x) = (x - 1)/[(x - 3)(x + 4)]$
(f) $f(x) = (x - 5)(2 - x)/[(x + 2)(x - 1)]$
(g) $f(x) = (x^2 - 1)/(x^2 - 9)$
(h) $f(x) = (x^2 - 4x + 4)/(x^2 + 2x - 3)$
(i) $f(x) = x + 1/x$
(j) $f(x) = 1 - 2/x^3$
(k) $f(x) = (x^3 - 3x^2)/x$
(l) $f(x) = (x^2 + 6x + 9)/(x + 3)$
(m) $f(x) = (x - 1)(x + 2)/[(x - 1)(x - 3)]$
(n) $f(x) = (\sqrt{x} - 2)/(x - 4)$
(o) $f(x) = (x - 1)/(\sqrt{x} - 1)$

5. Suppose that $f(1) = 5$ and $f(3) = -1$ for a certain continuous function f.
   (a) What conclusion can you draw about the possibility that the graph of f crosses the x axis between $x = 1$ and $x = 3$?
   (b) Is your conclusion in part (a) still valid if f is not continuous? Explain.

6. Suppose that f is continuous and that $f(a)$ and $f(b)$ have different signs.
   (a) What conclusion can you draw about the possibility that the graph of f crosses the x axis between $x = a$ and $x = b$?
   (b) Is your conclusion in part (a) still valid if f is not continuous? Explain.

7. Suppose that f is continuous and that the only values of x in the interval $a \leq x \leq b$ for which $f(x) = 0$ are $x = a$ and $x = b$.
   (a) What conclusion can you draw about the sign of $f(x)$ in the interval $a < x < b$?
   (b) Is your conclusion in part (a) still valid if f is not continuous? Explain.

8. Some texts give the following definition of continuity: A function f is continuous at $x = a$ if
   (i) $f(a)$ is defined (i.e., a is in the domain of f), and
   (ii) $\lim_{x \to a} f(x) = f(a)$.

Examine the functions in Examples 4.1, 4.2, 4.3, and 4.5 and use this definition of continuity to explain why these functions are not continuous at their discontinuities.

5. POLYNOMIALS

A *polynomial* is a function of the form

$$f(x) = a_0 + a_1 x + a_2 x^2 + \cdots + a_n x^n$$

where n is a non-negative integer and $a_0, a_1, \ldots, a_n$ are real numbers. If $a_n \neq 0$, the integer n is said to be the *degree* of the polynomial. For example, the function $f(x) = 3x^5 - 6x^2 + 7$ is a polynomial of degree 5. Many practical problems lead to polynomials. For instance, the cost function $f(x) = 300x^2 + 2{,}410x + 4{,}849$ in Example 1.9 and the profit function $f(x) = (120 - x)(x - 20)$ in Example 1.5 are both polynomials of degree 2.

Any number c for which $f(c) = 0$ is said to be a *root* of f. You may recall from algebra that if c is a root of a polynomial f then the linear term $x - c$ is a *factor* of f. That is, we can write

$$f(x) = (x - c)g(x)$$

where g is itself a polynomial. For example, 2 is a root of the polynomial $f(x) = x^2 - x - 2$ [since $f(2) = 0$], and $f(x)$ can be factored as $f(x) = (x - 2)(x + 1)$. This correspondence between roots and factors implies that a polynomial of degree n can have at most n roots. (If a polynomial had $n + 1$ roots, for example, its factorization would contain $n + 1$ linear terms and so its degree would be at least $n + 1$.) Of course, a polynomial of degree n may have fewer than n roots. For example, the second-degree polynomial $f(x) = x^2 + 1$ has no roots whatsoever. (Do you see why?)

Polynomials whose roots are known are particularly easy to sketch. A simple procedure for sketching such polynomials is illustrated in Example 5.1. It is based on the fact that polynomials are continuous functions whose graphs are unbroken curves.

Example 5.1

Sketch the graph of the polynomial $f(x) = x^3 - 3x^2 - x + 3$.

Solution

We begin by factoring the polynomial f as follows:

$$f(x) = (x - 1)(x + 1)(x - 3)$$

From this factored form we can immediately identify all the roots of f, namely, $x = 1$, $x = -1$, and $x = 3$. The graph of f must therefore pass through the points $(1, 0)$, $(-1, 0)$, and $(3, 0)$. These are known as the x *intercepts* of f and are the points at which the graph intersects the x axis. We begin the sketch by plotting these x intercepts (Figure 36).

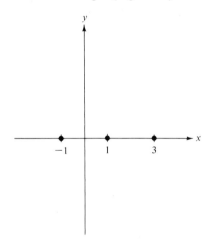

Figure 36

Next, we consider the behavior of the graph between adjacent intercepts. Since f is continuous and its graph an unbroken curve, there is no way that $f(x)$ can change sign between two adjacent x intercepts. Hence the portion of the graph between the intercepts $(1, 0)$ and $(3, 0)$ must consist of a curve that joins these intercepts and lies either entirely above the x axis as in Figure 37a or entirely below it

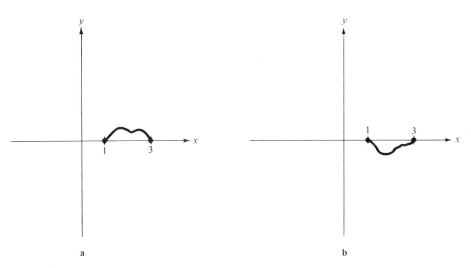

Figure 37

as in Figure 37b. We can settle the matter by simply plotting one point on the graph whose x coordinate lies between 1 and 3. Since any such point will do, we make the computation as simple as possible by taking $x = 2$. Since $f(2) = -3$, the point $(2, -3)$ is on the graph and so the curve lies *below* the axis as in Figure 37b.

Similarly, since $f(0) = 3$, the point $(0, 3)$ is on the graph of f and so the curve lies *above* the x axis between $x = -1$ and $x = 1$. With this information, we can now draw a rough sketch (Figure 38) of the portion of the graph between $x = -1$ and $x = 3$.

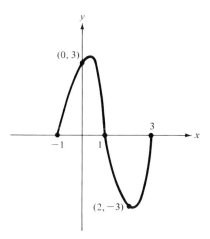

Figure 38

To complete the sketch, we must determine the behavior of the graph to the right of $x = 3$ and to the left of $x = -1$. If x is a large positive number, the corresponding value of our function $f(x) = x^3 - 3x^2 - x + 3$ is also large and positive. Indeed, as x increases without bound, so does $f(x)$. This is because the term x^3 in the expression for $f(x)$ grows more quickly than the remaining terms of lower degree.

Similarly, if x is a negative number whose absolute value is large, the absolute value of $f(x)$ is again large. This time, however, $f(x)$ is negative (since the dominant term x^3 is negative), and so $f(x)$ decreases without bound as x decreases without bound.

By incorporating this new information into our previous picture, we finally obtain the sketch shown in Figure 39.

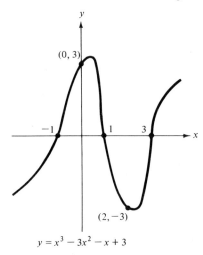

$y = x^3 - 3x^2 - x + 3$

Figure 39

Of course, this sketch is only a rough approximation of the actual graph of the polynomial. We have not determined, for example, the precise location of the peak that occurs between $x = -1$ and $x = 1$. This information, along with other details about the shape of the graph will become available in Chapter 2 where differential calculus will be applied to curve sketching. □

To simplify our discussion of polynomials, we extend the limit notation that was introduced in Section 4.

LIMITS

If $f(x)$ increases (or decreases) without bound as the variable x increases without bound, we say that the *limit* of $f(x)$ as x *approaches plus infinity* is plus (or minus) infinity and we write

$$\lim_{x \to +\infty} f(x) = +\infty \qquad \text{if } f(x) \text{ increases without bound}$$

and

$$\lim_{x \to +\infty} f(x) = -\infty \quad \text{if } f(x) \text{ decreases without bound}$$

Similarly, if $f(x)$ increases (or decreases) without bound as the variable x decreases without bound, we write

$$\lim_{x \to -\infty} f(x) = +\infty \quad \text{if } f(x) \text{ increases without bound}$$

and

$$\lim_{x \to -\infty} f(x) = -\infty \quad \text{if } f(x) \text{ decreases without bound}$$

The geometric interpretation of these limit statements is illustrated in Figure 40.

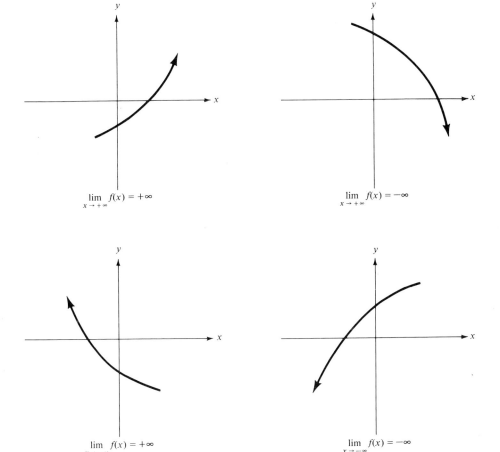

Figure 40

When we analyzed the function $f(x) = x^3 - 3x^2 - x + 3$ in Example 5.1, we examined the term x^3 to conclude that $\lim_{x \to +\infty} f(x) = +\infty$ and $\lim_{x \to -\infty} f(x) = -\infty$. In general, the limits of a polynomial are determined by its term of highest degree, since this is the term that will be increasing or decreasing most rapidly as x approaches $+\infty$ or $-\infty$.

Example 5.2

Find $\lim_{x \to +\infty} f(x)$ and $\lim_{x \to -\infty} f(x)$ if $f(x) = 1 + x^2 - x^3 - 3x^4$.

Solution

The term of highest degree is $-3x^4$ which decreases without bound as x approaches both $+\infty$ and $-\infty$. Hence, $\lim_{x \to +\infty} f(x) = -\infty$ and $\lim_{x \to -\infty} f(x) = -\infty$. □

Let us now summarize our procedure for sketching polynomials.

SKETCHING POLYNOMIALS

Step 1. Find the x intercepts.
Step 2. Determine the behavior of the graph between adjacent x intercepts.
Step 3. Evaluate $\lim_{x \to +\infty} f(x)$ and $\lim_{x \to -\infty} f(x)$.
Step 4. Sketch the graph using these clues.

To further illustrate this procedure, we consider one more example.

Example 5.3

Sketch the polynomial $f(x) = -x^3 + 3x - 2$.

Solution

The polynomial can be factored as

$$f(x) = -(x - 1)^2(x + 2)$$

Thus, the x intercepts are $(1, 0)$ and $(-2, 0)$. Since $f(0) = -2$, the curve lies below the x axis between these intercepts. Finally, since $-x^3$ is the term of highest degree, it follows that $\lim_{x \to +\infty} f(x) = -\infty$ and $\lim_{x \to -\infty} f(x) = +\infty$.
A sketch of f based on these observations is drawn in Figure 41. □

Our procedure for sketching polynomials has obvious limitations. When the factors are not known, explicit computation of the x intercepts may be impossible. Indeed, some polynomials have no x intercepts at all. Let us see how we might graph a polynomial without knowing its roots.

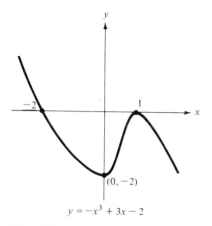

$y = -x^3 + 3x - 2$

Figure 41

Example 5.4

Sketch the polynomial $f(x) = x^3 - 9x + 1$.

Solution

Since we see no obvious factors, we begin with a short table of values.

x	-3	-2	-1	0	1	2	3
$f(x)$	1	11	9	1	-7	-9	1

We also observe that

$$\lim_{x \to +\infty} f(x) = +\infty \quad \text{and} \quad \lim_{x \to -\infty} f(x) = -\infty$$

This information suggests that the graph of f resembles the curve in Figure 42. □

Notice that although we were unable to locate the x intercepts in Example 5.4 we are now assured that precisely three exist. For example, there must be one somewhere between $x = 0$ and $x = 1$, since $f(0)$ is positive, $f(1)$ is negative, and the graph of f is unbroken. Similarly, the graph must cross the x axis between $x = 2$ [where $f(x)$ is negative] and $x = 3$ [where $f(x)$ is positive]. And finally, there is a third x intercept somewhere to the left of $x = -3$. This is because $f(-3)$ is positive, while the limit of $f(x)$ as x approaches minus infinity is *minus* infinity.

There is another way to obtain the sketch of the function in Example 5.4 that is based on the concept of vertical translation. We illustrate this technique in the next example.

56 FUNCTIONS AND GRAPHS

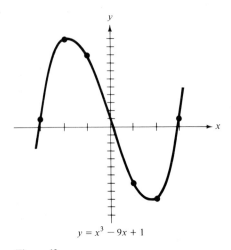

Figure 42

Example 5.5

Sketch the polynomial $f(x) = x^3 - 9x + 1$.

Solution

We consider the related function $g(x) = x^3 - 9x$. This function, which can be factored as

$$g(x) = x(x - 3)(x + 3)$$

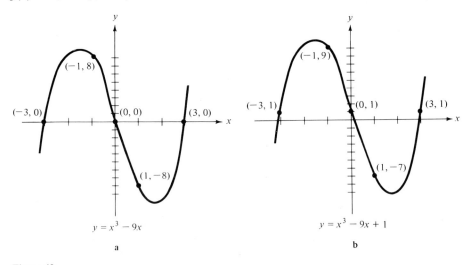

Figure 43

has x intercepts at $(0, 0)$, $(3, 0)$, and $(-3, 0)$ and can be easily sketched (Figure 43a). Since $f(x) = g(x) + 1$, the graph of f (Figure 43b) is obtained simply by raising the graph of g by 1 unit. □

In practical problems, graphs, or even rough sketches of graphs, often reveal significant information that may be obscured by verbal descriptions of the situation and concealed in the corresponding algebraic formulas. To illustrate this point, let us return to the profit problem we examined in Example 1.5.

Example 5.6

A manufacturer can produce cassette tape recorders at a cost of $20 each. He estimates that if he sells them for x dollars apiece he will be able to sell approximately $120 - x$ tape recorders a month. Express the manufacturer's monthly profit as a function of the selling price x and graph this profit function. How do the major geometric features of the graph reflect the underlying economic situation?

Solution

Let $f(x)$ denote the monthly profit if the tape recorders are sold for x dollars apiece. In Example 1.5 we found that

$$f(x) = (120 - x)(x - 20)$$

The graph of this polynomial is sketched in Figure 44.

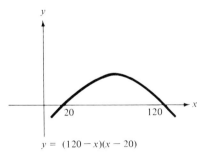

$y = (120 - x)(x - 20)$

Figure 44

There is an x intercept at $(20, 0)$, because the profit will be zero when the selling price x is equal to the cost, 20. The fact that the graph is rising for a while when x is greater than 20 reflects the fact that, as the price increases, the profit per recorder increases. However, as the price increases, fewer people are willing to buy the product, and there is a point at which the profit is maximal, after which it decreases and the graph begins to fall. The problem of determining the selling price that will generate the maximal profit is clearly of great interest to the manufacturer. We will calculate this optimal selling price in Chapter 2, using the tools of calculus. Indeed, a fundamental use of differential calculus is the solution of optimization problems of this sort. □

Problems

1. For each of the following polynomials, find the x intercepts, compute $\lim_{x \to +\infty} f(x)$ and $\lim_{x \to -\infty} f(x)$, and sketch a graph.
   (a) $f(x) = (x + 1)(x - 2)$
   (b) $f(x) = (x - 2)(x + 3)(x - 1)$
   (c) $f(x) = (x - 2)^2(x + 3)(x - 1)$
   (d) $f(x) = (1 - 2x)(x + 5)$
   (e) $f(x) = -x^2 + 2x + 3$
   (f) $f(x) = x^2 - 4x + 4$
   (g) $f(x) = x^3 + 4x^2 - x - 4$
   (h) $f(x) = -2x^2 + 4x$
   (i) $f(x) = x^3 + 3x^2$
   (j) $f(x) = (x - 1)^3$
   (k) $f(x) = (x^2 - 9)^3$
   (l) $f(x) = (x^2 - 1)(x^2 - 4)$
   (m) $f(x) = (x - 2)^3(2x + 1)^2$

2. Use suitable vertical translations in sketching the following polynomials.
   (a) $f(x) = x^2 + 3x + 4$
   (b) $f(x) = -2x^2 + 4x + 2$
   (c) $f(x) = x^3 + x^2 - 2x - 6$
   (d) $f(x) = x^3 - x^2 + 5$

3. Sketch the following polynomials. What conclusions can you draw about the location of the x intercepts?
   (a) $f(x) = x^2 + 3x + 1$
   (b) $f(x) = 1 + (x - 2)^2$
   (c) $f(x) = x^2 - 4x + 2$
   (d) $f(x) = x^2 - 2x + 3$
   (e) $f(x) = x^3 - 2x^2 + x - 5$

4. A manufacturer can produce waterbed frames at a cost of $10 each. He estimates that if he sells them for x dollars each he will be able to sell approximately $50 - x$ frames a month. (See Problem 6 on page 9.)
   (a) Express the manufacturer's monthly profit as a function of the selling price x and sketch this profit function.
   (b) Give an economic explanation for the x intercepts (50, 0) and (10, 0).
   (c) Use your graph to estimate the selling price at which the manufacturer will maximize his profit. (In Problem 14 on page 101 you will compute this optimal price exactly, using techniques of differential calculus.)

5. A college bookstore can obtain the book *Social Groupings of the American Dragonfly* from the publisher at a cost of $3 per book. The bookstore estimates that it can sell 200 copies at a price of $15 and that it will be able to sell 10 more copies for each 50-cent reduction in the price. (See Problem 7 on page 9.)
   (a) Express the store's total profit on this book as a function of the selling price x and sketch this profit function.

(b) Give an economic explanation for the x intercepts (3, 0) and (25, 0).
(c) Use your graph to estimate the price at which the booktsore should sell the book to maximize its profit. (In Problem 8 on page 151 you will compute this optimal price using differential calculus.)

6. A manufacturer is currently selling 2,000 lamps a month to retailers at a price of $2 per lamp. He estimates that for each 1-cent increase in the price he will sell 10 fewer lamps each month. The manufacturer's costs consist of a fixed overhead of $500 per month plus 40 cents a lamp for labor and materials. (See Problem 8 on page 9.)
   (a) Let x denote the price (in dollars) at which the manufacturer will sell the lamps, and express his monthly profit as a function of x.
   (b) Add 500 (representing the fixed overhead) to the profit function in part (a) and sketch the resulting function. Then, perform the appropriate translation to obtain a sketch of the profit function itself.
   (c) Use one of the graphs in part (b) to estimate the price that should be charged to maximize the profit. Does it matter which graph you use? Explain. (In Problem 10 on page 111 you will use differential calculus to compute the optimal price exactly.)

*7. Return to Problem 6. This time, let x denote the amount (in dollars) by which the price *exceeds* $2 (rather than the price itself).
   (a) Express the manufacturer's monthly profit as a function of x and sketch this function.
   (b) How is this graph related to the graph obtained in Problem 6? Explain this relationship in terms of the algebraic characterization of horizontal translation on page 16.

8. An open box is to be made from an 18- by 18-inch square piece of cardboard by removing a small square from each corner and folding up the flaps to form the sides. (See Problem 15 on page 11.)
   (a) Express the volume of the resulting box as a function of the length x of a side of the removed squares. Sketch this volume function.
   (b) For what values of x does this function represent the volume of the box?
   (c) Interpret the x intercepts in terms of the original practical problem.
   (d) Use your graph to estimate the dimensions of the box that has maximum volume. (In Problem 12 on page 151 you will compute the exact dimensions using differential calculus.)

9. (a) Sketch the polynomials $f(x) = (x + 3)(x - 1)^3$, $f(x) = (x + 3)(x - 1)^4$, $f(x) = (x - 3)(x - 1)^3$, and $f(x) = (x - 3)(x - 1)^4$. Observe the behavior of each graph near the intercept (1, 0).
   (b) A root c of a polynomial f is said to be a *root of order k* if k is the largest integer for which $(x - c)^k$ is a factor of $f(x)$. On the basis of your observations in part (a), what conclusions can you draw about the behavior of

the graph of a polynomial near an x intercept arising from a root of *even* order? What about an x intercept arising from a root of *odd* order?
*(c) Give a mathematical justification for your conclusions in part (b).

6. RATIONAL FUNCTIONS

A *rational function* is a quotient of two polynomials. For example, the functions $f(x) = (x^2 + 1)/(x - 2)$ and $f(x) = 1/x$ are both rational. Polynomials themselves may be thought of as rational functions having denominators equal to 1. We have already encountered rational functions arising from practical problems. For instance, the fencing function $f(x) = x + 6{,}400/x$ in Example 1.7 can be rewritten as $f(x) = (x^2 + 6{,}400)/x$ and hence is rational.

In general, rational functions are harder to graph than polynomials. There are two reasons for this. First, rational functions need not be continuous. In fact, as we have seen, a rational function will have a discontinuity whenever its denominator is zero. A technique for analyzing the behavior of rational functions near their discontinuities was developed in Section 4. (To refresh your memory, take another look at Example 4.4 in which this technique was illustrated.)

The second reason that rational functions are harder to graph than polynomials is that there is a greater variety of behavior possible as x approaches plus or minus infinity. Like polynomials, some rational functions increase or decrease without bound as x approaches infinity. There are others, however, that "level off." Consider the simple function $f(x) = 1/x$ sketched in Figure 45.

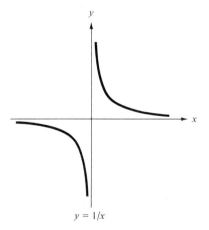

$y = 1/x$

Figure 45

As x approaches plus infinity, $1/x$ decreases and approaches zero. That is, the graph of f eventually approaches the horizontal line $y = 0$ as x increases without bound. This situation may be described compactly using limit notation as follows:

$$\lim_{x \to +\infty} f(x) = 0$$

Similarly, $1/x$ approaches zero as x decreases without bound, and we may write

$$\lim_{x \to -\infty} f(x) = 0$$

Analogous limit notation can be used to describe more general situations.

LIMITS

If $f(x)$ gets closer and closer to some number b as x increases without bound, we say that b is the *limit of $f(x)$ as x approaches plus infinity* and we write

$$\lim_{x \to +\infty} f(x) = b$$

If $f(x)$ gets closer and closer to some number c as x decreases without bound, we say that c is the *limit of $f(x)$ as x approaches minus infinity* and we write

$$\lim_{x \to -\infty} f(x) = c$$

These limit statements are illustrated in Figure 46. Note that the graph of f may cross the horizontal line that it eventually approaches.

The evaluation of this type of limit is particularly easy if the rational function is of the form $f(x) = a/x^n$, where a is a constant and n is a positive integer. For example, if $f(x) = 5/x^3$ and x increases without bound, then the denominator x^3 also increases without bound while the numerator remains fixed. Hence $f(x)$ approaches zero. Similarly, if x decreases without bound, the corresponding values of the function $f(x) = 5/x^3$ again approach zero. This argument can be applied without change to any rational function of this special form. Hence, we may conclude that if $f(x) = a/x^n$, where a is a constant and n is a positive integer, then $\lim_{x \to +\infty} f(x) = 0$ and $\lim_{x \to -\infty} f(x) = 0$.

For more complicated rational functions, the values of these limits may be less obvious. Try, for example, to evaluate $\lim_{x \to +\infty} f(x)$ if $f(x) = (53x^2 + 71x + 25)/(x^3 - 75x^2 - 56)$. This problem seems difficult because both the numerator and the denominator approach infinity as x does.

Fortunately, there is a simple trick by means of which we can reduce complicated problems like this to simple ones involving functions of the form a/x^n.

FUNCTIONS AND GRAPHS

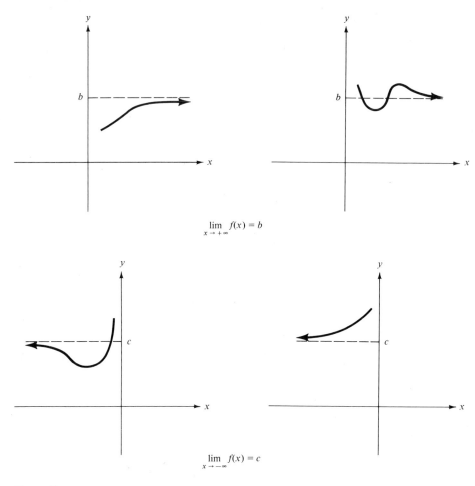

Figure 46

COMPUTATION OF $\lim_{x \to \infty} f(x)$ **FOR RATIONAL FUNCTIONS**

Compare the degrees of the numerator and denominator of the rational function f, and let k be the smaller of these two numbers. Divide both numerator and denominator of f by x^k and then examine the behavior of the new numerator and denominator as x approaches $+\infty$ and $-\infty$.

Let us apply this technique to some concrete examples.

Example 6.1

Find $\lim_{x \to +\infty} f(x)$ and $\lim_{x \to -\infty} f(x)$ if $f(x) = (53x^2 + 71x + 25)/(x^3 - 75x^2 - 56)$.

Solution

Since the degree of the numerator is 2 and the degree of the denominator is 3, we divide numerator and denominator by x^2 to get

$$f(x) = \frac{53 + 71/x + 25/x^2}{x - 75 - 56/x^2}$$

As x approaches $+\infty$, the terms $71/x$ and $25/x^2$ in the numerator approach zero, while 53 remains unchanged. Hence the numerator approaches 53. In the denominator, $56/x^2$ approaches zero, 75 remains unchanged, and x approaches $+\infty$. Hence the denominator approaches $+\infty$.

Since the numerator approaches 53 and the denominator approaches $+\infty$, the quotient $f(x)$ must approach zero. That is, $\lim_{x \to +\infty} f(x) = 0$.

Similarly, if x approaches $-\infty$, the numerator approaches 53, while the denominator approaches $-\infty$, and again the quotient approaches zero. That is, $\lim_{x \to -\infty} f(x) = 0$. □

Think of applying the reasoning used in Example 6.1 to an arbitrary rational function whose numerator is of smaller degree than its denominator. What can you say in general about the limit of such a rational function as x approaches $+\infty$ and $-\infty$? (For the answer, see Problem 15a.)

In the next example the degree of the numerator is greater than the degree of the denominator.

Example 6.2

Find $\lim_{x \to +\infty} f(x)$ and $\lim_{x \to -\infty} f(x)$ if $f(x) = (x^5 - 2x)/(3x^3 + 1)$.

Solution

We divide numerator and denominator by x^3 to get

$$f(x) = \frac{x^2 - 2/x^2}{3 + 1/x^3}$$

As x approaches $+\infty$, the numerator also approaches $+\infty$, while the denominator approaches 3. Hence, $\lim_{x \to +\infty} f(x) = +\infty$. As x approaches $-\infty$, the numerator approaches $+\infty$, while the denominator again approaches 3. Hence, $\lim_{x \to -\infty} f(x) = +\infty$. □

Do you see how to generalize Example 6.2 to an arbitrary rational function whose numerator is of larger degree than its denominator? (The appropriate generalization is given in Problem 15b.)

Finally, let us consider an example of a rational function whose numerator and denominator have the same degree. As you read the solution, think about the general case (Problem 15c).

Example 6.3

Find $\lim_{x \to +\infty} f(x)$ and $\lim_{x \to -\infty} f(x)$ if $f(x) = (3x^5 + x^2 - 2x - 4)/(2x^5 + 5x - 1)$.

Solution

In this case we divide numerator and denominator by x^5 to get

$$f(x) = \frac{3 + 1/x^3 - 2/x^4 - 4/x^5}{2 + 5/x^4 - 1/x^5}$$

As x approaches either $+\infty$ or $-\infty$, the terms $1/x^3$, $2/x^4$, $5/x^4$, $4/x^5$, and $1/x^5$ all approach zero, so $\lim_{x \to +\infty} f(x) = \frac{3}{2}$, and $\lim_{x \to -\infty} f(x) = \frac{3}{2}$. □

We can now describe an efficient procedure for sketching rational functions.

SKETCHING RATIONAL FUNCTIONS

Step 1. Factor the numerator and denominator (if possible).
Step 2. Find the x intercepts. (These are the values of x for which the numerator is zero.)
Step 3. Find all values of x for which f has a discontinuity. (These are the values of x for which the denominator is zero.) For each such value $x = a$, compute $\lim_{x \to a^+} f(x)$ and $\lim_{x \to a^-} f(x)$.
Step 4. Compute $\lim_{x \to +\infty} f(x)$ and $\lim_{x \to -\infty} f(x)$.
Step 5. Sketch the graph using these clues.

The next two examples illustrate the ease with which rational functions can be graphed using this procedure.

Example 6.4

Sketch the rational function $f(x) = (x^2 - 2x)/(x + 1)$.

Solution

The function can be factored as

$$f(x) = \frac{x(x-2)}{x+1}$$

The numerator is zero when $x = 0$ and when $x = 2$. Hence the x intercepts are $(0, 0)$ and $(2, 0)$. Discontinuities occur when the denominator is zero; in this case, when $x = -1$. We begin the sketch by plotting the x intercepts and indicating the presence of the discontinuity by a vertical dotted line (Figure 47).

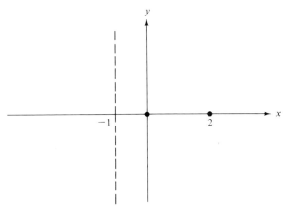

Figure 47

Next we determine the behavior of the graph on either side of the discontinuity. Since $\lim_{x \to -1^+} f(x) = +\infty$ and $\lim_{x \to -1^-} f(x) = -\infty$, we can continue the sketch as shown in Figure 48.

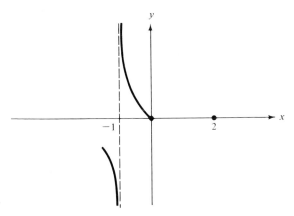

Figure 48

Between the intercepts (0, 0) and (2, 0), the function has no discontinuities, and so the graph must lie either entirely above the x axis or entirely below it. Since $f(1) = -\frac{1}{2}$, we conclude that the graph lies below the x axis when $0 < x < 2$. This is shown in Figure 49.

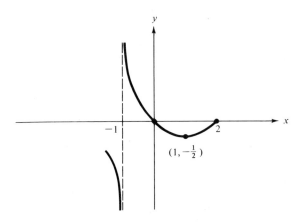

Figure 49

Finally, since $\lim\limits_{x \to +\infty} f(x) = +\infty$ and $\lim\limits_{x \to -\infty} f(x) = -\infty$, we may complete the sketch as shown in Figure 50.

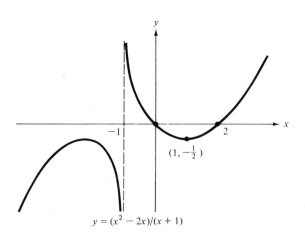

$y = (x^2 - 2x)/(x + 1)$

Figure 50

Example 6.5

Sketch the rational function $f(x) = (3x^2 - 9x)/(x^2 - 4)$.

Solution

After factorization we get

$$f(x) = \frac{3x(x-3)}{(x-2)(x+2)}$$

The x intercepts are $(0, 0)$ and $(3, 0)$. This time there are two discontinuities: when $x = 2$ and $x = -2$. We compute limits for each:

$$\lim_{x \to 2^+} f(x) = -\infty \qquad \lim_{x \to 2^-} f(x) = +\infty$$

and

$$\lim_{x \to -2^+} f(x) = -\infty \qquad \lim_{x \to -2^-} f(x) = +\infty$$

Finally,

$$\lim_{x \to +\infty} f(x) = 3 \quad \text{and} \quad \lim_{x \to -\infty} f(x) = 3$$

These observations lead to the sketch in Figure 51.

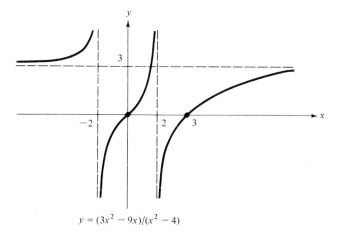

$y = (3x^2 - 9x)/(x^2 - 4)$

Figure 51

In our final example, we return to the practical problem we examined in Example 1.7.

Example 6.6

A rectangular lot is to be fenced off on three sides. The area of the lot is to be 3,200 square yards. Express the length of the fencing as a function of the length x of the unfenced side and sketch a graph.

Solution

We found in Example 1.7 that the amount of fencing required is given by the function $f(x) = x + 6{,}400/x$ for $x > 0$. This is a rational function, since it can be rewritten as

$$f(x) = \frac{x^2 + 6{,}400}{x}$$

Since $x^2 + 6{,}400$ is always positive, f has no x intercepts. There is, however, a discontinuity when $x = 0$, and the corresponding limits are

$$\lim_{x \to 0^+} f(x) = +\infty \quad \text{and} \quad \lim_{x \to 0^-} f(x) = -\infty$$

Moreover,

$$\lim_{x \to +\infty} f(x) = +\infty \quad \text{and} \quad \lim_{x \to -\infty} f(x) = -\infty$$

The corresponding graph is sketched in Figure 52. Of course, only when x is positive does this graph represent the amount of required fencing.

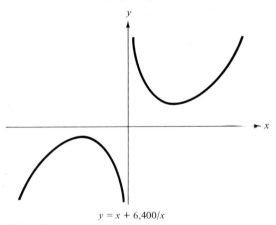

$y = x + 6{,}400/x$

Figure 52

Many of the geometric features of this graph reflect aspects of the original fencing problem. For example, if x is small, the corresponding lot would have to be quite long to have an area of 3,200 square yards (Figure 53a), and a considerable

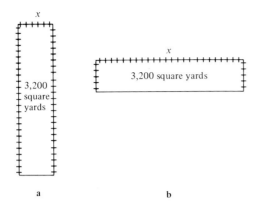

Figure 53

amount of fencing would be required. This is reflected by the fact that $\lim_{x \to 0^+} f(x) = +\infty$. Similarly, if x is very large (Figure 53b), the amount of fencing required is again large. In fact, $\lim_{x \to +\infty} f(x) = +\infty$.

It is also clear from the graph that there is some length x for which $f(x)$, the total length of the fencing, is a minimum. This optimal value of x will be computed in Chapter 2, using the tools of differential calculus. □

Problems

1. For each of the following rational functions, locate the discontinuities and use limit notation to describe the behavior of the function near each discontinuity.
   (a) $f(x) = 1/(x - 1)^2$
   (b) $f(x) = 3/(x + 2)^3$
   (c) $f(x) = (x - 2)/(x + 1)$
   (d) $f(x) = x/[(x - 2)(x + 1)]$
   (e) $f(x) = (x^2 + x - 2)/[x(x + 3)^2]$
   (f) $f(x) = (x^2 + 1)/(x^2 - 1)$
   (g) $f(x) = x + 1/x$
   (h) $f(x) = x - 1/x^2$

2. For each of the following rational functions, find $\lim_{x \to +\infty} f(x)$ and $\lim_{x \to -\infty} f(x)$.
   (a) $f(x) = 1/x^2$
   (b) $f(x) = (x + 1)/(2 - x)$
   (c) $f(x) = x^4/(x^2 - 1)$
   (d) $f(x) = x^5/(x^2 - 1)$
   (e) $f(x) = (9x + 5)/(x^2 - 12)$
   (f) $f(x) = (5 - x^3)/(3x^3 + 2x + 1)$
   (g) $f(x) = (x^4 - 9x^3)/(12x^3 + 5)$
   (h) $f(x) = (x + 1)(x - 2)/(x + 3)$
   (i) $f(x) = (x + 1)(x - 2)/(2x^2 - 9)$
   (j) $f(x) = (x - x^4)/(1 - 3x + x^4)$
   (k) $f(x) = 5x^3 - 2x + 1$
   (l) $f(x) = x + 1/x$
   (m) $f(x) = 1 - 2/x^3$
   (n) $f(x) = x^2 - 1/x$

3. For each of the functions f sketched below, determine $\lim_{x \to +\infty} f(x)$ and $\lim_{x \to -\infty} f(x)$, locate the discontinuities, and use limit notation to describe the behavior of f near each discontinuity.

(a)

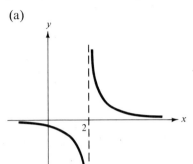

(b)

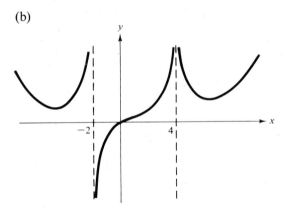

(c)

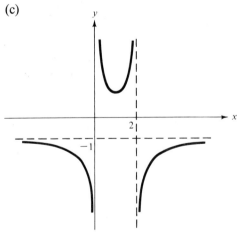

4. Sketch the graph of a function f that has all of the following properties:
   (a) The x intercepts of f are $(-1, 0)$ and $(1, 0)$.
   (b) The discontinuities of f occur when $x = -3$ and $x = 2$.
   (c) $\lim_{x \to -3^+} f(x) = +\infty$ $\lim_{x \to -3^-} f(x) = -\infty$
       $\lim_{x \to 2^+} f(x) = +\infty$ $\lim_{x \to 2^-} f(x) = +\infty$
   (d) $\lim_{x \to +\infty} f(x) = 0$ $\lim_{x \to -\infty} f(x) = 0$

5. Sketch the graph of a function f that has all of the following properties:
   (a) The only x intercept of f is $(0, 0)$.
   (b) The discontinuities of f occur when $x = 1$ and $x = -1$.
   (c) $\lim_{x \to 1^+} f(x) = -\infty$ $\lim_{x \to 1^-} f(x) = +\infty$
       $\lim_{x \to -1^+} f(x) = -\infty$ $\lim_{x \to -1^-} f(x) = +\infty$
   (d) $\lim_{x \to +\infty} f(x) = -\infty$ $\lim_{x \to -\infty} f(x) = +\infty$

6. Sketch the graph of a function f that has all of the following properties:
   (a) The x intercepts of f are $(-1, 0)$ and $(2, 0)$.
   (b) The discontinuities of f occur when $x = 0$ and $x = -2$.
   (c) $\lim_{x \to 0^+} f(x) = -\infty$ $\lim_{x \to 0^-} f(x) = -\infty$
       $\lim_{x \to -2^+} f(x) = +\infty$ $\lim_{x \to -2^-} f(x) = +\infty$
   (d) $\lim_{x \to +\infty} f(x) = 3$ $\lim_{x \to -\infty} f(x) = 3$

7. Sketch the following rational functions.
   (a) $f(x) = 1/x^3$
   (b) $f(x) = 1/(x-2)^2$
   (c) $f(x) = (1-x)/(x+1)$
   (d) $f(x) = (3x+6)/(x-3)$
   (e) $f(x) = (x-2)/(x+3)^2$
   (f) $f(x) = (x-2)/[(x+3)(x-1)]$
   (g) $f(x) = (x+2)/[(x+3)(x-1)]$
   (h) $f(x) = (x+5)/[(x+3)(x-1)]$
   (i) $f(x) = (x+1)(x-2)/[(x+2)(x-3)]$
   (j) $f(x) = x^2(x-4)/[(x-3)(x+2)]$
   (k) $f(x) = (5x^2 - 20x)/[(x-3)(x+2)]$
   (l) $f(x) = (x^3 - 9x)/(x^2 + x - 2)$
   (m) $f(x) = (x^2 + 2x - 3)/(x^2 - 4x + 4)$
   (n) $f(x) = (x^2 - 2x + 1)/(x^2 - 4x + 4)$
   (o) $f(x) = (3x + 5)/(x^2 - 4)$
   (p) $f(x) = 1 + 1/x$
   (q) $f(x) = x - 1/x$
   (r) $f(x) = 1 - 4/x^2$

(s) $f(x) = x + 27/x^2$
(t) $f(x) = 1 + x + 1/x$

8. A city recreation department is planning to build a rectangular playground having an area of 3,600 square yards. The playground will be surrounded by a fence costing $3 per yard.
    (a) Express the cost of enclosing the playground as a function of the length of one of its sides, and sketch this cost function.
    (b) What happens to the cost as the length of the side gets very small? Why? How is this reflected in the graph?
    (c) What happens to the cost as the length of the side gets very large? Why? How is this reflected in the graph?
    (d) Use your graph to estimate the dimensions of the playground for which the total cost of the fencing will be least. (In Problem 5 on page 151 you will compute these optimal dimensions using differential calculus.)

9. A closed box with a square base is to have a volume of 250 cubic feet. The material for the top and bottom of the box costs $2 per square foot, and the material for the sides costs $1 per square foot. (See Problem 16 on page 11.)
    (a) Express the cost of constructing the box as a function of the length of its base, and sketch a graph of this function.
    (b) What happens to the construction cost as the length of the base gets very small? Why? How is this reflected in the graph?
    (c) What happens to the construction cost as the length of the base gets very large? Why? How is this reflected in the graph?
    (d) Use your graph to estimate the dimensions of the box whose construction cost is minimal. (In Problem 10 on page 151 you will compute these optimal dimensions using differential calculus.)

*10. A plastics firm has received an order from the city recreation department to manufacture 8,000 special Styrofoam kickboards for its summer swimming program. The firm owns several machines, each of which can produce 30 kickboards an hour. The cost of setting up the machines to produce these particular kickboards is $20 per machine. Once the machines have been set up, the operation is fully automated and can be supervised by a single foreman earning $4.80 per hour. (See Problem 17 on page 11.)
    (a) Express the cost of producing the 8,000 kickboards as a function of the number of machines used, and sketch a graph of this function.
    (b) For what values of the independent variable does your function have a practical interpretation?
    (c) Use your graph to estimate the number of machines that should be used to minimize the cost. (In Problem 21 on page 153 you will compute this optimal number using differential calculus.)

*11. Find explicit formulas for rational functions whose graphs have the same properties as those in Problem 3.

*12. Find an explicit formula for a rational function that has the same properties as the one in Problem 4.

*13. Find an explicit formula for a rational function that has the same properties as the one in Problem 5.

*14. Find an explicit formula for a rational function that has the same properties as the one in Problem 6.

*15. Consider the general rational function

$$f(x) = \frac{a_0 + a_1 x + \cdots + a_n x^n}{b_0 + b_1 x + \cdots + b_m x^m}$$

where neither a_n nor b_m is zero.
  (a) If $n < m$, show that $\lim_{x \to +\infty} f(x) = 0$ and $\lim_{x \to -\infty} f(x) = 0$.
  (b) If $n > m$, show that $\lim_{x \to +\infty} f(x) = +\infty$ or $-\infty$ and $\lim_{x \to -\infty} f(x) = +\infty$ or $-\infty$. Give examples of actual functions to illustrate that all four possible combinations can occur.
  (c) If $n = m$, show that $\lim_{x \to +\infty} f(x) = a_n/b_m$ and $\lim_{x \to -\infty} f(x) = a_n/b_m$.

*16. Suppose f is a rational function for which $\lim_{x \to +\infty} f(x) = a$ (where a is finite). Must it also be true that $\lim_{x \to -\infty} f(x) = a$? Explain.

17. (a) Sketch the rational functions $f(x) = (x + 3)/(x - 2)^2$, $f(x) = (x + 3)/(x - 2)^3$, $f(x) = (x - 3)/(x - 2)^2$, and $f(x) = (x - 3)/(x - 2)^3$. Observe the behavior of each graph near the discontinuity at $x = 2$.
  (b) A rational function f is said to have a *discontinuity of order k* at $x = c$ if k is the largest integer for which $(x - c)^k$ is a factor of the denominator of f (and $x - c$ is not a factor of the numerator). On the basis of your observations in part (a), what conclusions would you draw about the behavior of the graph of a rational function near a discontinuity of *even* order? What about a discontinuity of *odd* order?
  (c) Compare your answers in part (b) to your answers in Problem 9b on page 59.
  *(d) Give a mathematical justification for your conclusions in part (b).

7. INTERSECTIONS OF GRAPHS

In many applications it is important to determine when two functions have the same value. For example, economists studying price trends of a particular commodity are often interested in finding the market price at which the supply of the commodity and the consumer demand for it are equal.

FUNCTIONS AND GRAPHS

In geometric terms, two functions $f(x)$ and $g(x)$ will be equal for those values of x for which the graphs of f and g intersect (Figure 54). To find these values of x algebraically, we must solve the equations $y = f(x)$ and $y = g(x)$ simultaneously. Consider the following examples.

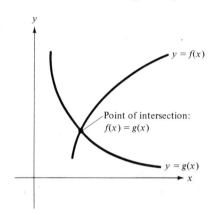

Figure 54

Example 7.1

Find the points of intersection of the graphs of the functions $f(x) = 2x$ and $g(x) = x^2$.

Solution

We equate the expressions for $f(x)$ and $g(x)$ and solve the resulting equation for x. In particular, we get

$$2x = x^2 \quad \text{or} \quad x(x - 2) = 0$$

from which we conclude that the graphs intersect when $x = 0$ and $x = 2$.

To compute the y coordinates of the two points of intersection, we may substitute these values of x into either function, f or g. Using $y = f(x)$ we find that $y = 0$ when $x = 0$ and $y = 4$ when $x = 2$. Hence, the points of intersection are $(0, 0)$ and $(2, 4)$.

As a final check, we could substitute the two values of x into the other equation, $y = g(x)$, to verify that we get the same values for y.

The two graphs are drawn in Figure 55. □

Example 7.2

Where do the lines $2y + 5x = 2$ and $6y - x = 12$ intersect?

7. INTERSECTIONS OF GRAPHS

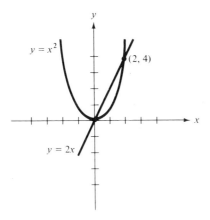

Figure 55

Solution

We could solve each of the equations for y, equate the resulting expressions, and then solve for x. However, there is a more efficient approach. We multiply both sides of the first equation by 3 to get

$6y + 15x = 6$

from which we subtract the second equation

$6y - x = 12$

This gives

$16x = -6$

or

$x = -\frac{3}{8}$

The corresponding value of y can be computed by solving either equation for y and then substituting $x = -\frac{3}{8}$. We get $y = \frac{31}{16}$ and conclude that $(-\frac{3}{8}, \frac{31}{16})$ is the point of intersection (Figure 56). □

An important economic application of these techniques arises in connection with the *law of supply and demand*. Suppose that p denotes the market price of a certain commodity, and that $S(p)$ is the number of units of this commodity that manufacturers are willing to produce when the market price is p dollars. In most cases, this *supply function* $S(p)$ increases as the price p increases. Let $D(p)$ be the corresponding *demand function*. This means that $D(p)$ is the number of units of the commodity that consumers are willing to buy when the market price is p dollars. In general, $D(p)$ decreases as p increases. A typical pair of supply and demand functions

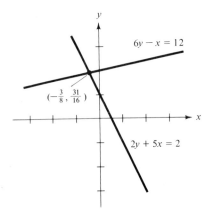

Figure 56

is sketched in Figure 57. (The letter q used to label the vertical axis stands for *quantity*.)

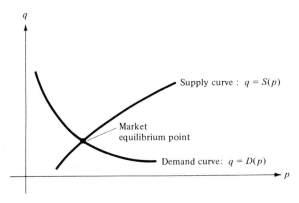

Figure 57

Actually, an economist's graph of these functions would not look quite like Figure 57. When dealing with supply and demand curves, economists usually depart from mathematical tradition and use the horizontal axis to represent the dependent variable q, and the vertical axis for the independent variable p.

The point of intersection of the supply and demand curves is called the point of *market equilibrium*. The p coordinate of this point (the *equilibrium price*) is the market price at which supply equals demand. That is, the price at which there will be neither a surplus nor a shortage of the commodity.

The law of supply and demand asserts that, in a situation of pure competition, a commodity will tend to be sold at its equilibrium price. The theory is that, if the commodity were sold for more than the equilibrium price, there would be an unsold surplus on the market and retailers would tend to lower their prices. On the other hand, if the commodity were sold for less than the equilibrium price, the demand would exceed the supply and retailers would be inclined to raise their prices.

Example 7.3

Find the equilibrium price and the corresponding number of units supplied and demanded if the supply function for a certain commodity is $S(p) = p^2 + 3p - 70$ and the demand function is $D(p) = 410 - p$.

Solution

We equate $S(p)$ and $D(p)$ and solve for p:

$$p^2 + 3p - 70 = 410 - p$$
$$p^2 + 4p - 480 = 0$$
$$(p - 20)(p + 24) = 0$$
$$p = 20 \text{ and } p = -24$$

Since only positive values of p are meaningful in this practical problem, we conclude that the equilibrium price is $20. The corresponding supply and demand are equal, and we use the simpler demand equation to compute this amount:

$$D(20) = 410 - 20 = 390$$

Thus 390 units are supplied (and demanded) when the market is in equilibrium.

The supply and demand curves are sketched in Figure 58. Notice that the supply curve crosses the p axis when $p = 7$. (Verify this.) What is the economic interpretation of this fact?

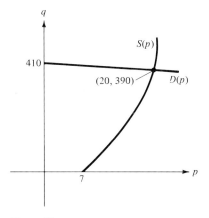

Figure 58

Many of the elementary practical problems that involve simultaneous equations can be rephrased and solved without any explicit reference to intersection or simultaneous equations. The two points of view are illustrated in the following example.

Example 7.4

Two jets bound for Los Angeles leave New York 30 minutes apart. The first travels 550 miles per hour, while the second goes 650 miles per hour. At what time will the second plane pass the first?

Solution 1

Our strategy will be to represent the position of each plane as a function of time, and then to determine the time at which the two positions are the same. That is, we will find the intersection of the graphs of the two position functions.

Let x denote the number of hours since the second plane left New York, and y the distance from New York after x hours.

The second plane travels for x hours at 650 miles per hour, and so its distance from New York is simply $y = 650x$. The first plane, having left New York $\frac{1}{2}$ hour earlier, has been in the air for $x + \frac{1}{2}$ hours, traveling at 550 miles per hour. Its distance from New York is therefore $y = 550(x + \frac{1}{2})$, or $y = 550x + 275$.

To determine the time at which the planes meet, we equate the two expressions for y and solve for x.

$650x = 550x + 275$

$100x = 275$

or

$x = \frac{11}{4}$

That is, 2 hours and 45 minutes after it leaves New York, the second plane catches up with the first.

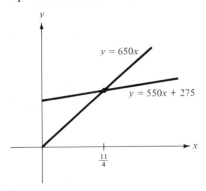

Figure 59

Figure 59 shows the graphs of the two position functions that intersect when $x = \frac{11}{4}$.

Solution 2

An equivalent strategy is to represent the distance *between* the planes as a function of time, and then to determine the time at which this distance function is zero.

As before, we let x denote the number of hours since the second plane left New York, but this time we let y be the corresponding distance *between* the planes. When $x = 0$, the first plane has already gone 275 miles. After that, the second plane begins to catch up, reducing the distance by 100 miles each hour. Hence the distance between the planes is given by the linear function

$y = 275 - 100x$

Setting y equal to zero we get

$100x = 275$

or

$x = \frac{11}{4}$

That is, after 2 hours and 45 minutes, the distance between the planes is zero. Figure 60 shows the linear function representing the distance between the planes.

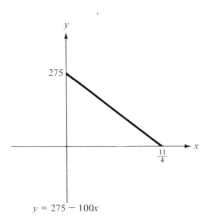

$y = 275 - 100x$

Figure 60

Notice that although the notion of intersection was not explicitly used in the second solution, we were eventually led to the same equation, $100x = 275$, that resulted from the simultaneous equations in the first solution.

Problems

1. Find the points of intersection (if any) of each of the following pairs of curves and sketch the curves.
   (a) $y = 3x + 5$ and $y = -x + 3$
   (b) $y = x^2$ and $y = 6 - x$
   (c) $y = x^2 - x$ and $y = x - 1$
   (d) $y = x^3 - 6x^2$ and $y = -x^2$
   (e) $y = x^3$ and $y = x^2$
   (f) $y = x^3$ and $y = -x^3$
   (g) $y = x^2 + 2$ and $y = x$
   (h) $y = 1/x$ and $y = x^2$
   (i) $y = 1/x^2$ and $y = 2$
   (j) $y = 1/x$ and $y = 1/x^2$
   (k) $y = 1/x^2$ and $y = -x^2$

2. Find the points of intersection (if any) of each of the following pairs of lines and graph the lines.
   (a) $y = 5x - 2$ and $y = -2x + 5$
   (b) $y - 3x + 10 = 0$ and $y - x + 2 = 0$
   (c) $2y - x - 6 = 0$ and $y + 5x - 3 = 0$
   (d) $3y + x = 0$ and $y - 3x = 0$
   (e) $y - 3x + 10 = 0$ and $2y - 6x + 5 = 0$
   (f) $y - 3x + 10 = 0$ and $2y - 6x + 20 = 0$
   (g) $2x + 3y + 4 = 0$ and $3x - 2y - 4 = 0$

3. Find the point (if any) that lies on all three of the lines $y = x - 1$, $y = 3x - 5$, and $3y - x - 2 = 0$.

4. Find the point (if any) that lies on all three of the lines $y = x - 1$, $y = 3x - 5$, and $3y - x - 1 = 0$.

5. A furniture manufacturer can sell dining-room tables for $70 apiece. His costs include a fixed overhead of $8,000 plus a production cost of $30 per table.
   (a) Express the manufacturer's revenue as a function of the number of tables he sells.
   (b) Express the manufacturer's cost as a function of the number of tables he sells.
   (c) Graph the revenue and cost functions and find the point of intersection of the two graphs. What is the economic significance of this point?

6. Suppose the furniture manufacturer in Problem 5 can sell his tables for $80 apiece. How many tables must he sell before he breaks even?

7. During the summer, a group of students builds kayaks in a converted garage. The rental for the garage is $600 for the summer, and the materials needed

to build a kayak cost $25. How many kayaks must the students sell at $175 apiece to break even?

8. The hero of a popular spy story has escaped from the headquarters of an international diamond smuggling ring in the tiny Mediterranean country of Azusa. Our hero, driving a stolen milk truck at 46 miles per hour, has a 40-minute headstart on his pursuers who are chasing him in a Ferrari going 96 miles per hour. The distance from the smugglers' headquarters to the border, and freedom, is 58.8 miles. Will he make it?

9. When electric blenders are sold for p dollars apiece, manufacturers are willing to supply $p^2/10$ blenders to local retailers. On the other hand, the local consumers will buy a total of $60 - p$ blenders at this price. At what market price will the manufacturers' supply of electric blenders be equal to the consumers' demand for the blenders? How many blenders will be sold at this price?

10. The supply and demand functions for a certain commodity are $S(p) = p - 10$ and $D(p) = 5{,}600/p$, respectively.
    (a) Find the equilibrium price and the corresponding number of units supplied and demanded.
    (b) Draw the supply and demand curves on the same graph.
    (c) Where does the supply function cross the p axis? What is the economic significance of this point?
    (d) Compute $\lim_{p \to 0^+} D(p)$. Interpret your answer in economic terms.

*11. Find an expression for the equilibrium price if the supply function for a certain commodity is $S(p) = p^2 + ap + c$ (where $a \geq 0$ and $c \leq 0$) and the demand function is $D(p) = mp + b$ (where $m < 0$ and $b > 0$).

*12. Suppose that the supply and demand functions for a certain commodity are $S(p) = ap + b$ and $D(p) = cp + d$, respectively.
    (a) What can you say about the signs of the coefficients a, b, c, and d if the supply and demand curves are oriented as shown in the following diagram?

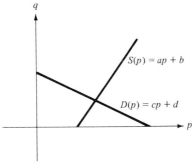

(b) Express the equilibrium price in terms of the coefficients a, b, c, and d.

(c) Use your answer in part (b) to determine what happens to the equilibrium price as a increases. Explain your answer in economic terms.
(d) Use your answer in part (b) to determine what happens to the equilibrium price as d increases. Explain your answer in economic terms.

*13. At exactly the same moment that an American oceanliner is departing from New York bound for Le Havre, a French liner is leaving Le Havre en route to New York. The distance between New York and Le Havre is 3,640 miles, and the Amercan ship, traveling at a constant speed, will cover this distance in 5 days. The French vessel, also sailing at a constant speed, will take 7 days to cross the ocean. How far from New York will the ships be when they pass each other?

8. REVIEW PROBLEMS

1. For each of the following functions, specify the domain and compute the values indicated.
   (a) $f(x) = x^2 - 2x + 5$; $f(-2), f(0), f(2)$
   (b) $g(t) = (t-1)/(t^2 - 4)$; $g(0), g(1), g(3)$
   (c) $h(x) = \sqrt{(x-1)(x+2)}$; $h(-2), h(2), h(3)$
   (d) $f(t) = \begin{cases} t & \text{if } 0 \le t \le 4 \\ 5 & \text{if } 4 < t \le 5 \\ 1-t & \text{if } t > 5 \end{cases}$ $[f(4), f(5), f(6)]$

2. For each of the following functions $f(x)$, find functions $h(x)$ and $g(u)$ such that $f(x) = g(h(x))$.
   (a) $f(x) = (x^3 - 2x^2 + 5)^6$
   (b) $f(x) = \sqrt{x^4 + 2x^2 + 1}$
   (c) $f(x) = 3(x+1)^5 - 2(x+1)^3 - 7(x+1) + 2$

3. Find the composite function $g(h(x))$ and specify its domain.
   (a) $g(u) = u^2$, $h(x) = 2x + 1$
   (b) $g(u) = 1/u$, $h(x) = x - 1$
   (c) $g(u) = \sqrt{u}$, $h(x) = 3 - x$
   (d) $g(u) = u^2 + 2u + 1$, $h(x) = x + 1$

4. Graph the function $f(x) = 1/x^3$. Then, without additional computation, graph
   (a) $f(x) = -3 + 1/x^3$
   (b) $f(x) = 1/(x-2)^3$
   (c) $f(x) = 3 - 1/(x+2)^3$

5. Find the number c for which the curve $y = 3x^2 - 2x + c$ passes through the point $(2, 4)$.

6. Find the slope and y intercept of each of the following lines and draw a graph.
   (a) $y = 2x + 1$
   (b) $3x + 2y = 6$
   (c) $x/2 + y/5 = 1$

7. Write equations for the lines with the following properties.
   (a) Through $(1, -2)$ with slope -3
   (b) Through $(6, -3)$ and $(-1, 2)$
   (c) Through $(2, 1)$ and parallel to the line $3y + 6x = 5$

8. Graph the function
$$f(x) = \begin{cases} 1/x^2 & \text{if } x < 0 \\ x & \text{if } 0 \leq x \leq 4 \\ 5 & \text{if } x > 4 \end{cases}$$
   Locate the discontinuities of f and use limit notation to describe the behavior of f near these discontinuities.

9. Locate the discontinuities of the following functions and use limit notation to describe the behavior of the functions near each of their discontinuities.
   (a) $f(x) = -1/(x - 2)^2$
   (b) $f(x) = (x - 1)(x + 2)/[(x - 3)(x + 1)]$
   (c) $f(x) = (\sqrt{x - 3})/(x - 9)$
   (d) $f(x) = (x^2 - x)/(x^2 - 3x + 2)$

10. For each of the following polynomials, find the x intercepts, compute $\lim_{x \to +\infty} f(x)$ and $\lim_{x \to -\infty} f(x)$, and sketch a graph.
    (a) $f(x) = x(x - 3)(x + 1)$ (b) $f(x) = (x - 1)^2(x + 2)(x + 1)$
    (c) $f(x) = x^4 - x^2$ (d) $f(x) = (x^2 - 4)^3$

11. Use suitable vertical translations in sketching the following polynomials.
    (a) $f(x) = x^2 - 3x + 5$ (b) $f(x) = x^3 - 2x^2 + 4$

12. Sketch the graph of a function f that has all of the following properties.
    (a) The only x intercept of f is $(0, 0)$.
    (b) The discontinuities of f occur when $x = -2$ and $x = 3$.
    (c) $\lim_{x \to -2^+} f(x) = -\infty$ $\lim_{x \to -2^-} f(x) = +\infty$
    $\lim_{x \to 3^+} f(x) = +\infty$ $\lim_{x \to 3^-} f(x) = +\infty$
    (d) $\lim_{x \to +\infty} f(x) = 2$ $\lim_{x \to -\infty} f(x) = 2$

13. For each of the following rational functions, find $\lim_{x \to +\infty} f(x)$ and $\lim_{x \to -\infty} f(x)$.
    (a) $f(x) = (3x^4 + 5x^3 - 7x + 1)/(2x^2 + 6x - 5)$

(b) $f(x) = (2x^2 + 6x - 5)/(3x^4 + 5x^3 - 7x + 1)$
(c) $f(x) = (5x^4 + 6x - 5)/(3x^4 + 5x^3 - 7x + 1)$

14. Sketch the following rational functions.
    (a) $f(x) = (x - 2)(x + 3)/(x^2 - 4x)$
    (b) $f(x) = 2x/[(x - 2)(x + 3)]$
    (c) $f(x) = (x - 1)^2(x + 3)/(x + 1)^2$
    (d) $f(x) = 1 - 1/x^3$

15. Find the points of intersection (if any) of each of the following pairs of curves and sketch the curves.
    (a) $y = 5x - 14$ and $y = 4 - x$
    (b) $3x + y - 4 = 0$ and $x - 2y + 3 = 0$
    (c) $y = x^2 - 1$ and $y = 1 - x^2$
    (d) $y = x^2$ and $y = 15 - 2x$

16. In 1970, the telegraph rate for interstate telegrams was $2.75 for 15 words or less, plus 9 cents for each additional word up to 50 words, and 6 cents a word thereafter.
    (a) Express the cost of sending a telegram as a function of its length. For what values of the independent variable does your function have a practical interpretation?
    (b) Compute the cost of sending a 60-word telegram.
    (c) Graph the function obtained in part (a).
    (d) Your graph should consist of three line segments. Compute the slope of each of these segments.

17. The following table is taken from the 1972 federal income tax rate schedule for single taxpayers.

IF THE TAXABLE INCOME IS:		THE INCOME TAX IS:	
Over ...	but not over ...		of the excess over ...
$8,000	$10,000	$1,590 + 25%	$8,000
$10,000	$12,000	$2,090 + 27%	$10,000
$12,000	$14,000	$2,630 + 29%	$12,000
$14,000	$16,000	$3,210 + 31%	$14,000

    (a) Compute the tax an individual will pay if his taxable income is $11,500.
    (b) Use this table to express an individual's income tax as a function of his taxable income x for $8,000 < x \le 16,000$.
    (c) Graph this income tax function. Does the function have any discontinuities?

(d) Your graph in part (c) should consist of four line segments. Compute the slope of each segment. What happens to these slopes as the taxable income increases?

18. A moving van, carrying the furniture of a recently promoted junior executive, leaves Chicago at noon, bound for the executive's new home in Cleveland, 345 miles away. Forty minutes later, the executive and his family leave Chicago in their station wagon. If the station wagon travels at a constant speed of 60 miles per hour and the truck goes at a constant speed of 50 miles per hour, when will the executive catch up with his furniture?

19. The consumer demand for a certain commodity is $-200p + 12{,}000$ units per month when the market price is p dollars per unit.
    (a) Graph this demand function.
    (b) Where does the graph cross the p axis? What is the economic significance of this point?
    (c) Express the consumers' *total monthly expenditure* for the commodity as a function of p. (The total monthly expenditure is the amount of money consumers will spend purchasing the commodity each month.)
    (d) Graph the total expenditure function.
    (e) Compare the graphs of the demand and expenditure functions. Discuss the economic significance of the p intercepts of the expenditure function.
    (f) Use your graph to estimate the market price that will result in maximal consumer expenditure.

20. An open box with a square base is to be built for $48. The sides of the box will cost $3 per square foot, and the base will cost $4 per square foot.
    (a) Express the volume of the box as a function of the length of its base.
    (b) Graph this volume function.
    (c) Where does your graph cross the positive x axis? Explain the significance of these points in terms of the original practical problem.

21. A manufacturer can produce radios at a cost of $2 apiece. He is currently selling 4,000 radios a month to retailers at a price of $5 per radio, and he estimates that for each 50-cent increase in his price he will sell 200 fewer radios each month.
    (a) Express the manufacturer's monthly profit as a function of the price at which he sells the radios.
    (b) Graph this profit function.
    (c) Use your graph to estimate the price that should be charged to maximize the profit.

22. A rectangular poster contains 25 square inches of print surrounded by margins of 2 inches on each side and 4 inches on the top and bottom.

(a) Express the total area of the poster (printing plus margins) as a function of the width of the printed portion.
(b) Graph this area function.
(c) Compute the limit of your function as the independent variable approaches zero from the right. Interpret your answer in terms of the original practical problem.
(d) Compute the limit of your function as the independent variable approaches plus infinity. Interpret your answer in terms of the original practical problem.

Differentiation

In this chapter, consideration of a simple optimization problem leads us to a related geometric problem; that of computing the slope of a line that is tangent to the graph of a function. The technique we shall develop to solve this geometric problem is called differentiation *and turns out to be a remarkably powerful mathematical tool. Not only does it give us useful information about the shape of graphs, but it can be used to solve a wide range of practical optimization problems and also has important applications to situations involving the rate at which one quantity changes with respect to another.*

1. INTRODUCTION

In many practical situations, we are faced with the problem of finding the maximum or minimum value of a particular function. In Chapter 1, for instance, we encountered such a problem in connection with the profit function in Example 5.6. Recall that a manufacturer's profit f was related to the price x at which he sold his tape recorders by the function $f(x) = (120 - x)(x - 20)$. This function is sketched again in Figure 1.

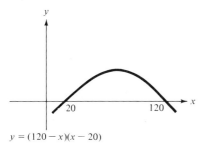

$y = (120 - x)(x - 20)$

Figure 1

The graph suggests that there is an optimal selling price x at which the manufacturer's profit will be maximal. In geometric terms, the optimal price is the x coordinate of the *peak* of the graph. To determine this optimal value, then, we must pinpoint the location of the peak. It looks as if the peak might occur at $x = 70$, midway between the x intercepts. Before we can check this assertion mathematically, we shall need a mathematical description of peaks.

In this relatively simple example, the peak can be completely characterized in terms of lines that are tangent to the graph. In particular, the peak is the only point on the graph at which the tangent line is horizontal; that is, at which the slope of the tangent line is zero. To the left of the peak, the slope of the tangent is positive. To the right of the peak, the slope is negative. But just at the peak itself, the curve levels off and the slope of its tangent line is zero (Figure 2).

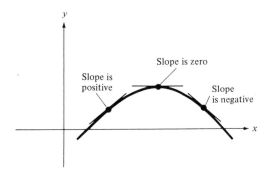

Figure 2

If only we had a device for computing slopes of tangent lines, we could finally solve our optimization problem. Such a device is the *derivative*, which we shall now introduce. Our development of the derivative will not be based on a *formal* definition of the notion of a tangent line. Instead, we shall rely on our intuitive understanding that the tangent to a curve at a point is the line that indicates the "direction" of the curve at that point.

2. THE DERIVATIVE

Our goal is to solve the following general problem: Given a point $(x, f(x))$ on the graph of a function f, compute the slope of the line that is tangent to the graph at this point (Figure 3).

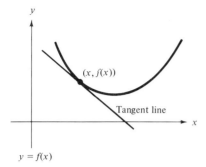

Figure 3

Recall that in calculating the slope of a line we use *two* points (Figure 4). In particular,

$$\text{Slope} = \frac{\Delta y}{\Delta x} = \frac{y_2 - y_1}{x_2 - x_1}$$

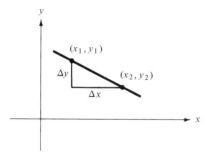

Figure 4

Unfortunately, in the present situation, we know the coordinates of only *one* point on the tangent line, namely, the point of tangency $(x, f(x))$. Hence, direct computation of the slope is impossible, and we are forced to adopt an indirect approach.

The strategy will be to approximate the tangent line by closely related lines whose slopes can be computed directly. In particular, we will consider lines joining the given point $(x, f(x))$ to neighboring points on the graph of f. These lines, called *secant lines*, are good approximations to the actual tangent line, provided the neighboring point is quite close to the given point (Figure 5).

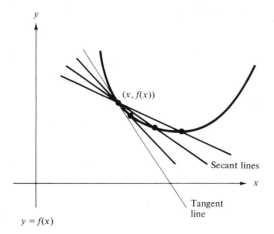

Figure 5

We can make the slope of the secant line as close as we like to the slope of the actual tangent by choosing our neighboring point sufficiently close to the given point $(x, f(x))$. Hence, we should be able to determine the slope of the tangent by first computing the slope of a corresponding secant line, and then studying the behavior of this slope as the neighboring point gets closer and closer to the given point.

To compute the slope of a secant line (Figure 6), we first label the coordinates of the neighboring point. As in Chapter 1, Section 3, we denote the change in the x coordinate by the symbol Δx. Hence the x coordinate of the neighboring point is $x + \Delta x$. Since the point lies on the graph of f, its y coordinate must be $f(x + \Delta x)$.

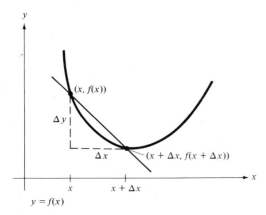

Figure 6

2. THE DERIVATIVE

Thus Δy, the change in the y coordinate, is $f(x + \Delta x) - f(x)$, and the slope of the secant line can be computed as follows:

$$\text{Slope of the secant line} = \frac{\Delta y}{\Delta x}$$

$$= \frac{f(x + \Delta x) - f(x)}{\Delta x}$$

Remember that this quotient is not the slope of the *tangent* line, but only an approximation to it. The smaller we take Δx, however, the closer the point $(x + \Delta x, f(x + \Delta x))$ is to the given point $(x, f(x))$, and the better the approximation becomes. The slope of the tangent line will be the number that this quotient approaches as Δx approaches zero. That is,

$$\text{Slope of tangent} = \lim_{\Delta x \to 0} \frac{f(x + \Delta x) - f(x)}{\Delta x}$$

Before looking at a specific example, let us summarize these observations.

TO FIND THE SLOPE OF A TANGENT LINE

To find the slope of the line that is tangent to the graph of f at the point $(x, f(x))$, we first form the quotient

$$\frac{f(x + \Delta x) - f(x)}{\Delta x}$$

This quotient is the slope of a secant line joining $(x, f(x))$ and a neighboring point $(x + \Delta x, f(x + \Delta x))$. As Δx approaches zero, this slope approaches the slope of the actual tangent line. Thus,

$$\text{Slope of tangent} = \lim_{\Delta x \to 0} \frac{f(x + \Delta x) - f(x)}{\Delta x}$$

Consider the following example.

Example 2.1

Compute the slope of the line that is tangent to the graph of the function $f(x) = x^2$ at the point (2, 4).

Solution

A sketch of f, the given point $(x, f(x)) = (2, 4)$, and a corresponding secant line are shown in Figure 7. Since the x coordinate of the given point is 2, it follows that

DIFFERENTIATION

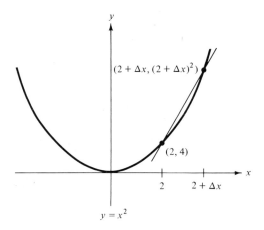

Figure 7

the x coordinate of the neighboring point is $2 + \Delta x$, and the y coordinate of this point is $f(2 + \Delta x) = (2 + \Delta x)^2$. Hence,

Slope of secant line $= \dfrac{f(2 + \Delta x) - f(2)}{\Delta x}$

$= \dfrac{(2 + \Delta x)^2 - 4}{\Delta x}$

This quotient can be simplified as follows:

$\dfrac{(2 + \Delta x)^2 - 4}{\Delta x} = \dfrac{4 + 4\Delta x + (\Delta x)^2 - 4}{\Delta x}$

$= \dfrac{4\Delta x + (\Delta x)^2}{\Delta x}$

$= 4 + \Delta x$

Hence,

Slope of secant line $= 4 + \Delta x$

As Δx approaches zero, the value of $4 + \Delta x$ gets closer and closer to 4. That is,

$\lim\limits_{\Delta x \to 0} (4 + \Delta x) = 4$

Hence, at the point (2, 4), the slope of the tangent is 4.

[To convince yourself that this answer is reasonable, carefully draw a graph of the curve $y = x^2$ and its tangent line at (2, 4).] □

For practice, let us try one more calculation of this sort.

Example 2.2

Compute the slope of the line that is tangent to the graph of the function $f(x) = x^2$ at the point (3, 9).

Solution

The graph in Figure 8 suggests that the tangent at (3, 9) is steeper than the one at (2, 4), so we expect our answer to be larger than that obtained in Example 2.1.

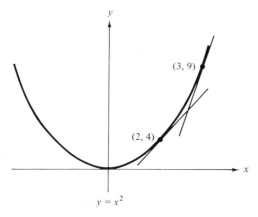

Figure 8

In this example, the given point is (3, 9), the neighboring point is $(3 + \Delta x, (3 + \Delta x)^2)$, and the slope of the corresponding secant line is

$$\frac{f(3 + \Delta x) - f(3)}{\Delta x} = \frac{(3 + \Delta x)^2 - 9}{\Delta x}$$
$$= 6 + \Delta x$$

Since

$$\lim_{\Delta x \to 0} (6 + \Delta x) = 6$$

it follows that the slope of the tangent is 6 at the point (3, 9). □

You may have noticed that the computation in Example 2.2 was quite similar to that in Example 2.1. If so, you may have wondered whether these repetitive computations have to be performed individually for each point on the graph. Fortunately, they do not. If we do the computation once, representing the given point algebraically as $(x, f(x))$, we obtain a formula into which we can substitute numerical values for x to calculate the slope of the tangent at particular points $(x, f(x))$. The next example illustrates this approach for the function $f(x) = x^2$.

Example 2.3

Let $f(x) = x^2$. Derive a formula that expresses the slope of the tangent at $(x, f(x))$ as a function of x.

Solution

We represent the given point $(x, f(x))$ as (x, x^2), and the neighboring point $(x + \Delta x, f(x + \Delta x))$ as $(x + \Delta x, (x + \Delta x)^2)$. The slope of the corresponding secant line is then

$$\frac{f(x + \Delta x) - f(x)}{\Delta x} = \frac{(x + \Delta x)^2 - x^2}{\Delta x}$$

$$= \frac{x^2 + 2x\,\Delta x + (\Delta x)^2 - x^2}{\Delta x}$$

$$= 2x + \Delta x$$

As Δx approaches zero, the expression $2x + \Delta x$ gets closer and closer to $2x$. Thus,

$$\lim_{\Delta x \to 0} (2x + \Delta x) = 2x$$

and we conclude that at the point $(x, f(x))$ the slope of the tangent is $2x$.

For example, at the point $(2, 4)$, $x = 2$, and so the slope of the tangent is $2(2) = 4$. This agrees with our calculation in Example 2.1. At $(3, 9)$, $x = 3$, and so the slope of the tangent is $2(3) = 6$, precisely the answer we obtained in Example 2.2. □

In Example 2.3, we started with a function f and derived a related function that expressed the slope of its tangent at $(x, f(x))$ in terms of the coordinate x. This derived function is known as the *derivative of f* and is frequently denoted by the symbol f'. In Example 2.3, we discovered that the derivative of x^2 is $2x$; that is, we found that if $f(x) = x^2$ then $f'(x) = 2x$.

THE DERIVATIVE

The *derivative f'* of a function f is defined as follows:

$$f'(x) = \lim_{\Delta x \to 0} \frac{f(x + \Delta x) - f(x)}{\Delta x}$$

The derivative is a function that expresses the slope of the tangent to the graph of f at $(x, f(x))$ in terms of the coordinate x.

Consider the following example.

Example 2.4

Suppose $f(x) = 1/x$.

(a) Compute the derivative $f'(x)$.
(b) Find the slope of the line that is tangent to the graph of f when $x = 2$.
(c) Find the equation of the line that is tangent to the graph when $x = 2$.

Solution

(a) $f'(x) = \lim_{\Delta x \to 0} \dfrac{1/(x + \Delta x) - 1/x}{\Delta x}$

$= \lim_{\Delta x \to 0} \dfrac{x - (x + \Delta x)}{x \Delta x (x + \Delta x)}$

$= \lim_{\Delta x \to 0} \dfrac{-1}{x^2 + x \Delta x}$

As Δx approaches zero, so does $x \Delta x$, and thus the quotient $-1/(x^2 + x \Delta x)$ approaches $-1/x^2$. Hence,

$$f'(x) = -\dfrac{1}{x^2}$$

(b) To find the slope of the tangent when $x = 2$, we compute the corresponding value of the derivative $f'(x)$. In particular,

(Slope of tangent when $x = 2$) $= f'(2) = -\tfrac{1}{4}$

(c) In general, to find the equation of a line, we need to know its slope and a point through which it passes. An identifiable point through which the tangent line passes is the point of tangency itself; in this case, the point on the graph whose x coordinate is 2. The y coordinate of this point is simply $f(2) = \tfrac{1}{2}$. Thus the tangent line in question is the straight line through $(2, \tfrac{1}{2})$ having slope $-\tfrac{1}{4}$. Its equation is

$$y - \tfrac{1}{2} = -\tfrac{1}{4}(x - 2)$$

or

$$y = -\tfrac{1}{4}x + 1$$

The graph of f and the tangent line are sketched in Figure 9. □

It may be instructive to examine more closely the limit process involved in the computation of derivatives. When we evaluate the limit

$$\lim_{\Delta x \to 0} \dfrac{f(x + \Delta x) - f(x)}{\Delta x}$$

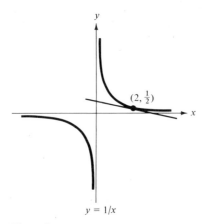

Figure 9

we are actually thinking of the quotient $[f(x + \Delta x) - f(x)]/\Delta x$ as a function of the variable Δx. Viewed in this way, the quotient has a discontinuity when $\Delta x = 0$. In general, *both* the numerator and the denominator of this quotient will approach zero as Δx approaches zero. This was the case, for instance, in Example 2.3, where

$$\frac{f(x + \Delta x) - f(x)}{\Delta x} = \frac{(x + \Delta x)^2 - x^2}{\Delta x}$$

and again in Example 2.4, where

$$\frac{f(x + \Delta x) - f(x)}{\Delta x} = \frac{1/(x + \Delta x) - 1/x}{\Delta x}$$

In such situations, the quotient must be simplified before the limit can be found. In Example 2.3, for instance, we rewrote the quotient as

$$\frac{f(x + \Delta x) - f(x)}{\Delta x} = 2x + \Delta x$$

before we took the limit. And in Example 2.4, we worked with the simplified quotient

$$\frac{f(x + \Delta x) - f(x)}{\Delta x} = \frac{-1}{x^2 + x\,\Delta x}$$

We first encountered limits of this sort in Chapter *1*, Section 4. Take another look at Example 4.6 in Chapter *1* and see if you can now identify the limit in that example as the value of a certain derivative. (*Hint:* Let $\Delta x = x - 1$.)

The derivative is often denoted by symbols other than f'. If the function is expressed in terms of y instead of $f(x)$, the symbol dy/dx (suggesting slope, $\Delta y/\Delta x$) is frequently used instead of $f'(x)$. For example, as an alternative to the statement

If $f(x) = x^2$, then $f'(x) = 2x$

we could write

If $y = x^2$, then $\dfrac{dy}{dx} = 2x$

Sometimes the two notations are combined, as in the statement

If $f(x) = x^2$, then $\dfrac{df}{dx} = 2x$

These statements may be condensed by omitting reference to y and f altogether. For example, the compact notation

$$\frac{d}{dx}(x^2) = 2x$$

may be used to state that the derivative of x^2 is $2x$.

Some functions fail to have a derivative at certain points. Consider the function

$$f(x) = \begin{cases} x & \text{if } x \leq 1 \\ 1 & \text{if } x > 1 \end{cases}$$

shown in Figure 10. At the point (1, 1), the tangent line cannot be uniquely determined; it might be the diagonal line $y = x$, or the horizontal line $y = 1$, or

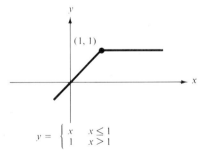

Figure 10

perhaps something in between. Because of this ambiguity, the derivative of f (the slope of the tangent) cannot be defined for $x = 1$.

A function whose derivative is defined at a certain point is said to be *differentiable* at that point. No function can be differentiable at a point of discontinuity. However, as we have just seen, the absence of a discontinuity does not guarantee the existence of a derivative. Practically all of the functions we will encounter in this text will be differentiable at most points. Polynomials, for example, are differentiable everywhere. Rational functions fail to have derivatives only at points of discontinuity. The function $f(x) = 1/x$ is typical. In Example 2.4 we found that

$$\frac{d}{dx}\left(\frac{1}{x}\right) = -\frac{1}{x^2}$$

When $x = 0$, this derivative is undefined. This reflects the fact that the tangent when $x = 0$ is, in a sense, the y axis, a vertical line whose slope is undefined. (See Figure 11.) For all other values of x, however, $-1/x^2$ can be computed and the derivative exists.

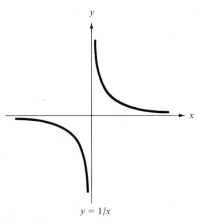

$y = 1/x$

Figure 11

To conclude this section, let us return to the optimization problem with which we began the discussion of differential calculus back in Section 1.

Example 2.5

A manufacturer can produce cassette tape recorders at a cost of $20 each. He estimates that if he sells them for x dollars apiece he will be able to sell approximately $120 - x$ tape recorders a month. Determine the selling price x that will generate the most profit for the manufacturer.

Solution

Let $f(x)$ denote the monthly profit if the tape recorders are sold for x dollars apiece. Then, as we have already seen,

$$f(x) = (120 - x)(x - 20) = -x^2 + 140x - 2{,}400$$

The graph of f is sketched once again in Figure 12.

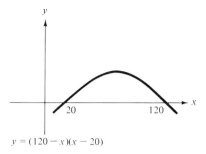

$y = (120 - x)(x - 20)$

Figure 12

We wish to find the value of x for which the profit $f(x)$ is maximal. This will be the value of x for which the slope of the tangent is zero. Since this slope is expressed by the derivative, we begin by computing $f'(x)$.

$$f'(x) = \lim_{\Delta x \to 0} \frac{f(x + \Delta x) - f(x)}{\Delta x}$$

$$= \lim_{\Delta x \to 0} \frac{[-(x + \Delta x)^2 + 140(x + \Delta x) - 2{,}400] - [-x^2 + 140x - 2{,}400]}{\Delta x}$$

$$= \lim_{\Delta x \to 0} \frac{-(\Delta x)^2 - 2x\, \Delta x + 140\, \Delta x}{\Delta x}$$

$$= \lim_{\Delta x \to 0} (-\Delta x - 2x + 140)$$

$$= -2x + 140$$

The optimal value of x is the one for which $f'(x) = 0$. That is,

$$-2x + 140 = 0$$

or

$$x = 70$$

Thus the manufacturer's profit will be maximal if he sells his tape recorders for $70 apiece. Moreover, if he does sell the recorders at this optimal price, his profit will be $f(70) = 2{,}500$ dollars. □

Problems

1. For each of the following functions, compute the derivative and find the slope of the tangent line at the specified point.
   (a) $f(x) = 5x - 3$; (2, 7)
   (b) $y = x^2 - 1$; (−1, 0)
   (c) $f(x) = 2x^2 - 3x + 5$; (0, 5)
   (d) $y = x^3 - 1$; (2, 7)
   (e) $f(x) = x + 1/x$; $(-3, -\frac{10}{3})$
   (f) $y = 1/x^2$; $(\frac{1}{2}, 4)$
   (g) $f(x) = \sqrt{x}$; (9, 3)

2. For each of the functions in Problem 1, write an equation for the line tangent to the graph at the specified point.

3. Let $f(x) = x^2$.
   (a) Compute the slope of the secant line joining the points on the graph of f whose x coordinates are $x = -2$ and $x = -\frac{19}{10}$.
   (b) Use calculus to compute the slope of the line that is tangent to the graph when $x = -2$, and compare this slope with your answer in part (a).
   (c) Sketch the graph of f, the secant line, and the tangent line.

4. Let $f(x) = 2x^2 - 8$.
   (a) Compute the slope of the secant line joining the points on the graph of f whose x coordinates are $x = 3$ and $x = \frac{31}{10}$.
   (b) Use calculus to compute the slope of the line that is tangent to the graph when $x = 3$, and compare this slope with your answer in part (a).
   (c) Sketch the graph of f, the secant line, and the tangent line.

5. (a) Compute the derivative of the linear function $y = 3x - 2$.
   (b) Write the equation of the tangent line at $(-1, -5)$.
   (c) Explain how your answers to parts (a) and (b) could have been obtained from geometric considerations, with no calculation whatsoever.

6. (a) Compute the derivatives of the functions $y = x^2$ and $y = x^2 - 3$, and account geometrically for their similarity.
   (b) Without further computation, find the derivative of the function $y = x^2 + 5$.

*7. (a) Compute the derivatives of the functions $y = x^2$ and $y = (x - 1)^2$, and account geometrically for their similarity.
   (b) Without further computation, find the derivative of the function $y = (x + 4)^2$.

8. (a) Compute the derivative of the function $y = x^2 + 3x$.
   (b) Compute the derivatives of the functions $y = x^2$ and $y = 3x$ separately.
   (c) How is the answer in part (a) related to the answers in part (b)?

9. (a) Sketch the graph of the function $f(x) = x^2 - 3x$.
   (b) Use calculus to determine the precise location of the lowest point on the graph.

10. (a) Sketch the graph of the function $y = 1 - x^2$.
    (b) Use calculus to determine the precise location of the highest point on the graph.

11. (a) Graph the function $f(x) = x^3 - x^2$.
    (b) Determine the values of x for which the derivative of f is equal to zero. What happens to the graph at the corresponding points?

12. What can you conclude about the graph of a function between $x = a$ and $x = b$ if its derivative is positive whenever $a \leq x \leq b$?

*13. Sketch the graph of a function f whose derivative has all of the following properties.
   (a) $f'(x) > 0$ when $x < 1$ and when $x > 5$
   (b) $f'(x) < 0$ when $1 < x < 5$
   (c) $f'(1) = 0$ and $f'(5) = 0$

14. A manufacturer can produce waterbed frames at a cost of $10 each. He estimates that if he sells them for x dollars apiece he will be able to sell approximately $50 - x$ frames a month. (See Problem 4 on page 58.)
   (a) Express the manufacturer's monthly profit as a function of the selling price x and sketch this profit function.
   (b) Use calculus to determine the selling price at which the manufacturer will maximize his profit. What is the maximum profit?

15. (a) Compute the derivatives of the functions $y = x$, $y = x^2$, and $y = x^3$.
    (b) Examine your answers in part (a). Can you detect a pattern? What do you think the derivative of $y = x^4$ is? How about the derivative of $y = x^{27}$?

*16. (a) Compute the derivatives of the functions $y = 1/x$, $y = 1/x^2$, and $y = 1/x^3$.
    (b) Examine your answers in part (a). Can you detect a pattern? What do you think the derivative of $y = 1/x^4$ is? How about the derivative of $y = 1/x^{27}$?
    (c) Can you see a connection between your answers to this problem and your answers to Problem 15? (*Hint*: Think of $1/x^n$ as x^{-n}.)

*17. (a) Compute the derivative of $y = \sqrt{x}$.
    (b) Is your answer to part (a) consistent with the pattern discovered in Problems 15 and 16? Can you guess what the derivative of $y = \sqrt[3]{x}$ is?

3. TECHNIQUES OF DIFFERENTIATION

To compute the derivative of a function, we first write an expression for the slope of a secant line, then simplify this expression algebraically, and finally take its limit as Δx approaches zero. For all but the simplest functions, this process is tedious and time-consuming. Fortunately, there are some shortcuts. In fact, on the basis of one key observation and a few general rules, the derivatives of most useful functions can be painlessly computed.

The key observation concerns the derivatives of power functions $f(x) = x^n$. In Example 2.3, we found that

$$\frac{d}{dx}(x^2) = 2x$$

In Problem 15 on page 101, you also computed the derivatives of x and x^3. If your calculations were correct, you found that

$$\frac{d}{dx}(x) = 1$$

and

$$\frac{d}{dx}(x^3) = 3x^2$$

If you rewrote the first of these two equations as

$$\frac{d}{dx}(x^1) = 1x^0$$

you might have noticed a pattern. When n is 1, 2, or 3, the derivative of x^n is obtained by first reducing the power of x by 1 and then multiplying by the original power. That is, for $n = 1, 2,$ and 3,

$$\frac{d}{dx}(x^n) = nx^{n-1}$$

This formula remains true for other values of n as well. For example, using computations similar to those for x^2 and x^3, one can establish this formula for any positive integer n. A general proof that the formula holds for all positive integers is outlined in Problem 14.

The formula also holds when n is a negative integer. In Problem 16 on page 101, you should have discovered that

$$\frac{d}{dx}\left(\frac{1}{x}\right) = -\frac{1}{x^2}$$

$$\frac{d}{dx}\left(\frac{1}{x^2}\right) = -\frac{2}{x^3}$$

and

$$\frac{d}{dx}\left(\frac{1}{x^3}\right) = -\frac{3}{x^4}$$

The pattern emerges if we rewrite these equations as follows:

$$\frac{d}{dx}(x^{-1}) = -1x^{-2}$$

$$\frac{d}{dx}(x^{-2}) = -2x^{-3}$$

and

$$\frac{d}{dx}(x^{-3}) = -3x^{-4}$$

The fact that the formula $d/dx(x^n) = nx^{n-1}$ holds for all negative integers can be proved by an argument that combines the formula for positive integers with a rule for differentiating quotients. This proof is outlined in Problem 16.

Actually, the formula holds for *any* value of n whatsoever. A proof of this fact, based on properties of exponential and logarithmic functions, will be given in Chapter 3.

DIFFERENTIATION OF POWER FUNCTIONS

For any number n,

$$\frac{d}{dx}(x^n) = nx^{n-1}$$

Consider the following example.

Example 3.1

Differentiate (find the derivative of) each of the following functions.

(a) $y = x^{27}$
(b) $y = 1/x^{27}$
(c) $y = \sqrt{x}$
(d) $y = 1/\sqrt[3]{x}$

Solution

In each case, we first write the function as a power function and then apply our formula.

(a) $\dfrac{d}{dx}(x^{27}) = 27x^{26}$

(b) $\dfrac{d}{dx}\left(\dfrac{1}{x^{27}}\right) = \dfrac{d}{dx}(x^{-27}) = -27x^{-28} = -\dfrac{27}{x^{28}}$

(c) $\dfrac{d}{dx}(\sqrt{x}) = \dfrac{d}{dx}(x^{1/2}) = \dfrac{1}{2}x^{-1/2} = \dfrac{1}{2\sqrt{x}}$

(d) $\dfrac{d}{dx}\left(\dfrac{1}{\sqrt[3]{x}}\right) = \dfrac{d}{dx}(x^{-1/3}) = -\dfrac{1}{3}x^{-4/3} = \dfrac{-1}{3\sqrt[3]{x^4}}$ □

In addition to the formula for differentiating power functions, we shall need a few general rules for differentiating certain combinations of functions. In the statements of these rules, f and g will denote differentiable functions.

CONSTANT MULTIPLE RULE

For any constant c,

$$\frac{d}{dx}(cf) = c\frac{df}{dx}$$

That is, the derivative of a constant times a function is equal to the constant times the derivative of that function.

The use of this rule is illustrated in the following example.

Example 3.2

Differentiate the function $y = 5x^2$.

Solution

We already know that $d/dx(x^2) = 2x$. According to the constant multiple rule, it follows that

$$\frac{d}{dx}(5x^2) = 5\frac{d}{dx}(x^2) = 10x \qquad \square$$

The constant multiple rule expresses the geometric fact that the graph of the function $y = cf(x)$ is c times as steep as the graph of the function $y = f(x)$. To verify this rule mathematically, let us consider a secant line on the graph of the function $y = cf(x)$ (Figure 13). Note that the y coordinate corresponding to x is $cf(x)$ and that the y coordinate corresponding to $x + \Delta x$ is $cf(x + \Delta x)$. Hence,

$$\text{Slope of secant} = \frac{cf(x + \Delta x) - cf(x)}{\Delta x}$$

$$= c\frac{f(x + \Delta x) - f(x)}{\Delta x}$$

As Δx approaches zero, the quotient $[f(x + \Delta x) - f(x)]/\Delta x$ approaches df/dx, the derivative of f, while the constant c remains unchanged. Hence,

$$\frac{d}{dx}(cf) = \lim_{\Delta x \to 0} c\frac{f(x + \Delta x) - f(x)}{\Delta x}$$

$$= c\frac{df}{dx}$$

and the constant multiple rule is proved.

3. TECHNIQUES OF DIFFERENTIATION

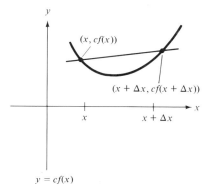

Figure 13 $y = cf(x)$

One easy consequence of this rule is worth noting.

THE DERIVATIVE OF A CONSTANT

For any constant c,

$$\frac{d}{dx}(c) = 0$$

That is, the derivative of a constant function is zero.

To see this, just write c as cx^0 so that

$$\frac{d}{dx}(c) = \frac{d}{dx}(cx^0) = c\frac{d}{dx}(x^0) = 0$$

This result should come as no surprise. The graph of the constant function $y = c$ is a horizontal straight line, and its slope is zero.

Our next rule states that a sum can be differentiated term by term.

SUM RULE

$$\frac{d}{dx}(f+g) = \frac{df}{dx} + \frac{dg}{dx}$$

That is, the derivative of a sum is the sum of the individual derivatives.

Consider the following example.

Example 3.3

Differentiate the function $y = x^2 + 3x$.

Solution

We know that $d/dx(x^2) = 2x$ and that $d/dx(3x) = 3$. According to the sum rule, it follows that

$$\frac{d}{dx}(x^2 + 3x) = \frac{d}{dx}(x^2) + \frac{d}{dx}(3x)$$
$$= 2x + 3 \qquad \square$$

Using the sum rule, we can now differentiate any polynomial.

Example 3.4

Differentiate the polynomial $f(x) = 5x^3 - 4x^2 + 12x - \frac{1}{2}$.

Solution

According to the sum rule, f can be differentiated term by term. That is,

$$f'(x) = \frac{d}{dx}(5x^3) + \frac{d}{dx}(-4x^2) + \frac{d}{dx}(12x) + \frac{d}{dx}\left(-\frac{1}{2}\right)$$
$$= 15x^2 - 8x + 12 \qquad \square$$

To prove the sum rule, we consider a secant line on the graph of the function $f + g$ (Figure 14). The y coordinate corresponding to x is $f(x) + g(x)$, and the y coordinate corresponding to $x + \Delta x$ is $f(x + \Delta x) + g(x + \Delta x)$. Hence,

$$\text{Slope of secant} = \frac{[f(x + \Delta x) + g(x + \Delta x)] - [f(x) + g(x)]}{\Delta x}$$

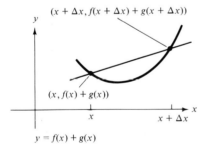

Figure 14

We can rewrite this expression as the sum of two quotients, one involving f and the other g. In particular,

$$\text{Slope of secant} = \frac{f(x + \Delta x) - f(x)}{\Delta x} + \frac{g(x + \Delta x) - g(x)}{\Delta x}$$

3. TECHNIQUES OF DIFFERENTIATION

To get the derivative of $f + g$, we now let Δx approach zero. The limit of the first quotient is simply df/dx, the derivative of f, while the limit of the second quotient is dg/dx. Hence,

$$\frac{d}{dx}(f + g) = \frac{df}{dx} + \frac{dg}{dx}$$

and the sum rule is proved.

Suppose you want to differentiate the product $f(x) = x^2(3x + 1)$. You might be tempted to differentiate the factors x^2 and $3x + 1$ separately, and then multiply your answers. That is, since $d/dx(x^2) = 2x$ and $d/dx(3x + 1) = 3$, you might conclude that $f'(x) = 6x$. However, this answer is wrong. To see this, let us rewrite the function as $f(x) = 3x^3 + x^2$. Now it is clear that the derivative $f'(x)$ is $9x^2 + 2x$, not $6x$. Thus the derivative of a product is *not* the product of the individual derivatives. The correct rule for differentiating products is slightly more complicated.

PRODUCT RULE

$$\frac{d}{dx}(fg) = f\frac{dg}{dx} + g\frac{df}{dx}$$

The use of this rule is illustrated in the next example.

Example 3.5

Differentiate the function $y = x^2(3x + 1)$.

Solution

According to the product rule,

$$\frac{d}{dx}[x^2(3x + 1)] = x^2 \frac{d}{dx}(3x + 1) + (3x + 1)\frac{d}{dx}(x^2)$$

$$= x^2(3) + (3x + 1)(2x)$$

$$= 9x^2 + 2x \qquad \square$$

The product rule tells us how to differentiate the product of any two functions, while the constant multiple rule tells us how to differentiate those special products in which one of the factors is constant. It is easy to see that these two rules are consistent. According to the product rule,

$$\frac{d}{dx}(cf) = c\frac{df}{dx} + f\frac{dc}{dx}$$

But the derivative dc/dx is zero (since c is a constant), and so this equation reduces to

$$\frac{d}{dx}(cf) = c\frac{df}{dx}$$

which is the answer predicted by the constant multiple rule.

Let us prove the product rule. A secant line on the graph of the function fg is shown in Figure 15.

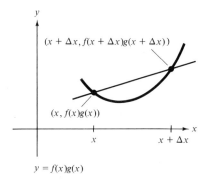

Figure 15

Its slope is expressed by the equation

$$\text{Slope of secant} = \frac{f(x + \Delta x)g(x + \Delta x) - f(x)g(x)}{\Delta x}$$

This quotient does not simplify as readily as the corresponding one in the proof of the sum rule. In fact, we have to resort to an algebraic trick. We rewrite the quotient by subtracting and then adding the quantity $f(x + \Delta x)g(x)$ to the numerator. We get

$$\text{Slope of secant} = \frac{f(x + \Delta x)g(x + \Delta x) - f(x + \Delta x)g(x)}{\Delta x}$$

$$+ \frac{f(x + \Delta x)g(x) - f(x)g(x)}{\Delta x}$$

$$= f(x + \Delta x)\frac{g(x + \Delta x) - g(x)}{\Delta x} + g(x)\frac{f(x + \Delta x) - f(x)}{\Delta x}$$

The derivative $d/dx(fg)$ is the limit of this expression as Δx approaches zero. But as Δx approaches zero, $f(x + \Delta x)$ approaches $f(x)$ and $g(x)$ remains unchanged, while $[g(x + \Delta x) - g(x)]/\Delta x$ approaches dg/dx and $[f(x + \Delta x) - f(x)]/\Delta x$ approaches df/dx. Hence,

$$\frac{d}{dx}(fg) = f\frac{dg}{dx} + g\frac{df}{dx}$$

and the product rule is proved.

The derivative of a quotient is *not* the quotient of the individual derivatives. (Consider the function $y = 1/x$, for example.) Here is the correct rule.

QUOTIENT RULE

$$\frac{d}{dx}\left(\frac{f}{g}\right) = \frac{g \, df/dx - f \, dg/dx}{g^2}$$

As our first illustration, let us apply the quotient rule to a function whose derivative can be computed easily by another method as well.

Example 3.6

Differentiate the function $y = 1/x^2$.

Solution

According to the quotient rule,

$$\frac{d}{dx}\left(\frac{1}{x^2}\right) = \frac{x^2 \, d/dx(1) - 1 \, d/dx(x^2)}{(x^2)^2}$$

$$= \frac{x^2(0) - 2x}{x^4}$$

$$= -\frac{2}{x^3}$$

Notice that this is precisely the answer that the rule for differentiating power functions would have given. In this case, of course, the power function rule is much more efficient than the quotient rule. In practice, one would not use the cumbersome quotient rule in such a simple example. □

With the quotient rule, we can now differentiate any rational function. Consider the following typical example.

Example 3.7

Differentiate the rational function $f(x) = (x^2 + 2x - 21)/(x - 3)$.

Solution

$$f'(x) = \frac{(x-3) \, d/dx(x^2 + 2x - 21) - (x^2 + 2x - 21) \, d/dx(x - 3)}{(x-3)^2}$$

$$= \frac{(x-3)(2x+2) - (x^2 + 2x - 21)}{(x-3)^2}$$

$$= \frac{x^2 - 6x + 15}{(x-3)^2}$$

□

The proof of the quotient rule is similar to that of the product rule. An outline of the proof is given in Problem 15.

We have developed techniques that allow us to differentiate a large class of useful functions. There are still, however, many simple functions that cannot be handled easily by our present methods. For example, although the function $f(x) = (3x^2 - 2x + 1)^{12}$ can be differentiated by techniques already developed, the resulting computations would be prohibitive. (Can you think of at least two methods that, in principle, could be used to differentiate this function?) There is one final rule, called the *chain rule*, that will significantly expand the class of functions that can be easily differentiated. Section 4 will be devoted to this important rule.

Problems

1. Differentiate the following functions.
   (a) $y = \frac{2}{5}x^5 - 4x^3 + 9x - 6$
   (b) $f(x) = x^{10} - 5x^4 - x + 2$
   (c) $g(u) = 2u - 3/u^2 + 5u^{1/3}$
   (d) $y = 3\sqrt{x} + 2/(5x^3) - \sqrt{2}$
   (e) $h(t) = -16t + 1/\sqrt{t} - t^{3/2} + 1/(3t)$
   (f) $f(x) = (5x^4 - 3x^2 + 12x - 6)(1 - 6x)$
   (g) $g(x) = \frac{1}{3}(x^4 - 3x^2 + 1)$
   (h) $y = (t^3 - t^2 + t - 1)(t^2 + 1)$
   (i) $g(x) = x/(x^2 - 2)$
   (j) $h(u) = (u^2 + 1)/(3u - u^3)$
   (k) $f(t) = 1/(t - 2)$
   (l) $y = (x^2 + 2x + 1)/3$
   (m) $h(x) = x^3/2 - 1/(2x)$
   (n) $f(x) = (2x + 1)(x^2 + 6x - 3)(1 - x^5)$
   (o) $g(u) = (u^2 + 2u + 1)(u + 5)/(2u - 1)$

2. For each of the following functions, find the slope and the equation of the tangent line at the specified point.
   (a) $y = x^5 - 3x^3 - 5x + 2$; $(1, -5)$
   (b) $f(x) = (x^2 + 1)(1 - x^3)$; $(0, 1)$
   (c) $y = (x + 1)/(x - 1)$; $(0, -1)$
   (d) $f(x) = 1 - 1/x + 2/\sqrt{x}$; $(4, \frac{7}{4})$

3. For each of the following functions, find the equation of the tangent line at the point $(x, f(x))$ for the specified value of x.
   (a) $f(x) = x^4 - 3x^3 + 2x^2 - 6$; $x = 2$
   (b) $f(x) = x - 1/x^2$; $x = 1$
   (c) $f(x) = (x^2 + 2)/(x^2 - 2)$; $x = -1$

4. (a) Differentiate the function $f(x) = 2x^2 - 5x - 3$.
   (b) Now factor this function as $f(x) = (2x + 1)(x - 3)$ and differentiate using the product rule. Compare your answers.

5. (a) Use the quotient rule to differentiate the function $f(x) = (2x - 3)/x^3$.
   (b) Now rewrite this function as $f(x) = x^{-3}(2x - 3)$ and differentiate using the product rule.
   (c) Show that your answers to parts (a) and (b) are the same.

6. Sketch the graph of the function $f(x) = x^2 - 4x - 5$ and use calculus to determine the lowest point on the graph.

7. Sketch the graph of the function $f(x) = 3 - 2x - x^2$ and use calculus to determine the highest point on the graph.

8. (a) Use calculus to find all the points on the graph of the function $f(x) = (x^2 + 3)/(x - 1)$ at which the tangent line is horizontal.
   (b) Sketch a graph showing f and its horizontal tangent lines.

*9. Find numbers a and b so that the lowest point on the graph of the function $f(x) = ax^2 + bx$ is $(3, -8)$.

10. A manufacturer is currently selling 2,000 lamps a month to retailers at a price of $2 per lamp. He estimates that for each 1-cent increase in the price he will sell 10 fewer lamps each month. The manufacturer's costs consist of a fixed overhead of $500 per month plus 40 cents a lamp for labor and materials. (See Problem 6 on page 59.)
    (a) Express the manufacturer's monthly profit as a function of the price at which he sells the lamps, and sketch this profit function.
    (b) Use calculus to determine the selling price at which the manufacturer's profit will be maximal. What is the maximal profit?

*11. Prove the following generalization of the product rule:

$$\frac{d}{dx}(fgh) = fg\frac{dh}{dx} + fh\frac{dg}{dx} + gh\frac{df}{dx}$$

12. Find numbers a, b, and c, so that the graph of the function $f(x) = ax^2 + bx + c$ will have x intercepts at $(0, 0)$ and $(5, 0)$, and a tangent line with slope 1 when $x = 2$.

*13. Find the equations of all the tangent lines to the graph of the function $f(x) = x^2 - 4x + 25$ that pass through the origin $(0, 0)$.

*14. Suppose n is a positive integer and $f(x) = x^n$. Prove that $f'(x) = nx^{n-1}$. (Hint: To begin, rewrite the quotient $[f(x + \Delta x) - f(x)]/\Delta x$ using the fact that

$$f(x + \Delta x) = (x + \Delta x)^n$$

$$= x^n + \frac{n}{1!}x^{n-1}\Delta x + \frac{n(n-1)}{2!}x^{n-2}(\Delta x)^2$$

$$+ \cdots + nx(\Delta x)^{n-1} + (\Delta x)^n$$

Then let Δx approach zero.)

*15. Prove the quotient rule:

$$\frac{d}{dx}\left(\frac{f}{g}\right) = \frac{g\,df/dx - f\,dg/dx}{g^2}$$

(*Hint*: Show that the slope of the usual secant line is

$$\frac{1}{\Delta x}\left[\frac{f(x+\Delta x)}{g(x+\Delta x)} - \frac{f(x)}{g(x)}\right] = \frac{g(x)f(x+\Delta x) - f(x)g(x+\Delta x)}{g(x+\Delta x)g(x)\Delta x}$$

Before letting Δx approach zero, rewrite this quotient using the trick of subtracting and adding $g(x)f(x)$ to the numerator.)

*16. Suppose n is a *negative* integer and $f(x) = x^n$. Prove that $f'(x) = nx^{n-1}$. (*Hint*: Use the corresponding result for positive integers together with the quotient rule.)

4. THE CHAIN RULE

At the end of Section 3, we encountered the function $f(x) = (3x^2 - 2x + 1)^{12}$ and observed that we had no simple method for computing its derivative. The rule we shall now develop will provide us with an efficient technique for differentiating this function and many others like it.

We begin our discussion by considering powers of functions; that is, functions of the form $f(x) = [h(x)]^n$. For example, $f(x) = (3x^2 - 2x + 1)^{12}$ is such a function, since it can be written as $f(x) = [h(x)]^{12}$, where $h(x) = 3x^2 - 2x + 1$. Similarly, the function $f(x) = \sqrt{(x+1)/(x^2-2)}$ can be expressed as $f(x) = [h(x)]^n$, where $h(x) = (x+1)/(x^2-2)$ and $n = \frac{1}{2}$.

Recalling the rule

$$\frac{d}{dx}(x^n) = nx^{n-1}$$

for differentiating power functions, we might suspect that $d/dx(h^n) = nh^{n-1}$. For example, if $f(x) = (2x^4 - x)^3$, we might be tempted to conclude that $f'(x) = 3(2x^4 - x)^2$. Unfortunately, this answer is wrong. To see this, let us expand $(2x^4 - x)^3$ and rewrite our function as $f(x) = 8x^{12} - 12x^9 + 6x^6 - x^3$. Now it is clear that the derivative $f'(x)$ is $96x^{11} - 108x^8 + 36x^5 - 3x^2$, and not $3(2x^4 - x)^2$.

Actually, our incorrect conjecture about the derivative of h^n turns out to have been a fairly good guess. According to the correct rule (called the *chain rule for powers*), we differentiate h^n by first computing nh^{n-1} and then multiplying this expression by the derivative dh/dx of the function h.

CHAIN RULE FOR POWERS

For any number n,

$$\frac{d}{dx}(h^n) = nh^{n-1}\frac{dh}{dx}$$

We illustrate the use of this rule in the following two examples.

Example 4.1

Differentiate the function $f(x) = (2x^4 - x)^3$.

Solution

As we have seen, one way to do this problem is to rewrite the function as

$$f(x) = 8x^{12} - 12x^9 + 6x^6 - x^3$$

and then differentiate this polynomial term by term to get

$$f'(x) = 96x^{11} - 108x^8 + 36x^5 - 3x^2$$

But see how much easier it is to use the chain rule. According to the chain rule,

$$f'(x) = 3(2x^4 - x)^2 \frac{d}{dx}(2x^4 - x)$$

$$= 3(2x^4 - x)^2(8x^3 - 1)$$

And the answer even comes out in factored form! □

Example 4.2

Differentiate the function $f(x) = \sqrt{(x+1)/(x^2 - 2)}$.

Solution

We rewrite the function as

$$f(x) = \left(\frac{x+1}{x^2 - 2}\right)^{1/2}$$

Then, according to the chain rule,

$$f'(x) = \frac{1}{2}\left(\frac{x+1}{x^2-2}\right)^{-1/2} \frac{d}{dx}\left(\frac{x+1}{x^2-2}\right)$$

Now we use the quotient rule to compute $d/dx[(x + 1)/(x^2 - 2)]$:

$$\frac{d}{dx}\left(\frac{x+1}{x^2-2}\right) = \frac{(x^2-2)(1) - (x+1)(2x)}{(x^2-2)^2}$$

$$= \frac{-x^2 - 2x - 2}{(x^2-2)^2}$$

Finally, we substitute this expression into our equation for $f'(x)$ to get

$$f'(x) = \frac{1}{2}\left(\frac{x+1}{x^2-2}\right)^{-1/2}\left[\frac{-x^2-2x-2}{(x^2-2)^2}\right]$$

$$= -\frac{x^2 + 2x + 2}{2(x+1)^{1/2}(x^2-2)^{3/2}} \qquad \square$$

The general proof of the chain rule is rather subtle. It will be discussed later in this section. For certain special cases, however, elementary proofs based on the product rule can be given. Consider the following two examples.

Example 4.3

Prove the chain rule for powers for the integer $n = 2$. That is, prove

$$\frac{d}{dx}(h^2) = 2h\frac{dh}{dx}$$

Solution

According to the product rule,

$$\frac{d}{dx}(h^2) = \frac{d}{dx}(h \cdot h) = h\frac{dh}{dx} + h\frac{dh}{dx} = 2h\frac{dh}{dx} \qquad \square$$

By combining the product rule with the chain rule for $n = 2$, we can now prove the chain rule for $n = 3$.

Example 4.4

Prove

$$\frac{d}{dx}(h^3) = 3h^2\frac{dh}{dx}$$

Solution

We think of h^3 as $h \cdot h^2$ and apply the product rule.

$$\frac{d}{dx}(h^3) = \frac{d}{dx}(h \cdot h^2) = h\frac{d}{dx}(h^2) + h^2\frac{dh}{dx}$$

Now we evaluate $d/dx(h^2)$ using the chain rule for $n = 2$. We get

$$\frac{d}{dx}(h^2) = 2h\frac{dh}{dx}$$

which we substitute into our equation for $d/dx(h^3)$ to obtain

$$\frac{d}{dx}(h^3) = 2h^2\frac{dh}{dx} + h^2\frac{dh}{dx} = 3h^2\frac{dh}{dx} \qquad \square$$

Similar computations can be used to derive the rule for $n = 4$ from the rule for $n = 3$, for $n = 5$ from $n = 4$, and so on. The chain rule for negative integers can be obtained from the rule for the corresponding positive integers via the quotient rule. A short proof of this is outlined in Problem 10.

You may have noticed that the chain rule for powers is actually a rule for differentiating certain composite functions. In particular, $[h(x)]^n$ can be thought of as $g(h(x))$, where $g(u) = u^n$. (To refresh your memory, reread the discussion of composite functions in Chapter 1, Section 1.) Notice that $n[h(x)]^{n-1}$ is just $g'(h(x))$. Hence the chain rule for powers may be rephrased as follows. If $g(u) = u^n$, then

$$\frac{d}{dx}[g(h(x))] = g'(h(x))h'(x)$$

It turns out that this formula remains valid even when g is not a power function. The general form of the chain rule is an assertion of this fact.

CHAIN RULE

For any differentiable functions h and g,

$$\frac{d}{dx}[g(h(x))] = g'(h(x))h'(x)$$

Let us apply the chain rule to an example in which g is not a power function.

Example 4.5

Find $d/dx[g(h(x))]$ if $g(u) = u^3 - 3u^2 + 1$ and $h(x) = x^2 + 2$.

Solution

Let us first find the derivative without using the chain rule, by computing $g(h(x))$ explicitly. Since

$$g(h(x)) = (x^2 + 2)^3 - 3(x^2 + 2)^2 + 1$$
$$= x^6 + 6x^4 + 12x^2 + 8 - 3x^4 - 12x^2 - 12 + 1$$
$$= x^6 + 3x^4 - 3$$

it follows that

$$\frac{d}{dx}[g(h(x))] = 6x^5 + 12x^3 = 6x^3(x^2 + 2)$$

Now let us try the chain rule. Notice the superior efficiency and elegance of this solution.

Since $g'(u) = 3u^2 - 6u$ and $h'(x) = 2x$, the chain rule gives

$$\frac{d}{dx}[g(h(x))] = g'(h(x))h'(x)$$
$$= [3(x^2 + 2)^2 - 6(x^2 + 2)](2x)$$
$$= 6x^3(x^2 + 2) \qquad \square$$

The chain rule is especially convenient when we want to calculate the derivative of a composite function for a particular value of the independent variable. Consider the following example.

Example 4.6

Find $f'(3)$ if $f(x) = g(h(x))$, where $g(u) = 2\sqrt{u^2 + u - 18}$ and $h(x) = 20 - 6/x$.

Solution

According to the chain rule,

$f'(3) = g'(h(3))h'(3)$

Since $g'(u) = (2u + 1)/\sqrt{u^2 + u - 18}$ and $h(3) = 18$, it follows that
$g'(h(3)) = g'(18) = \frac{37}{18}$
Moreover, $h'(x) = 6/x^2$, so
$h'(3) = \frac{2}{3}$
Hence,
$f'(3) = \frac{37}{18}(\frac{2}{3}) = \frac{37}{27} \qquad \square$

Observe that we were able to compute $f'(3)$ in Example 4.6 without actually finding an explicit algebraic formula for $f'(x)$. You will become convinced of the efficiency of this technique if you now write out the formula for $f'(x)$.

It is often convenient to write the chain rule in a somewhat different form. If we let $y = g(h(x))$, then $y = g(u)$, where $u = h(x)$. Thus,

$$g'(h(x)) = g'(u) = \frac{dy}{du}$$

$$h'(x) = \frac{du}{dx}$$

and

$$\frac{d}{dx}[g(h(x))] = \frac{dy}{dx}$$

Hence the chain rule can be rewritten as

$$\frac{dy}{dx} = \frac{dy}{du}\frac{du}{dx}$$

In this form, the chain rule is particularly easy to remember. If we pretend that the derivatives dy/du and du/dx are quotients and cancel du, we can reduce the expression on the right-hand side of the equation to the expression dy/dx on the left-hand side.

Let us see how the computation in Example 4.5 would look with this notation.

Example 4.7

Find dy/dx if $y = u^3 - 3u^2 + 1$ and $u = x^2 + 2$.

Solution

By the chain rule,

$$\frac{dy}{dx} = \frac{dy}{du}\frac{du}{dx} = (3u^2 - 6u)(2x)$$

But $u = x^2 + 2$, so

$$\frac{dy}{dx} = [3(x^2 + 2)^2 - 6(x^2 + 2)](2x) = 6x^3(x^2 + 2) \qquad \square$$

We conclude this section with a "proof" of the chain rule. Actually, the argument we are about to give is not a rigorous proof, because at one crucial point we commit the mathematical error of dividing by a quantity that may be zero. (As you read the "proof," see if you can spot this error.) Fortunately, it is possible to modify our argument to produce a completely rigorous proof of the chain rule. However, the rigorous proof is subtle and rather technical, and will not be given here. It can be found in more advanced calculus texts.

Let $f(x) = g(h(x))$. Our goal is to show that $f'(x) = g'(h(x))h'(x)$. To compute $f'(x)$, we consider the quotient

$$\frac{f(x + \Delta x) - f(x)}{\Delta x} = \frac{g(h(x + \Delta x)) - g(h(x))}{\Delta x}$$

We rewrite this quotient by multiplying its numerator and denominator by the quantity $h(x + \Delta x) - h(x)$. After regrouping the terms we get

$$\frac{f(x + \Delta x) - f(x)}{\Delta x} = \frac{g(h(x + \Delta x)) - g(h(x))}{h(x + \Delta x) - h(x)} \frac{h(x + \Delta x) - h(x)}{\Delta x}$$

The first quotient on the right-hand side of this equation will look more familiar if we let $\Delta h = h(x + \Delta x) - h(x)$. Then, $h(x + \Delta x) = h(x) + \Delta h$, and so

$$\frac{f(x + \Delta x) - f(x)}{\Delta x} = \frac{g(h(x) + \Delta h) - g(h(x))}{\Delta h} \frac{h(x + \Delta x) - h(x)}{\Delta x}$$

We get the derivative of f by letting Δx approach zero. The quotient $[h(x + \Delta x) - h(x)]/\Delta x$ will approach $h'(x)$, the derivative of h. Moreover, as Δx approaches zero, so does Δh. (Why?) Hence $[g(h(x) + \Delta h) - g(h(x))]/\Delta h$ approaches $g'(h(x))$, and so

$$f'(x) = g'(h(x))h'(x)$$

which is precisely the formula we were seeking.

Were you able to spot the flaw in this "proof"? It occurred toward the beginning of the argument, when we multiplied the numerator and denominator of a quotient by $h(x + \Delta x) - h(x)$. This quantity can conceivably be zero, even though Δx is not.

Problems

1. Differentiate the following functions.
   (a) $f(x) = (3x^4 - 7x^2 + 9)^5$
   (b) $h(t) = (t^5 - 4t^3 - 7)^8$
   (c) $y = \sqrt{5x^6 - 12}$
   (d) $g(t) = (5t^9 - 6t + 2)^{-3}$
   (e) $h(u) = 1/\sqrt{5u^3 + 2}$
   (f) $y = [(t + 1)/(t - 1)]^{2/3}$
   (g) $f(x) = \sqrt[3]{x^2/(3x - 5)}$
   (h) $g(u) = (3u + 9)^{12}(1 - u^2)^6$
   (i) $f(t) = (t^2 - t + 2)\sqrt{t^5 - 1}$
   (j) $y = (x^2 - x + 2)^9\sqrt{x^5 - 1}$
   (k) $h(x) = (3x + 1)^5/(x^2 - 2)^4$
   (l) $f(x) = \sqrt{(x^2 + 1)^5 - 7x}$

2. For each of the following functions, find the equation of the tangent line at the point $(x, f(x))$ for the specified value of x.
   (a) $f(x) = (3x^2 + 1)^2$; $x = -1$
   (b) $f(x) = (x - 1)/\sqrt{5x + 4}$; $x = 1$
   (c) $f(x) = (x^2 - 3)^5(2x - 1)^3$; $x = 2$
   (d) $f(x) = [(x + 1)/(x - 1)]^3$; $x = 3$

3. Find $f'(x)$ if $f(x) = g(h(x))$.
   (a) $g(u) = u^4 + 3u^3 - u^2$, $h(x) = 3x + 2$
   (b) $g(u) = \sqrt{u}$, $h(x) = x^2 + 2x - 3$

(c) $g(u) = 1/u$, $h(x) = 3x^2 + 5$
(d) $g(u) = u^2$, $h(x) = 1/(x-1)$
(e) $g(u) = u^2 + u - 2$, $h(x) = 1/x$

4. Find dy/dx.
   (a) $y = 4u^5 - 5u^4 + 3$, $u = x^2 + x + 1$
   (b) $y = \sqrt{u}$, $u = 1/x^2$
   (c) $y = 1/u^2$, $u = x^2 + 1$
   (d) $y = 1/\sqrt{u}$, $u = x^2 - 9$
   (e) $y = u^2 + u - 2$, $u = 2/x$

5. Compute $f'(x)$ for the specified values of x if $f(x) = g(h(x))$.
   (a) $g(u) = u^5 - 3u^2 + 6u - 5$, $h(x) = x^2 - 1$; $x = 1$
   (b) $g(u) = \sqrt{u}$, $h(x) = x^2 - 2x + 6$; $x = 3$
   (c) $g(u) = 1/u^2$, $h(x) = 1 - \sqrt{x+1}$; $x = 15$
   (d) $g(u) = u^2 - 12u + 3$, $h(x) = 1/x^2$; $x = \frac{1}{2}$

6. Compute dy/dx for the specified values of x.
   (a) $y = 3u^4 - 4u + 5$, $u = x^3 - 2x - 5$; $x = 2$
   (b) $y = 2u^3 - u^2 + 5$, $u = \sqrt{x+1}$; $x = 8$
   (c) $y = 1/u$, $u = 3 - 1/x^2$; $x = \frac{1}{2}$
   (d) $y = \sqrt{u+1}$, $u = 1 + x + 1/x$; $x = 1$

7. Differentiate the function $f(x) = (3x^2 - 9x)^2$ by two different methods, first using the chain rule and then the product rule. Compare your answers.

8. Sketch the graph of the function $f(x) = (x - 4)^4$, and use calculus to determine the lowest point on the graph.

9. (a) Use calculus to find all the points on the graph of the function $f(x) = x^2(x - 3)^4$ at which the tangent line is horizontal.
   (b) Sketch a graph showing f and its horizontal tangent lines.

*10 Use the chain rule for positive powers to prove the chain rule for negative powers. That is, prove that

$$\frac{d}{dx}(h^n) = nh^{n-1}\frac{dh}{dx}$$

for negative numbers n. (*Hint*: To begin, let $m = -n$, write h^n as $1/h^m$, and apply the quotient rule.)

5. IMPLICIT DIFFERENTIATION

For all the functions we have encountered so far, we have been able to solve for the dependent variable in terms of the independent variable. That is, we have been able to find equations, such as $y = (x^2 + 1)/(2x - 1)$, in which the dependent variable y is given *explicitly* by an expression involving the independent variable x.

Sometimes, practical problems lead to equations that are not written in this convenient explicit form; equations, such as $x^2y^3 - 6 = 5y^3 + x$ or $x^2y + 2y^3 = 3x + 2y$, for example. Because they have not been solved for y, these equations are said to define y *implicitly* as functions of x.

Suppose you have an equation that defines y implicitly as a function of x, and you want to find the derivative dy/dx. For instance, you may be interested in the slope of a line that is tangent to the curve consisting of all points (x, y) that satisfy the given equation. One approach might be to solve the equation for y explicitly and then differentiate using the techniques we have already developed. Unfortunately, it is not always possible to find y explicitly. For example, there is no obvious way to solve for y in the equation $x^2y + 2y^3 = 3x + 2y$. Moreover, even when an explicit formula for y can be found, it is often quite complicated and cannot be differentiated easily. For example, if we solve the equation $x^2y^3 - 6 = 5y^3 + x$ for y, we get $y = \sqrt[3]{(x+6)/(x^2-5)}$. The computation of dy/dx for this function would be tedious, involving both the chain rule and the quotient rule.

Fortunately, there is a simple technique, based on the chain rule, that allows us to compute dy/dx without solving for y explicitly. This technique, called *implicit differentiation*, is illustrated in the following example.

Example 5.1

Find dy/dx if $x^2y + 2y^3 = 3x + 2y$.

Solution

The given equation defines y implicitly as a function of x. Let f denote this (unknown) function. That is, let $y = f(x)$. We rewrite our equation, replacing y by $f(x)$, to get

$$x^2 f(x) + 2[f(x)]^3 = 3x + 2f(x)$$

We now differentiate both sides of this equation term by term.

$$\frac{d}{dx}[x^2 f(x)] + \frac{d}{dx}\{2[f(x)]^3\} = \frac{d}{dx}(3x) + \frac{d}{dx}[2f(x)]$$

By the product rule,

$$\frac{d}{dx}[x^2 f(x)] = 2xf(x) + x^2 f'(x)$$

By the chain rule,

$$\frac{d}{dx}\{2[f(x)]^3\} = 6[f(x)]^2 f'(x)$$

Also,

$$\frac{d}{dx}(3x) = 3 \quad \text{and} \quad \frac{d}{dx}[2f(x)] = 2f'(x)$$

Substituting these expressions into our differentiated equation we get

$$2xf(x) + x^2 f'(x) + 6[f(x)]^2 f'(x) = 3 + 2f'(x)$$

Since $f(x) = y$ and $f'(x) = dy/dx$, we can rewrite this equation as

$$2xy + x^2 \frac{dy}{dx} + 6y^2 \frac{dy}{dx} = 3 + 2\frac{dy}{dx}$$

Finally, we solve this equation for the derivative dy/dx to get

$$\frac{dy}{dx} = \frac{3 - 2xy}{x^2 + 6y^2 - 2} \qquad \square$$

Notice that the formula for the derivative in Example 5.1 contains both the independent variable x and the dependent variable y. As we shall see in Example 5.3, unfamiliar formulas of this type can be just as useful as the more conventional formulas that contain only the independent variable x.

Implicit differentiation is not really as cumbersome a process as the solution in Example 5.1 might suggest. For instance, there is no need to introduce a new letter f to stand for the dependent variable y. You should be able to differentiate the given equation directly. Just keep in mind that y is really a function of x, and remember to use the chain rule when it is appropriate.

Let us rewrite our solution to Example 5.1 in this more compact and efficient way.

Example 5.2

Find dy/dx if $x^2 y + 2y^3 = 3x + 2y$.

Solution

We think of y as a function of x and differentiate both sides of the given equation with respect to x. We get

$$2xy + x^2 \frac{dy}{dx} + 6y^2 \frac{dy}{dx} = 3 + 2\frac{dy}{dx}$$

We now solve this equation for dy/dx to obtain the formula

$$\frac{dy}{dx} = \frac{3 - 2xy}{x^2 + 6y^2 - 2} \qquad \square$$

In our final example, we use implicit differentiation to find the slope of a certain tangent line.

Example 5.3

Find the slope of the line that is tangent to the curve $x^2y^3 - 6 = 5y^3 + x$ when $x = 2$.

Solution

We first differentiate the given equation implicitly to find a formula for dy/dx. We get

$$2xy^3 + 3x^2y^2 \frac{dy}{dx} = 15y^2 \frac{dy}{dx} + 1$$

or

$$\frac{dy}{dx} = \frac{1 - 2xy^3}{3x^2y^2 - 15y^2}$$

The desired slope is the value of the derivative dy/dx when $x = 2$. Before we can compute this slope, we need to know the y coordinate of the point of tangency. To find this value of y, we substitute $x = 2$ into the original equation and solve. We get

$$4y^3 - 6 = 5y^3 + 2$$
$$y^3 = -8$$

or

$$y = -2$$

Finally, we substitute $x = 2$ and $y = -2$ into the formula for dy/dx to compute the desired slope:

$$\text{Slope} = \frac{1 - 2(2)(-2)^3}{3(2)^2(-2)^2 - 15(-2)^2} = -\frac{11}{4} \qquad \square$$

For practice, do Example 5.3 again. This time, solve for y explicitly and then compute the derivative using the chain rule and the quotient rule. Which method is easier?

Problems

1. Find dy/dx by implicit differentiation.
   (a) $xy = 1$
   (b) $x + 1/y = 2xy^2$
   (c) $2xy^2 - x^2y = y^2 - 3x + 1$
   (d) $x - 1/(y + 1)^2 = 2xy$
   (e) $(2xy + y^3)^2 = 4y + 3$
   (f) $(x + 1)/y = y^2 + 1$

2. Find the slope of the line that is tangent to the given curve for the specified value of x.
   (a) $x^2 = y^3; x = 8$
   (b) $1/x - 1/y = 2; x = \frac{1}{4}$
   (c) $x^2y^3 - 2xy = 6x + y + 1; x = 0$
   (d) $x^2/y + 1 = 3x; x = 1$

3. Find dy/dx in two ways: by implicit differentiation of the given equation, and by differentiation of an explicit formula for y. Show that the two answers are really the same.
   (a) $x^2 + y^3 = 12$
   (b) $xy + 2y = x^2$
   (c) $x + 1/y = 5$
   (d) $xy - x = y + 2$

4. Find the equations of all the lines that are tangent to the curve $x^2 + y^2 = 25$ when $x = 4$. Explain why there is more than one such line. (*Hint*: The curve is a circle.)

*5. Assume that the rule

$$\frac{d}{dx}(x^n) = nx^{n-1}$$

for differentiating power functions is true whenever n is an integer, and prove that the rule is true whenever n is a rational number. (*Hint*: Let $y = x^n$ and $n = p/q$, where p and q are integers. Rewrite the equation $y = x^{p/q}$ as $y^q = x^p$, and differentiate implicitly to get dy/dx.)

6. RELATIVE MAXIMA AND MINIMA

In Section 1, we considered a simple practical problem involving a manufacturer's profit. The graph of the profit function is sketched in Figure 16.

We observed that the maximum profit corresponded to the peak of the graph, and that this peak was the unique point on the graph at which the slope of the tangent

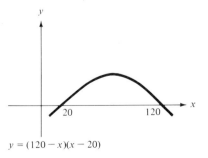

$y = (120 - x)(x - 20)$

Figure 16

was zero. In terms of derivatives, the x coordinate of the peak was the unique value of x for which $f'(x) = 0$.

The simplicity of this example is misleading. There is not, in general, a one-to-one correspondence between peaks of graphs and points at which derivatives are zero. It has probably already occurred to you, for example, that a point at which $f'(x) = 0$ might not be a peak, but the bottom of a valley instead. For example, the familiar function $f(x) = x^2$, shown in Figure 17, reaches its *lowest* point at $x = 0$ when the derivative $f'(x) = 2x$ is zero.

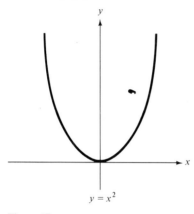

Figure 17

It is even possible for the derivative of a function to be zero at a point that is neither a peak nor a valley. Consider the function $f(x) = x^3$. Its derivative is $f'(x) = 3x^2$, which is zero when $x = 0$. The graph of f (Figure 18) "levels off" when $x = 0$, yet has neither a peak nor a valley at this point.

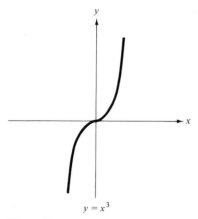

Figure 18

The situation is further complicated by the existence of functions that have peaks when $f'(x)$ is not zero. A simple example is the function

$$f(x) = \begin{cases} 1 - x & \text{if } x \geq 0 \\ 1 + x & \text{if } x < 0 \end{cases}$$

The graph of f (Figure 19) has a peak at (0, 1), although the derivative is not even defined at that point.

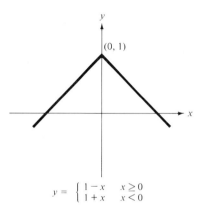

$$y = \begin{cases} 1 - x & x \geq 0 \\ 1 + x & x < 0 \end{cases}$$

Figure 19

The purpose of this section is to clarify the situation by developing a systematic procedure for locating and identifying peaks and valleys. We begin by defining the concepts of relative maximum and relative minimum. (Although these concepts are really quite simple, their formal definitions contain the technical mathematical term *open interval*. If you are unfamiliar with this term, or fail to see its significance in this context, you may ignore the formal definitions and turn directly to the intuitive descriptions of these concepts.)

RELATIVE MAXIMUM

A function f has a *relative maximum* at $x = a$ if $f(x) < f(a)$ for all values of x (except $x = a$) in some open interval containing a.

In other words, a relative maximum is a peak, a point that is higher than any nearby point on the graph. For example, the function sketched in Figure 20 has two relative maxima, one at $x = a$ and one at $x = c$.

Notice that a relative maximum need not be the highest point on the graph. It is maximal only in relation to nearby points.

RELATIVE MINIMUM

A function f has a *relative minimum* at $x = a$ if $f(x) > f(a)$ for all values of x (except $x = a$) in some open interval containing a.

A relative minimum is a point that is lower than any nearby point on the graph. For example, the function sketched in Figure 20 has two relative minima, one at $x = b$ and one at $x = d$.

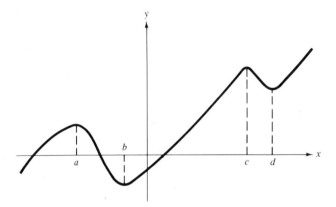

Figure 20

A relative minimum need not be the lowest point on the graph. It is minimal only in relation to nearby points. Indeed, in Figure 20, the relative minimum at $x = d$ is actually *higher* than the relative maximum at $x = a$.

We now introduce two concepts that will play a central role in the classification of relative maxima and minima.

FUNCTIONS INCREASING ON AN INTERVAL

A function f is *increasing* on an interval if, for x_1 and x_2 in the interval, $f(x_1) < f(x_2)$ whenever $x_1 < x_2$.

In other words, f is increasing if its graph is *rising* as x increases. The function in Figure 21 is increasing for $a < x < b$ and for $x > c$.

FUNCTIONS DECREASING ON AN INTERVAL

A function f is *decreasing* on an interval if, for x_1 and x_2 in the interval, $f(x_1) > f(x_2)$ whenever $x_1 < x_2$.

6. RELATIVE MAXIMA AND MINIMA

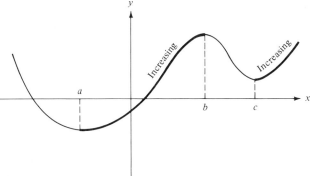

Figure 21

In other words, f is decreasing if its graph is *falling* as x increases. The function in Figure 22 is decreasing for $x < a$ and for $b < x < c$.

We can tell where a differentiable function is increasing or decreasing by checking the sign of its derivative. This is because the derivative is the slope of the tangent line. When the derivative is positive, the slope of the tangent is positive and the function is increasing (Figure 23a). When the derivative is negative, the slope of the tangent is negative and the function is decreasing (Figure 23b).

Here is a more precise statement of the situation.

GEOMETRIC SIGNIFICANCE OF THE SIGN OF THE DERIVATIVE

If $f'(x) > 0$ for each value of x in some interval, then f is increasing on that interval. If $f'(x) < 0$ for each value of x in some interval, then f is decreasing on that interval.

In the following two examples, we illustrate how one uses the sign of the derivative to determine where a function is increasing and where it is decreasing.

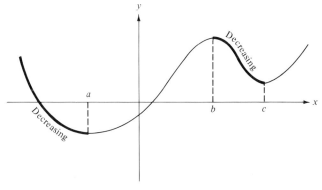

Figure 22

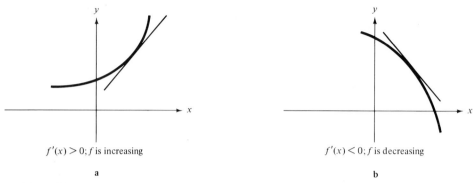

$f'(x) > 0; f$ is increasing

a

$f'(x) < 0; f$ is decreasing

b

Figure 23

Example 6.1

Determine where the function $f(x) = 2x^3 + 3x^2 - 12x - 7$ is increasing and where it is decreasing.

Solution

We begin by computing the derivative

$$f'(x) = 6x^2 + 6x - 12 = 6(x + 2)(x - 1)$$

The derivative is zero only when $x = -2$ and $x = 1$. We check the sign of the derivative for the other values of x.

When $x < -2$, both $x + 2$ and $x - 1$ are negative, so the derivative is positive. Thus f is increasing when $x < -2$.

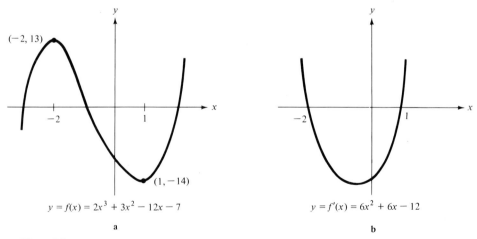

$y = f(x) = 2x^3 + 3x^2 - 12x - 7$

a

$y = f'(x) = 6x^2 + 6x - 12$

b

Figure 24

When $-2 < x < 1$, $x + 2$ is positive, while $x - 1$ is still negative. Hence the derivative is negative, and f is decreasing when $-2 < x < 1$.

Finally, when $x > 1$, both $x + 2$ and $x - 1$ are positive. Hence f is increasing when $x > 1$.

The graph of f is sketched in Figure 24a. The graph of the derivative f' is sketched beside it in Figure 24b. Comparison of these two graphs shows clearly the correlation between the sign of the derivative and the direction of the graph of f. Notice that f has a relative maximum when $x = -2$ and a relative minimum when $x = 1$. The derivative f' is zero for these two values of x. □

Example 6.2

Determine where the function $f(x) = x^{2/3}$ is increasing and where it is decreasing.

Solution

The derivative of f is $f'(x) = \frac{2}{3}x^{-1/3}$, which is positive for $x > 0$ and negative for $x < 0$. Hence f is increasing when $x > 0$ and decreasing when $x < 0$. The graph of f is shown in Figure 25a and the graph of f' beside it in Figure 25b. Notice that f has a relative minimum when $x = 0$, the value of x for which the derivative is undefined. This corresponds to the fact that when $x = 0$ the tangent to the graph of f is the y axis, whose slope is undefined.

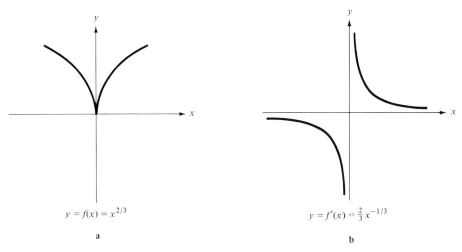

$y = f(x) = x^{2/3}$

a

$y = f'(x) = \frac{2}{3}x^{-1/3}$

b

Figure 25 □

As the two preceding examples suggest, knowledge of the intervals on which a function is increasing and decreasing leads to easy identification of its relative extrema. A relative maximum occurs when a function stops increasing and starts decreasing. That is, a relative maximum occurs when the sign of the derivative

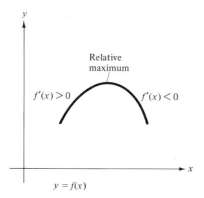

Figure 26

changes from positive to negative (Figure 26). A relative minimum occurs when a function stops decreasing and starts increasing. That is, a relative minimum occurs when the sign of the derivative changes from negative to positive (Figure 27).

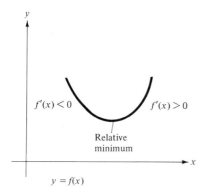

Figure 27

At the relative extremum itself, one of two things can happen. Either the derivative fails to exist, or it exists and is equal to zero.

If the derivative fails to exist, the relative extremum is "pointed." This is the case at $x = 0$ for the functions

$$f(x) = \begin{cases} 1 - x & \text{if } x \geq 0 \\ 1 + x & \text{if } x < 0 \end{cases}$$

and $f(x) = x^{2/3}$ (Figure 28).

If the derivative is zero, the curve "levels off" and has a horizontal tangent at the relative extremum. This was the case in Example 6.1 at the relative maximum $(-2, 13)$ and at the relative minimum $(1, -14)$.

The points at which the derivative of a function is zero or undefined are often called the *critical points* of the function. Every relative extremum is a critical point.

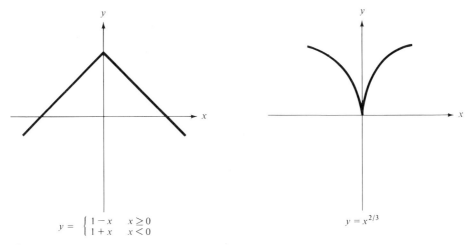

$$y = \begin{cases} 1-x & x \geq 0 \\ 1+x & x < 0 \end{cases} \qquad\qquad y = x^{2/3}$$

Figure 28

However [as the function $f(x) = x^3$ illustrates], not every critical point is a relative extremum.

The preceding observations suggest the following general procedure for finding relative maxima and minima.

TO IDENTIFY RELATIVE MAXIMA AND MINIMA

Step 1. Compute the derivative $f'(x)$.

Step 2. Find all the points $(a, f(a))$ for which $f'(a)$ is either zero or undefined. (That is, find the critical points.) These are the only possible points at which relative extrema may occur.

Step 3. Check the sign of the derivative near each of these points. If f' is positive to the left of $x = a$ and negative to the right of $x = a$, then $(a, f(a))$ is a relative maximum. If f' is negative to the left of $x = a$ and positive to the right of $x = a$, then $(a, f(a))$ is a relative minimum.

We apply this procedure in the following example.

Example 6.3

Find the relative maxima and minima of the function $f(x) = 3x^4 - 8x^3 + 6x^2 + 2$ and sketch the graph.

Solution

The derivative

$$f'(x) = 12x^3 - 24x^2 + 12x = 12x(x-1)^2$$

is zero when $x = 0$ and $x = 1$. Hence, relative extrema may occur at the corresponding points (0, 2) and (1, 3). We begin the sketch by plotting these two critical points (Figure 29a).

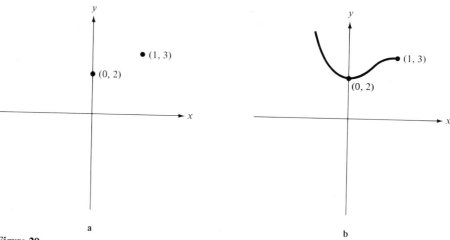

a
b
Figure 29

Next we check the sign of the derivative. If $x < 0$, then $f'(x) < 0$, and so f is decreasing; if $0 < x < 1$, then $f'(x) > 0$, and so f is increasing. Hence f has a relative minimum at (0, 2). This information has been added to the sketch in Figure 29b.

Finally, if $x > 1$, then $f'(x) > 0$, and so f is still increasing. Thus, although the graph levels off at (1, 3), f has neither a relative maximum nor a relative minimum there. We complete the sketch in Figure 30 by including this final observation.

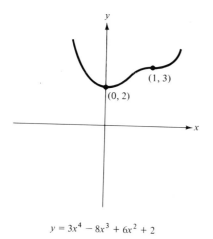

$y = 3x^4 - 8x^3 + 6x^2 + 2$

Figure 30

□

Example 6.4

Find the relative maxima and minima of the function $f(x) = x^2/(x-2)$ and sketch the graph.

Solution

$$f'(x) = \frac{(x-2)(2x) - x^2(1)}{(x-2)^2} = \frac{x(x-4)}{(x-2)^2}$$

This derivative is zero when $x = 0$ and $x = 4$, and is undefined when $x = 2$. In this case, $x = 2$ does not correspond to a relative extremum, since the function itself is undefined when $x = 2$. However, the critical points $(0, 0)$ and $(4, 8)$ may be relative extrema, and we test them by checking the sign of the derivative:

If $x < 0$, then $f'(x) > 0$, and so f is increasing.
If $0 < x < 2$, then $f'(x) < 0$, and so f is decreasing.
If $2 < x < 4$, then $f'(x) < 0$, and so f is decreasing.
If $x > 4$, then $f'(x) > 0$, and so f is increasing.

Thus f must have a relative maximum at $(0, 0)$ and a relative minimum at $(4, 8)$, as shown in Figure 31.

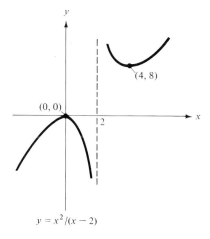

Figure 31

You may have been disturbed by the behavior of the graph in Example 6.4 near the discontinuity. According to our analysis of the derivative, the function was supposed to be decreasing on either side of the discontinuity, that is, for $0 < x < 2$ and for $2 < x < 4$. Yet the graph was much higher after $x = 2$ than before. This is not really contradictory. Since the value $x = 2$ does not belong to either of the intervals on which f is decreasing, nothing whatsoever is implied about the relation of the part of the graph before $x = 2$ to the part after $x = 2$.

In Example 6.3, we used information obtained from the derivative to assist us in sketching the graph of a function. In Example 6.4, on the other hand, we could have avoided the detailed analysis of the sign of the derivative had we first applied some elementary techniques of curve sketching. In particular, we might have observed that the function $f(x) = x^2/(x - 2)$ has an x intercept at $(0, 0)$ and a discontinuity when $x = 2$, that

$$\lim_{x \to 2^+} f(x) = +\infty \qquad \lim_{x \to 2^-} f(x) = -\infty$$

and that

$$\lim_{x \to +\infty} f(x) = +\infty \qquad \lim_{x \to -\infty} f(x) = -\infty$$

This would have led us to a rough sketch (Figure 32), which suggests that f has at least two relative extrema: a relative minimum somewhere to the right of $x = 2$ and a relative maximum at $(0, 0)$. This sketch, combined with the observation that $f'(x) = 0$ only when $x = 0$ and $x = 4$, would have told us immediately that $(0, 0)$ is the only relative maximum and that the only relative minimum occurs when $x = 4$.

When finding relative extrema, you can routinely follow the prescription on page 131. However, as the foregoing analysis suggests, you may save time and avoid tedium if you cultivate the art of combining calculus with elementary curve sketching techniques.

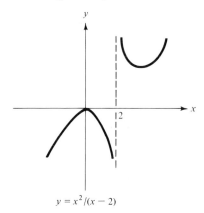

$y = x^2/(x - 2)$

Figure 32

Problems

1. Determine where each of the functions sketched below is increasing and where it is decreasing.

   (a)   (b)

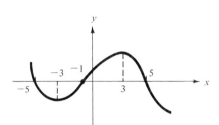
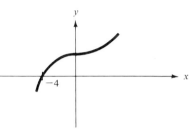

   (c)

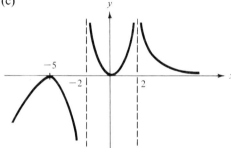

2. Sketch the graph of a function that has all of the following properties.
   (a) $f'(x) > 0$ when $x < -5$ (b) $f'(x) < 0$ when $-5 < x < 1$
   (c) $f'(x) > 0$ when $x > 1$ (d) $f(-5) = 4$ and $f(1) = -1$

3. Sketch the graph of a function that has all of the following properties.
   (a) $f'(x) < 0$ when $x < -1$
   (b) $f'(x) > 0$ when $-1 < x < 3$ and when $x > 3$
   (c) $f'(3) = 0$

4. Sketch the graph of a function that has all of the following properties.
   (a) $f'(x) > 0$ when $-1 < x < 3$ and when $x > 6$
   (b) $f'(x) < 0$ when $x < -1$ and when $3 < x < 6$
   (c) $f'(-1) = 0$ and $f'(6) = 0$
   (d) f' is undefined when $x = 3$

5. Sketch the graph of a function that has all of the following properties.
   (a) $f'(x) > 0$ when $x > 2$
   (b) $f'(x) < 0$ when $x < 0$ and when $0 < x < 2$
   (c) f has a discontinuity when $x = 0$

6. Determine where each of the following functions is increasing, where it is decreasing, and where its relative extrema occur. Sketch the graphs.
   (a) $f(x) = x^3 + 3x^2 + 1$
   (b) $f(x) = \frac{1}{3}x^3 - 9x + 2$
   (c) $f(x) = x^3 - 3x - 4$
   (d) $f(x) = x^5 - 5x^4 + 1$
   (e) $f(x) = 3x^5 - 5x^3$
   (f) $f(x) = x^3 + 3x^2 - 9x + 15$
   (g) $f(x) = 324x - 72x^2 + 4x^3$
   (h) $f(x) = 2x^3 + 6x^2 + 6x + 5$
   (i) $f(x) = 10x^6 + 24x^5 + 15x^4 + 3$
   (j) $f(x) = (x^2 - 1)^5$
   (k) $f(x) = (x^2 - 1)^4$
   (l) $f(x) = (x^3 - 1)^4$
   (m) $f(x) = x^2/(x - 1)$
   (n) $f(x) = 1/x^2$
   (o) $f(x) = 1/(x^2 - 9)$
   (p) $f(x) = x + 1/x$
   (q) $f(x) = 2x + 18/x + 1$
   (r) $f(x) = (x^2 - 3x)/(x + 1)$
   (s) $f(x) = 6x^2 + 12{,}000/x$
   (t) $f(x) = (x^2 + 4)/(x^2 - 4)$
   (u) $f(x) = (2x - 4)/(1 - x)$
   (v) $f(x) = 1 + x^{1/3}$
   (w) $f(x) = \sqrt{x^2 + 2x + 2}$

*7. (a) Find the relative minimum of the function $f(x) = (x - 1)(x - 5)$ and sketch the graph. Where is the minimum located in relation to the x intercepts?
   (b) Prove, in general, that the relative minimum of the function $f(x) = (x - a)(x - b)$ occurs at $(a + b)/2$, midway between the x intercepts of f.
   (c) Give an example of a function whose relative extrema do not occur midway between its x intercepts. (*Hint*: Try a polynomial of degree 3.)

8. A manufacturer can produce radios at a cost of $2 apiece. He is currently selling 4,000 radios a month to retailers at a price of $5 per radio, and he estimates that for each 50-cent increase in his price he will sell 200 fewer radios each month. (See Problem 21 on page 85.)
   (a) Express the manufacturer's monthly profit as a function of the price at which he sells the radios, and sketch this profit function.
   (b) Determine the price that should be charged to maximize the monthly profit. What is the maximal monthly profit?

9. Find constants a, b, and c so that the graph of the function $f(x) = ax^2 + bx + c$ has a relative maximum at $(5, 12)$ and crosses the y axis at $(0, 3)$.

*10. Write the equation of a function whose derivative is zero when $x = 2$, but that has neither a relative maximum nor a relative minimum when $x = 2$.

*11. Write the equation of a function that has a relative minimum when $x = 1$ and a relative maximum when $x = 4$.

*12. Find the largest and smallest values of the function $f(x) = x^3 + 3x^2 + 1$ on the interval $-3 \leq x \leq 2$.

7. ABSOLUTE MAXIMA AND MINIMA

In many practical optimization problems, we are interested in finding the largest and smallest values of a function on some interval; the *absolute* maximum and minimum on the interval. These absolute extrema may coincide with relative extrema, but not necessarily. For example, on the interval $a \leq x \leq b$, the function in Figure 33 attains its largest value at the relative maximum. However, its smallest value on the interval occurs at $x = a$ and not at a relative minimum.

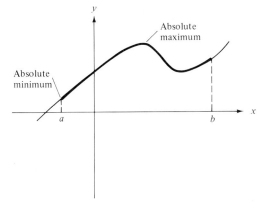

Figure 33

A *closed* interval is one that contains its endpoints. A function that is continuous throughout a *closed* interval $a \leq x \leq b$ attains an absolute maximum and an absolute minimum on that interval. An absolute extremum can occur either at a relative extremum in the interval or at an endpoint $x = a$ or $x = b$. The various possibilities are illustrated in Figure 34.

These observations lead to a simple procedure for locating absolute extrema.

TO FIND THE ABSOLUTE EXTREMA OF A FUNCTION ON AN INTERVAL

To find the absolute extrema of a continuous function f on a closed interval $a \leq x \leq b$, first locate all the critical points (the possible relative extrema) in the interval and compute the corresponding values of f. Then compute $f(a)$ and $f(b)$, the values of f at the endpoints. The largest and smallest of all these values are the absolute maximum and minimum, respectively.

This procedure is illustrated in the following example.

Example 7.1

Find the absolute maximum and minimum of the function $f(x) = 2x^3 + 3x^2 - 12x - 7$ on the interval $-3 \leq x \leq 0$.

138 DIFFERENTIATION

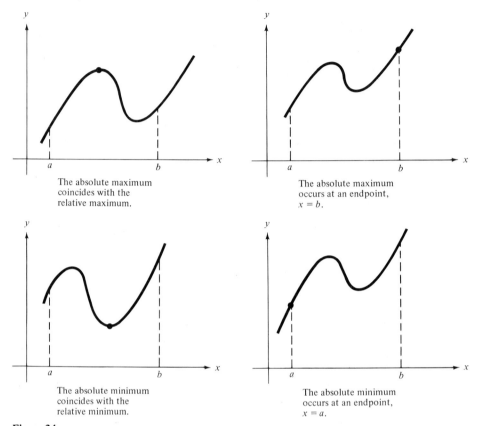

Figure 34

The absolute maximum coincides with the relative maximum.

The absolute maximum occurs at an endpoint, $x = b$.

The absolute minimum coincides with the relative minimum.

The absolute minimum occurs at an endpoint, $x = a$.

Solution

When we examined this function in Example 6.1, we found that it had a relative maximum when $x = -2$ and a relative minimum when $x = 1$. Of these, only the value $x = -2$ lies in the interval $-3 \leq x \leq 0$. Hence, $f(-2) = 13$ is one possibility for a relative extremum on this interval. The only other possibilities are the values at the endpoints $f(-3) = 2$ and $f(0) = -7$. Comparing these three values, we conclude that on the interval $-3 \leq x \leq 0$ the absolute maximum of f is 13, which occurs when $x = -2$, and the absolute minimum of f is -7, which occurs when $x = 0$.

Although we did not need a graph of the function to locate the extrema, a sketch (Figure 35) can help us visualize our results. □

If a function is not continuous at *all* points of a closed interval, it need not have an absolute maximum or minimum on that interval. Consider the following simple example.

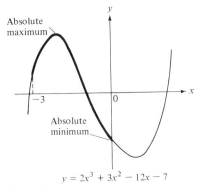

Figure 35

Example 7.2

Find the absolute maximum and minimum of the function $f(x) = 1/x^2$ on the interval $-1 \leq x \leq 2$.

Solution

This function (Figure 36) has no relative extrema. Hence the only possible absolute extrema on our interval occur at the endpoints $x = -1$ and $x = 2$. The minimum is clearly $f(2) = \frac{1}{4}$, but there is no maximum, since the function is undefined at $x = 0$ and increases without bound as x approaches zero. In this example, the discontinuity at $x = 0$ is responsible for the failure of the function to have an absolute maximum on $-1 \leq x \leq 2$. ☐

We conclude this section with a practical application of the theory of absolute extrema.

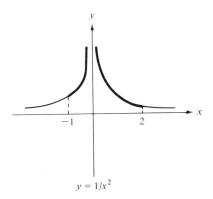

Figure 36

Example 7.3

For several weeks, a city traffic department has been recording the speed of freeway traffic flowing past a certain downtown exit. The data suggest that between the hours of noon and 7:00 P.M. on a normal weekday the speed of the traffic at the exit is approximately

$$\tfrac{1}{4}x^4 - \tfrac{10}{3}x^3 + \tfrac{27}{2}x^2 - 18x + 50 \quad \text{miles per hour}$$

where x denotes the number of hours after noon.

(a) At what time between the hours of noon and 3:00 P.M. is the traffic moving the fastest? How fast is it moving at this time?

(b) At what time between the hours of noon and 7:00 P.M. is the traffic moving the slowest? How fast is it moving at this time?

Solution

Let $f(x) = \tfrac{1}{4}x^4 - \tfrac{10}{3}x^3 + \tfrac{27}{2}x^2 - 18x + 50$. We want to find the absolute maximum of $f(x)$ on the interval $0 \leq x \leq 3$, and the absolute minimum of $f(x)$ on the interval $0 \leq x \leq 7$. We begin by finding the critical points of f. Since

$$f'(x) = x^3 - 10x^2 + 27x - 18$$
$$= (x - 1)(x - 3)(x - 6)$$

the critical points occur when $x = 1$, $x = 3$, and $x = 6$. The corresponding values of $f(x)$ are $f(1) = 42\tfrac{5}{12}$, $f(3) = 47\tfrac{3}{4}$, and $f(6) = 32$.

(a) To maximize $f(x)$ on the interval $0 \leq x \leq 3$, we compare $f(1)$, $f(3)$, and $f(0)$. Since $f(0) = 50$, we conclude that the traffic is moving the fastest at noon, when the speed is 50 miles per hour.

(b) To minimize $f(x)$ on the interval $0 \leq x \leq 7$, we compare $f(1)$, $f(3)$, $f(6)$, $f(0)$, and $f(7)$. We have already computed the first four of these values. Since $f(7) = 42\tfrac{5}{12}$, we conclude that the traffic is moving the slowest at 6:00 P.M., when the speed is 32 miles per hour.

The graph of f is sketched in Figure 37.

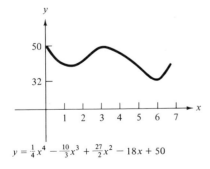

Figure 37

Problems

1. For each of the following functions, find the absolute maximum and absolute minimum (if any) on the specified interval.
   (a) $f(x) = x^3 + 3x^2 + 1$; $-3 \le x \le 2$
   (b) $f(x) = \frac{1}{3}x^3 - 9x + 2$; $0 \le x \le 2$
   (c) $f(x) = x^5 - 5x^4 + 1$; $0 \le x \le 5$
   (d) $f(x) = 3x^5 - 5x^3$; $-2 \le x \le 0$
   (e) $f(x) = 10x^6 + 24x^5 + 15x^4 + 3$; $-1 \le x \le 1$
   (f) $f(x) = (x^2 - 4)^5$; $-3 \le x \le 2$
   (g) $f(x) = x^2/(x - 1)$; $-2 \le x \le -\frac{1}{2}$
   (h) $f(x) = x^2/(x - 1)$; $1 \le x \le 4$
   (i) $f(x) = 1/x^2$; $x \ge 1$
   (j) $f(x) = x + 1/x$; $\frac{1}{2} \le x \le 3$
   (k) $f(x) = 1 + x^{1/3}$; $-8 \le x \le 27$
   (l) $f(x) = 1/(x^2 - 9)$; $0 \le x \le 2$
   (m) $f(x) = 1/(x^2 - 9)$; $x \ge 4$

2. Suppose that a certain civil rights organization, founded in 1960, had a membership of $100(2x^3 - 45x^2 + 264x)$ people x years later.
   (a) At what time between 1960 and 1974 was the membership of the organization largest? What was the membership at that time?
   (b) At what time between 1961 and 1974 was the membership of the organization the smallest? What was the membership at that time?
   (c) Sketch the membership function for $0 \le x \le 14$.
   (d) What happened to the membership between 1964 and 1971?

3. An all-news radio station has made a survey of the listening habits of local residents between the hours of 5:00 P.M. and midnight. The survey indicates that x hours after 5:00 P.M. on a typical weeknight, $-\frac{1}{4}x^3 + \frac{27}{8}x^2 - \frac{27}{2}x + 30$ percent of the local adult population is tuned in to the station.
   (a) At what time between 5:00 P.M. and midnight are the most people listening to the station?
   (b) At what time between 5:00 P.M. and midnight are the fewest people listening?

8. PRACTICAL OPTIMIZATION PROBLEMS

We have already seen examples in which differential calculus was applied to simple optimization problems arising from practical situations. In this section, we shall analyze a few additional problems more thoroughly to illustrate some of the techniques of successful problem solving.

You may be discouraged at first by the great variety of optimization problems to which calculus can be applied. You may wonder whether there can possibly be a single strategy that will apply both to a problem in which a businessman is to maximize his profit and to one in which a specified area is to be enclosed using the least amount of fencing.

Fortunately, these problems are not as unrelated as they may seem to be at first glance. In each, you are asked to maximize or minimize something. The first step in solving any such problem is to decide precisely what it is you are to optimize. Once you have identified this quantity, choose a letter to represent it. Some people are most comfortable using the standard letter f for this purpose. Others find it helpful to choose a letter more closely related to the quantity, such as R for revenue or A for area.

Remember that your goal is to represent the quantity to be optimized as a function of some other variable so that you can apply calculus. It is usually helpful to express the desired function in words before trying to symbolize it mathematically.

The next step is to choose an appropriate variable. Frequently the choice is obvious. In many business problems, for example, revenue, cost, and profit are most naturally regarded as functions of the number of items produced or sold. In other problems, however, you may be faced with a choice among several natural variables. Think ahead and try to choose the variable that leads to the simplest functional representation.

Now express the quantity to be optimized as a function of the variable you have chosen. In most problems, the function makes practical sense only when the variable lies in a certain interval. Once you have written the function and identified the appropriate interval, the difficult part is done and the rest is routine. To complete the problem, simply apply the familiar techniques of calculus to optimize your function on the specified interval.

Let us now apply this approach to a particular problem.

Example 8.1

A cable is to be run from a power plant on one side of a river 900 feet wide to a factory on the other side, 3,000 feet downstream. The cost of running the cable across the water is $5 per foot, while the cost over land is $4 per foot. What is the most economical route over which to run the cable?

Solution

As a preliminary step, we draw a diagram to help us visualize the situation (Figure 38). We are to find the most economical route over which to run the cable. That is, we are to *minimize* the *cost* of installing the cable. Let C denote this cost. Then,

$C = 5$ (number of feet of cable over water)

$\quad + 4$ (number of feet of cable over land)

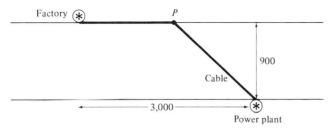

Figure 38

Notice that in drawing the diagram we have already made one assumption about the optimal route. We have assumed that the cable should be run in a *straight line* from the power plant to some point P on the opposite bank. (Why is this assumption justified?) Since we are supposed to describe the route, it will be convenient to choose a variable in terms of which we can easily locate the point P. There seem to be two reasonable choices for the variable x, as illustrated in Figures 39 and 40.

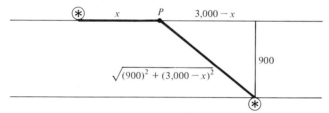

Figure 39

Before plunging into the computations, let us take a minute to choose the variable most advantageously. Consider the two possibilities.

Suppose x represents the distance from the factory to the point P (Figure 39). Then, the distance across the water from the power plant to the point P is $\sqrt{(900)^2 + (3{,}000 - x)^2}$, and we would have to minimize the function

$$C(x) = 5\sqrt{(900)^2 + (3{,}000 - x)^2} + 4x$$

On the other hand, suppose x represents the distance from P to the point directly across the river from the power plant (Figure 40). In this case we would be working

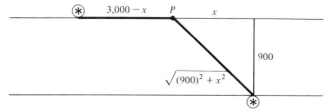

Figure 40

with the function

$$C(x) = 5\sqrt{(900)^2 + x^2} + 4(3{,}000 - x)$$

Which function is more attractive? Clearly the second one, since the term $3{,}000 - x$ is merely multiplied by 4, while in the first formula it is squared and appears under the square root sign. For this reason, we choose x as in Figure 40 and consider the cost function

$$C(x) = 5\sqrt{(900)^2 + x^2} + 12{,}000 - 4x$$

Since the distances x and $3{,}000 - x$ cannot be negative, this function is relevant only when $0 \leq x \leq 3{,}000$. Our goal, then, is to minimize the function $C(x)$ on this interval.

We compute the derivative

$$C'(x) = \frac{5x}{\sqrt{(900)^2 + x^2}} - 4$$

and set it equal to zero to get

$$\sqrt{(900)^2 + x^2} = \tfrac{5}{4}x$$

Squaring both sides of this equation and solving for x, we get

$$(900)^2 + x^2 = \tfrac{25}{16}x^2$$

$$x^2 = \tfrac{16}{9}(900)^2$$

$$x = \pm 1{,}200$$

Since $0 \leq x \leq 3{,}000$, we choose the positive value $x = 1{,}200$.

Finally, we should verify that $x = 1{,}200$ does indeed correspond to the absolute minimum of $C(x)$ on the interval $0 \leq x \leq 3{,}000$. Since the only critical point in this interval occurs when $x = 1{,}200$, we need only compare the function value at $x = 1{,}200$ with the values at the two endpoints. We find

$C(1{,}200) = 14{,}700$

$C(0) = 16{,}500$

and

$C(3{,}000) = 1{,}500\sqrt{109} > 15{,}000$

Hence we conclude that the installation cost is minimal if the cable reaches the opposite bank 1,200 feet downstream from the power plant. □

In the next example, the application of calculus to a practical problem gives an answer that makes no sense in the practical context. Additional analysis is necessary to obtain a meaningful solution.

Example 8.2

A bus company is willing to charter buses only to groups of 35 or more people. If a group contains exactly 35 people, each person pays $60. However, in larger groups, everybody's fare is reduced by 50 cents for each person in excess of 35. What size group will produce the largest revenue?

Solution

Since we are asked to maximize the revenue, let us begin by choosing the letter R to denote this revenue. Then,

$R =$ (number of people)(fare per person)

We could let x denote the number of people in the group. However, it is slightly more convenient to let x be the number of people in excess of 35. Then, $35 + x$ is the number of people in the group, and $60 - \frac{1}{2}x$ is the fare per person. Hence,

$$R(x) = (35 + x)(60 - \tfrac{1}{2}x)$$

which is graphed in Figure 41.

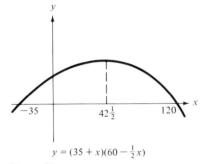

$y = (35 + x)(60 - \tfrac{1}{2}x)$

Figure 41

Since the derivative

$$R'(x) = -x + 42\tfrac{1}{2}$$

is zero when $x = 42\tfrac{1}{2}$, the function $R(x)$ attains its maximum value at $x = 42\tfrac{1}{2}$.

But x is a certain number of people, so $x = 42\tfrac{1}{2}$ cannot be the solution to the bus company's optimization problem. Indeed, the function $R(x)$ represents the revenue for *integer* values of x only. Hence we must find the largest value of $R(x)$ corresponding to an integer x. Calculus tells us where to look. R is increasing for all $x < 42\tfrac{1}{2}$ and decreasing for all $x > 42\tfrac{1}{2}$. Hence, the optimal integer value of x is either $x = 42$ or $x = 43$. Since

$$R(42) = 3{,}003$$

and

$R(43) = 3{,}003$

we conclude that the revenue will be maximal when the group contains either 42 or 43 extra people; that is, for groups of 77 or 78 people. □

The preceding example illustrates an important point. The graph of revenue as a function of x is actually a collection of discrete points corresponding to the integer values of x (Figure 42). Calculus cannot be used to study such a function. Therefore, we introduced a differentiable function

$R(x) = (35 + x)(60 - \tfrac{1}{2}x)$

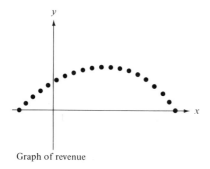

Graph of revenue

Figure 42

defined for all values of x, whose graph (Figure 43) "connected" the points in Figure 42. After applying differential calculus to this continuous mathematical model, we obtained a mathematical solution that was not the solution to the discrete practical problem, but that did suggest where to look for the practical solution.

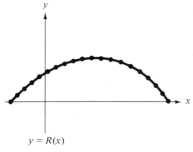

$y = R(x)$

Figure 43

In the next example, the quantity to be optimized is expressed most naturally in terms of *two* other variables. Fortunately, additional information in the problem allows us to write one of the variables in terms of the other, so that ultimately we have a function involving only a single variable.

Example 8.3

A farmer wishes to fence off a rectangular pasture along the bank of a river. The area of the pasture is to be 3,200 square yards, and no fencing is needed along the river bank. The farmer has 150 yards of fencing on hand. Is this sufficient fencing to complete the job?

Solution

This is really an optimization problem in disguise. To determine whether 150 yards of fencing is sufficient, we shall compare 150 with the *minimal* amount of fencing required.

Let F denote the amount of fencing needed to form three sides of the pasture. If we label the sides as shown in Figure 44, we see that

$F = x + 2y$

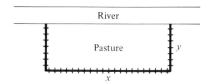

Figure 44

Since we want to express F in terms of a single variable, we look for an equation relating x and y. The fact that the area of the pasture is to be 3,200 square yards provides us with this equation. Specifically,

$xy = 3,200$

or

$y = \dfrac{3,200}{x}$

We can now substitute this into our previous expression for F to get

$F(x) = x + \dfrac{6,400}{x}$

We wish to minimize $F(x)$ for $x > 0$. Since

$F'(x) = 1 - \dfrac{6,400}{x^2}$

it follows that $F'(x) = 0$ when $x = \pm 80$. Only the positive value is of interest, so we take $x = 80$.

To verify that $x = 80$ actually minimizes F, we note that if $0 < x < 80$, then $F'(x) < 0$, and so F is decreasing. On the other hand, if $x > 80$, then $F'(x) > 0$, and so F is increasing (Figure 45). Hence $x = 80$ does indeed correspond to the minimum value of $F(x)$ for $x > 0$.

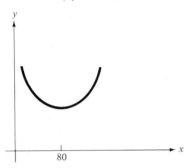

Figure 45

Finally, since $F(80) = 160$, we conclude that the 150 yards of fencing on hand is not sufficient. □

In the final example, we apply calculus to solve a typical inventory problem arising in business.

Example 8.4

A bicycle manufacturer who buys 6,000 tires a year from a distributor is trying to decide how often to order the tires. If he orders infrequently, each shipment will be large and the cost of storing the tires that are waiting to be used will be high. On the other hand, he must pay an ordering fee on each shipment for handling and transportation, so the more orders he places, the higher his ordering costs will be.

Suppose that the ordering fee is $20 per shipment, that the cost of storing one tire for a year is 96 cents, and that the bicycle manufacturer pays a fixed price of 25 cents per tire. Suppose, in addition, that the tires are used at a uniform rate throughout the year, and that each shipment arrives just as the previous shipment has been used up.

How many tires should the bicycle manufacturer order each time to minimize his costs? How often should he order the tires to minimize his costs?

Solution

Let C denote the bicycle manufacturer's total yearly cost for the tires. Then,

$C = $ storage cost $+$ ordering cost $+$ cost of the tires

8. PRACTICAL OPTIMIZATION PROBLEMS   149

We first represent the storage cost. Suppose there are N tires in each shipment. When a shipment arrives, all N tires are put into storage. The inventory then decreases linearly until there are no tires left, at which point the next shipment arrives (Figure 46a). The average number of tires in storage during the year is $N/2$, and the total yearly storage cost is the same as if $N/2$ tires were kept in storage for the entire year (Figure 46b). (This assertion, although reasonable, is not really

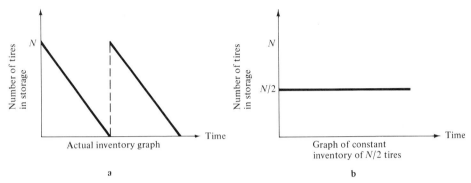

Figure 46

obvious and you have every right to be unconvinced. In Problem 15 on page 316, we shall prove this fact using integral calculus. For the present, however, let us accept it without further justification and proceed.) It follows that

Storage cost $= \dfrac{N}{2}$ (cost of storing one tire 1 year)

$= \dfrac{N}{2}(0.96)$

$= 0.48N$

The other two components of the total cost are easier to analyze.

Cost of tires $=$ (total number of tires ordered)(cost per tire)

$= 6{,}000(0.25)$

$= 1{,}500$

Also,

Ordering cost $=$ (ordering cost per shipment)(number of shipments)

Since 6,000 tires are ordered during the year and each shipment contains N tires, the number of shipments is $6{,}000/N$, and so

Ordering cost $= 20\left(\dfrac{6{,}000}{N}\right) = \dfrac{120{,}000}{N}$

Hence,

$$C(N) = 0.48N + \frac{120{,}000}{N} + 1{,}500$$

We wish to minimize this function on the interval $0 < N \leq 6{,}000$. We compute the derivative

$$C'(N) = 0.48 - \frac{120{,}000}{N^2}$$

which is zero if

$$N^2 = \frac{120{,}000}{0.48} = 250{,}000$$

or

$$N = \pm 500$$

We take the positive value $N = 500$.

It is easy to check that C is decreasing for $0 < N < 500$ and increasing for $N > 500$. Hence the value $N = 500$ does indeed minimize C on the interval $0 < N \leq 6{,}000$, and we conclude that the tires should be ordered in lots of 500 to minimize the cost.

Finally, since

$$\text{Number of shipments} = \frac{6{,}000}{N} = \frac{6{,}000}{500} = 12$$

it follows that the tires should be ordered 12 times a year (monthly) to minimize the cost. $\square$

Problems

1. There are 320 feet of fencing available to enclose a rectangular field. How should the fencing be used so that the enclosed area is as large as possible?

2. Prove that of all rectangles with a given perimeter the square has the largest area.

3. A city recreation department plans to build a rectangular playground, 3,600 square yards in area, and surround it by a fence. How can this be done using the least amount of fencing?

4. Prove that of all rectangles with a given area the square has the smallest perimeter.

5. A city recreation department is planning to build a rectangular playground having an area of 3,600 square yards. The playground will be surrounded by a fence costing $3 per yard. (See Problem 8 on page 72.)
   (a) How can this be done so that the total cost of the fencing will be minimal?
   (b) Is it a coincidence that the answer to part (a) is the same as the answer to Problem 3? Explain.

6. The product of two positive numbers is 128. The first number is added to the square of the second.
   (a) How small can this sum be?
   (b) How large can this sum be?

7. Find the number that exceeds its square by the greatest amount.

8. A college bookstore can obtain the book *Social Groupings of the American Dragonfly* from the publisher at a cost of $3 per book. The bookstore estimates that it can sell 200 copies at a price of $15 and that it will be able to sell 10 more copies for each 50-cent reduction in the price. At what price should the bookstore sell the books to maximize its profit? (See Problem 5 on page 58.)

9. During the summer, members of a local boys' club have been collecting used bottles that they plan to deliver to a glass company for recycling. So far, in 80 days, the boys have collected 24,000 pounds of glass for which the glass company currently offers a penny a pound. However, because bottles are accumulating faster than they can be recycled, the company plans to reduce by 1 cent each day the price it will pay for 100 pounds of used glass. Assume that the boys can continue to collect bottles at the same rate and that transportation costs make more than one trip to the glass company unfeasible. When is the most profitable time for the boys to conclude their summer project and deliver the bottles?

10. A closed box with a square base is to have a volume of 250 cubic feet. The material for the top and bottom of the box costs $2 per square foot, and the material for the sides costs $1 per square foot. Can the box be constructed for less than $300? (See Problem 9 on page 72.)

11. A carpenter has been asked to build an open box with a square base. The sides of the box will cost $3 per square foot, and the base will cost $4 per square foot. What are the dimensions of the box of maximal volume that can be constructed for $48.

12. An open box is to be made from an 18- by 18-inch square piece of cardboard by removing a small square from each corner and then folding up the flaps to form the sides. What are the dimensions of the box that has maximal volume? (See Problem 8 on page 59.)

*13. A wire, 20 inches long, is cut into two pieces. One of the pieces is bent to form a circle, and the other a square.
 (a) What is the smallest total area that can be enclosed in this way?
 (b) Can the total area be as large as 33 square inches?

14. A truck is 300 miles due east of a car, and is traveling west at a constant speed of 30 miles per hour. Meanwhile, the car is going north at 60 miles per hour. (See Problem 12 on page 10.) At what time will the car and truck be closest to each other? (*Hint*: You will simplify the calculation if you minimize the *square* of the distance between the car and truck, rather than the distance itself. Why is this simplification justified?)

*15. It is noon, and the hero of a popular spy story (the same fellow who escaped from the diamond smugglers in Problem 8 on page 81), is driving a Jeep through the sandy desert in the tiny principality of Alta Loma. He is 20 miles from the nearest point on a straight, paved road. Down the road 10 miles is a power station in which a band of international saboteurs has placed a time bomb set to explode at 12:50. The Jeep can travel 30 miles per hour in the sand, and 50 miles per hour on the paved road. If he arrives at the power station in the shortest possible time, how long will our hero have to defuse the bomb? (*Hint*: You may find it useful to know that $\sqrt{500} \approx 22.36$.)

16. A printer receives an order to produce a rectangular poster containing 25 square inches of print surrounded by margins of 2 inches on each side and 4 inches on the top and bottom. What are the dimensions of the smallest piece of paper he can use to make the poster? (*Hint*: An unwise choice of variables will make the calculations unnecessarily complicated.)

17. If the factory in Example 8.1 is only 1,000 feet downstream from the power plant, what is the most economical route over which to run the cable?

*18. For the summer, the company that is installing the cable in Example 8.1 has hired a young man with a Ph.D. in mathematics. The mathematician, recalling a problem from freshman calculus, asserts that no matter how far downstream the factory is located (beyond 1,200 feet), it would be most economical to have the cable reach the opposite bank 1,200 feet downstream from the power plant. The foreman, amused by the naïveté of his over-educated employee, retorts: "Any fool can see that if the factory is farther away the cable should reach the opposite bank farther downstream. It's just common sense." Who is right? And why?

19. Each year, the owner of Pierre's Gourmet Restaurant expects to sell 800 bottles of the popular white wine, Vouvrez. The cost of the wine is 85 cents per bottle, the ordering fee is $10 per shipment, and the cost of storing one bottle for a year is 40 cents. The wine is consumed at a uniform rate throughout the year,

and each shipment arrives just as the previous shipment has been used up.
(a) How many bottles should Pierre order in each shipment to minimize his costs?
(b) How often should he order the wine?
(c) How will the answers to parts (a) and (b) change if the cost of the wine is increased to 95 cents a bottle?

20. An electronics firm uses 600 cases of transistors each year. The cost of storing one case for a year is 90 cents, and the ordering fee is $30 per shipment. How frequently should the transistors be ordered to keep costs at a minimum? (Assume that the transistors are used at a uniform rate throughout the year and that each shipment arrives just as the previous shipment has been used up.)

21. A plastics firm has received an order from the city recreation department to manufacture 8,000 special Styrofoam kickboards for its summer swimming program. The firm owns 10 machines, each of which can produce 30 kickboards an hour. The cost of setting up the machines to produce these particular kickboards is $20 per machine. Once the machines have been set up, the operation is fully automated and can be supervised by a single foreman earning $4.80 per hour. (See Problem 10 on page 72.)
    (a) How many of the machines should be used to minimize the cost of producing the kickboards?
    (b) How much will the foreman earn during this production run if the optimal number of machines is used?

22. Suppose the plastics firm in Problem 21 has only six machines. How many of these should be used to minimize the cost?

*23. A manufacturer receives an order for Q items. He owns several machines, each of which can produce n items an hour. The cost of setting up a machine for the production run is s dollars. Once the machines have been set up, the operation is fully automated and can be supervised by a single foreman earning p dollars an hour.
    (a) Derive a formula for the number of machines that should be used to keep production costs as small as possible.
    (b) Prove that the production costs are minimal when the cost of setting up the machines is equal to the cost of running the machines (that is, the wages of the foreman).

24. An enterprising student at Rapid River State College has contracted to produce 150 candles in the shape of the college mascot, the otter. He plans to buy a quantity of reusable candle molds from a local metal works at $3 apiece, and then hire a freshman at $1.50 an hour to fill the molds with wax. It takes 3 hours to produce a single candle from a mold.

(a) How many molds should the student buy to keep his costs as small as possible?

(b) How much money will the freshman earn if the optimal number of molds is used?

25. In economics, the terms *marginal cost* and *marginal revenue* are used to denote the derivatives of the cost and revenue functions, respectively. An economic law states that profit is maximized when marginal revenue equals marginal cost.
   (a) Use the theory of relative extrema to explain why this law is true.
   (b) What assumptions about the shape of the profit graph are implicit in this law?

26. Suppose the total cost of manufacturing x items is given by the function $C(x) = 3x^2 + x + 48$.
   (a) What is the cost of manufacturing 20 items?
   (b) What is the cost of manufacturing the twentieth item?
   (c) What is the average cost per item of manufacturing 20 items?
   (d) Express the average manufacturing cost per item as a function of x.
   (e) For what value of x is the average cost the smallest?
   (f) For what value of x is the average cost equal to the marginal cost? Compare this value with your answer in part (e).
   (g) On the same axes, graph the total cost, marginal cost, and average cost functions.

*27. An economic law states that average cost is smallest when average cost equals marginal cost.
   (a) Prove this law. [*Hint*: If $C(x)$ represents the total cost of manufacturing x items, the average cost is $A(x) = C(x)/x$. Average cost will be smallest when $A'(x) = 0$. Use the quotient rule to compute $A'(x)$, then set $A'(x)$ equal to zero and the economic law will emerge.]
   (b) What assumptions about the shape of the average cost graph are implicit in this economic law?

9. THE SECOND DERIVATIVE

To solve most of the practical problems in this chapter, we have determined where an appropriate function was increasing, where it was decreasing, and where its relative extrema occurred. There are some practical problems, however, for which a more detailed analysis of the graph of a function is required. Consider the following example.

When a student writes a term paper, there may be a functional relationship, like the one shown in Figure 47, between the number of hours he has been working and the number of pages he has produced.

The graph reflects the fact that when a typical student first sits down to write, his

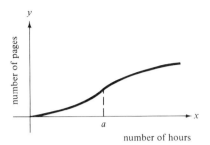

Figure 47

ideas take form gradually and his writing is frustratingly slow. As time goes on, his ideas begin to flow more freely, and he turns out pages at an increasing rate. There comes a time ($x = a$) at which he is working most efficiently, and after which fatigue sets in and his rate of production declines. This moment is known as the *point of diminishing returns*. It can be characterized geometrically as the value of x at which the curvature of the graph changes direction. In this particular example, the graph is turning in a counterclockwise direction before $x = a$, and in a clockwise direction after $x = a$.

Our primary goal in this section is to find a procedure for determining the direction in which a graph turns. The procedure will involve not merely the derivative of a function, but the function's *second derivative*, the derivative of the derivative. As an extra dividend, we will find that the second derivative can also be used to classify the critical points of functions. Let us begin by defining the second derivative.

THE SECOND DERIVATIVE

Suppose f' is the derivative of a function f. The *second derivative* of f is the derivative of f', and is denoted by f''.

The derivative f' is sometimes called the *first derivative* to further distinguish it from f''. If the function is denoted by y, the symbol d^2y/dx^2 is often used instead of f'' to denote the second derivative.

Let us try two simple examples.

Example 9.1

Compute the second derivative of the function $f(x) = 5x^4 - 3x^2 - 3x + 7$.

Solution

The first derivative is

$$f'(x) = 20x^3 - 6x - 3$$

To compute the second derivative, we simply differentiate again:

$$f''(x) = 60x^2 - 6$$

Example 9.2

Compute the second derivative of the function $y = x^2/(x-3)$.

Solution

$$\frac{dy}{dx} = \frac{2x(x-3) - x^2}{(x-3)^2} = \frac{x^2 - 6x}{(x-3)^2}$$

Hence,

$$\frac{d^2y}{dx^2} = \frac{(x-3)^2(2x-6) - 2(x^2-6x)(x-3)}{(x-3)^4}$$

$$= \frac{18}{(x-3)^3} \qquad \square$$

The next definition makes precise the notions of curvature with which we began this section.

CONCAVITY

A curve is said to be *concave downward* if its tangent turns in a clockwise direction as it moves along the curve from left to right.

A curve is said to be *concave upward* if its tangent turns in a counterclockwise direction as it moves along the curve from left to right.

The curve in Figure 48, for example, is concave upward for $x < a$ and concave downward for $x > a$.

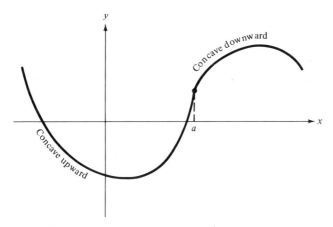

Figure 48

Notice that when the graph of a function $y = f(x)$ is concave upward the slope of its tangent increases as x increases (Figure 49a). On the other hand, when the graph is concave downward the slope of its tangent decreases as x increases (Figure 49b). Using these geometric observations, we can now determine the connection between concavity and second derivatives.

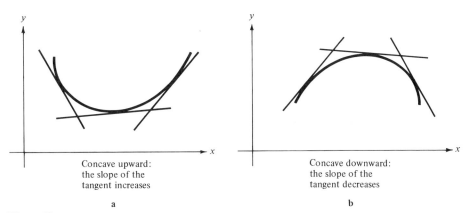

Concave upward: the slope of the tangent increases

a

Concave downward: the slope of the tangent decreases

b

Figure 49

Suppose the second derivative f'' is positive on an interval. This implies that the first derivative f' must be increasing on the interval. But f' is the slope of the tangent. Hence the slope of the tangent is increasing, and so the graph of f is concave upward on the interval. On the other hand, if f'' is negative on an interval, then f' is decreasing. This implies that the slope of the tangent is decreasing, and so the graph of f is concave downward on the interval.

Let us summarize these important observations. For convenience, we will be slightly imprecise and speak of the *concavity of a function*, when we actually mean the concavity of its graph.

GEOMETRIC SIGNIFICANCE OF THE SIGN OF THE SECOND DERIVATIVE

If $f''(x) > 0$ for each value of x in some interval, then f is concave upward on that interval.

If $f''(x) < 0$ for each value of x in some interval, then f is concave downward on that interval.

Do not confuse concavity with the concepts of functional increase and decrease. A function that is concave upward on an interval may be increasing or decreasing on that interval. Similarly, a function that is concave downward may be increasing or decreasing. The four possibilities are illustrated in Figure 50.

A point at which the concavity of a function changes is called an *inflection point*. The curve in Figure 48, for example, had an inflection point when $x = a$. In our

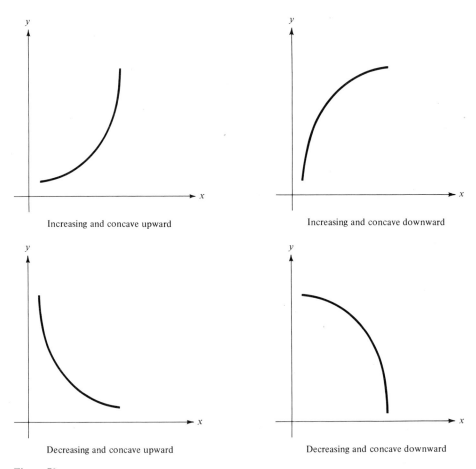

Figure 50

initial example on the student's writing efficiency, the inflection point corresponded to the point of diminishing returns. At an inflection point, the second derivative is either zero or undefined.

By combining the analysis of concavity with our other curve sketching techniques we can often obtain highly accurate graphs of functions. Consider the following example.

Example 9.3

Determine where the function $f(x) = \frac{1}{3}x^3 - x^2 - 3x + 4$ is increasing, decreasing concave upward, and concave downward. Find the relative extrema and inflection points and sketch the graph.

Solution

The first derivative

$$f'(x) = x^2 - 2x - 3 = (x-3)(x+1)$$

is zero when $x = 3$ and $x = -1$. The critical points are therefore $(-1, \frac{17}{3})$ and $(3, -5)$. Moreover:

If $x < -1$, then $f'(x) > 0$, and so f is increasing.
If $-1 < x < 3$, then $f'(x) < 0$, and so f is decreasing.
If $x > 3$, then $f'(x) > 0$, and so f is increasing.

Hence, $(-1, \frac{17}{3})$ is a relative maximum, and $(3, -5)$ is a relative minimum.

Now let us consider the second derivative

$$f''(x) = 2x - 2 = 2(x-1)$$

This is zero when $x = 1$. Moreover:

If $x < 1$, then $f''(x) < 0$, and so f is concave downward.
If $x > 1$, then $f''(x) > 0$, and so f is concave upward.

Since the concavity changes when $x = 1$, the point $(1, \frac{1}{3})$ is an inflection point.

We may use this information to draw a detailed sketch of f (Figure 51).

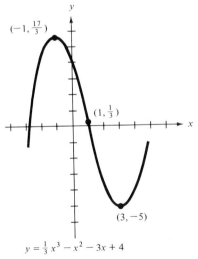

Figure 51

In the preceding example, the second derivative of f was zero at the inflection point $(1, \frac{1}{3})$. The next example illustrates that inflection points may occur where the second derivative is undefined.

Example 9.4

Find the inflection points of the function $f(x) = (x-1)^{5/3}$.

Solution

The first derivative

$$f'(x) = \tfrac{5}{3}(x-1)^{2/3}$$

is zero when $x = 1$ and is positive for all other values of x. Hence f is always increasing, and the critical point $(1, 0)$ is neither a relative maximum nor a relative minimum.

The second derivative is

$$f''(x) = \tfrac{10}{9}(x-1)^{-1/3}$$

which is undefined when $x = 1$, negative when $x < 1$, and positive when $x > 1$. Hence f is concave downward when $x < 1$, is concave upward when $x > 1$, and has an inflection point at $(1, 0)$. The graph of f is sketched in Figure 52.

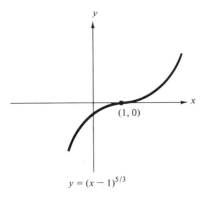

Figure 52

In the preceding example, a critical point coincided with an inflection point of the function. This phenomenon occurs again in the next example. Notice that neither of these critical points is a relative extremum. Indeed, a point at which f' equals zero can never be both a relative extremum and an inflection point. Do you see why?

Example 9.5

Determine where the function $f(x) = x^4 + 8x^3 + 18x^2 - 8$ is increasing, decreasing, concave upward, and concave downward. Find the relative extrema and inflection points and sketch the graph.

Solution

From the derivatives

$$f'(x) = 4x^3 + 24x^2 + 36x = 4x(x+3)^2$$

and

$$f''(x) = 12x^2 + 48x + 36 = 12(x+3)(x+1)$$

we can easily obtain the following information:

f is decreasing when $x < -3$ and when $-3 < x < 0$.

f is increasing when $x > 0$.

f is concave downward when $-3 < x < -1$.

f is concave upward when $x < -3$ and when $x > -1$.

The points $(0, -8)$ and $(-3, 19)$ are the critical points of f, but only $(0, -8)$, the relative minimum, is a relative extremum. There are two inflection points, $(-3, 19)$ and $(-1, 3)$. The graph of this function is sketched in Figure 53.

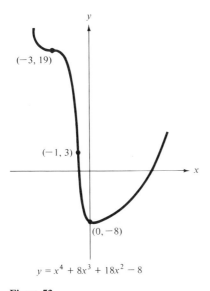

$y = x^4 + 8x^3 + 18x^2 - 8$

Figure 53

Were you able to justify the assertion that a point at which f' is zero cannot be both a relative extremum and an inflection point? Let us examine this question more carefully, for in the process we shall discover how the second derivative may be used to classify critical points. Recall that the graph of f levels off when f' is zero. That is, if $f'(a) = 0$, the graph of f near $(a, f(a))$ must behave in one of the four ways indicated in Figure 54. In Figures 54a and 54b, the point $(a, f(a))$ is a relative extremum but not an inflection point, while in Figures 54c and 54d the point $(a, f(a))$ is an inflection point but not a relative extremum.

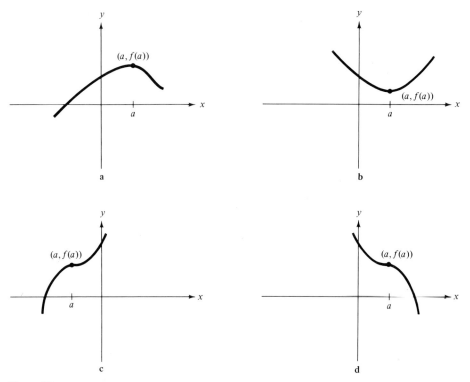

Figure 54

How do these pictures pertain to the use of the second derivative in classifying critical points? Suppose that f is a function whose first and second derivatives exist and that $f'(a) = 0$. Figure 54a suggests that if $(a, f(a))$ is a relative maximum then f must be concave downward on some interval containing a and $f''(a) \leq 0$. Similarly, Figure 54b suggests that if $(a, f(a))$ is a relative minimum then f must be concave upward on some interval containing a and $f''(a) \geq 0$. On the other hand, Figures 54c and 54d suggest that if $(a, f(a))$ is not a relative extremum then the concavity of f must change at a. That is, $(a, f(a))$ is an inflection point, and $f''(a)$ is either zero or undefined.

From these observations we may deduce the following simple test for classifying critical points.

9. THE SECOND DERIVATIVE

THE SECOND DERIVATIVE TEST

Suppose $f'(a) = 0$.

(a) If $f''(a) > 0$, then f has a relative minimum at $x = a$.

(b) If $f''(a) < 0$, then f has a relative maximum at $x = a$.

Example 9.6

Use the second derivative test to find the relative maxima and minima of the function $f(x) = 2x^3 + 3x^2 - 12x - 7$.

Solution

$$f'(x) = 6x^2 + 6x - 12 = 6(x + 2)(x - 1)$$

Hence the critical points of f are $(-2, 13)$ and $(1, -14)$. To test these points, we compute the second derivative

$$f''(x) = 12x + 6$$

We conclude that $(-2, 13)$ is a relative maximum, since $f''(-2) = -18 < 0$, and that $(1, -14)$ is a relative minimum, since $f''(1) = 18 > 0$. □

The function in the preceding example is the same one we analyzed using only the first derivative back in Example 6.1. Notice the relative ease with which we can now identify the extrema. Using the second derivative test, we compute f'' only at the critical points themselves; using the first derivative, we had to investigate the sign of f' over entire intervals.

There are, however, several disadvantages to the second derivative test. For many complicated functions, the tedious work involved in computing the second derivative may diminish the efficiency of the test. Moreover, the second derivative test is no help whatever in classifying critical points that arise when f' is undefined. [If $f'(a)$ is undefined, so is $f''(a)$.] Worst of all, if both $f'(a)$ and $f''(a)$ are zero, the second derivative test is inconclusive.

WARNING

If both $f'(a) = 0$ and $f''(a) = 0$, the point $(a, f(a))$ may be a relative maximum, a relative minimum, or an inflection point that is not a relative extremum.

We conclude this section with three simple examples illustrating what may happen when both f' and f'' are zero.

First, let $f(x) = x^4$. Then,

$$f'(x) = 4x^3 \quad \text{and} \quad f''(x) = 12x^2$$

Both f' and f'' are zero when $x = 0$. From the sign of the first derivative we conclude that f is decreasing for $x < 0$ and increasing for $x > 0$. Hence, $(0, 0)$ is a relative *minimum* (Figure 55).

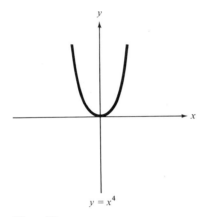

Figure 55

Next, let $f(x) = -x^4$. Again, $f'(0) = 0$ and $f''(0) = 0$, but this time f has a relative *maximum* when $x = 0$. The graph of f (Figure 56) is simply the reflection across the x axis of the graph in Figure 55.

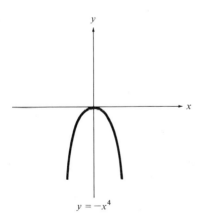

Figure 56

Finally, let $f(x) = x^3$. For this familiar function, $f'(0) = 0$ and $f''(0) = 0$, but the critical point $(0, 0)$ is neither a relative maximum nor a relative minimum (Figure 57).

9. THE SECOND DERIVATIVE

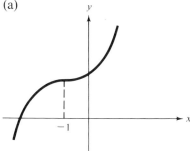

Figure 57

Problems

1. For each of the following functions, find the second derivative.
   (a) $f(x) = 5x^{10} - 6x^5 - 27x$
   (b) $y = \frac{2}{5}x^5 - 4x^3 + 9x - 6$
   (c) $f(x) = 1 - 1/x$
   (d) $y = 5\sqrt{x} - 3/x^2 + 5x^{-1/3}$
   (e) $f(x) = (x^3 + 1)^5$
   (f) $y = \sqrt{1 - x^2}$
   (g) $f(x) = 1/(x - 2)$
   (h) $y = (x + 1)/(x - 1)$

2. Determine where the functions sketched below are concave upward and where they are concave downward.

   (a)
   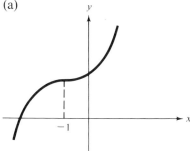

   (b)
   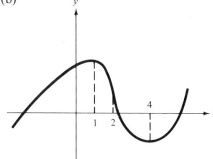

   (c)
   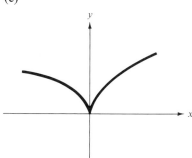

   (d)
   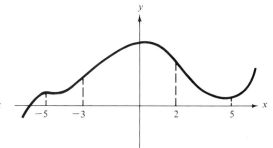

3. Sketch the graph of a function that has all of the following properties.
   (a) $f'(x) > 0$ when $x < -1$ and when $x > 3$
   (b) $f'(x) < 0$ when $-1 < x < 3$
   (c) $f''(x) < 0$ when $x < 2$
   (d) $f''(x) > 0$ when $x > 2$

4. Sketch the graph of a function that has all of the following properties.
   (a) $f'(x) > 0$ when $x < 2$ and when $2 < x < 5$
   (b) $f'(x) < 0$ when $x > 5$
   (c) $f'(2) = 0$
   (d) $f''(x) < 0$ when $x < 2$ and when $4 < x < 7$
   (e) $f''(x) > 0$ when $2 < x < 4$ and when $x > 7$

5. Sketch the graph of a function that has all of the following properties.
   (a) $f'(x) > 0$ when $x < 1$
   (b) $f'(x) < 0$ when $x > 1$
   (c) $f''(x) > 0$ when $x < 1$ and when $x > 1$
   What can you say about the derivative of f when $x = 1$?

6. Determine where each of the following functions is increasing, decreasing, concave upward, and concave downward. Graph the function, showing the relative extrema and inflection points.
   (a) $f(x) = x^3 + 3x^2 + 1$
   (b) $f(x) = \frac{1}{3}x^3 - 9x + 2$
   (c) $f(x) = \frac{1}{3}x^3 - x^2 + x$
   (d) $f(x) = x^4 - 4x^3 + 10$
   (e) $f(x) = x^5 - 5x$
   (f) $f(x) = (x - 2)^3$
   (g) $f(x) = (x - 2)^4$
   (h) $f(x) = (x^2 - 1)^3$
   (i) $f(x) = (x^2 - 1)^2$
   (j) $f(x) = x + 1/x$
   (k) $f(x) = x^2/(x - 3)$
   (l) $f(x) = 1 + 2x + 18/x$
   (m) $f(x) = (x^2 - 3x)/(x + 1)$
   (n) $f(x) = (x + 1)^{1/3}$
   (o) $f(x) = (x + 1)^{2/3}$
   (p) $f(x) = (x + 1)^{4/3}$
   (q) $f(x) = (x + 1)^{5/3}$
   (r) $f(x) = \sqrt{x^2 + 1}$

7. The derivative of a certain function is $f'(x) = x^2 - 4x$.
   (a) On what intervals is f increasing? Decreasing?
   (b) On what intervals is f concave upward? Concave downward?
   (c) Find the x coordinates of the relative extrema and inflection points of f.

8. The derivative of a certain function is $f'(x) = x^2 - 2x - 8$.
   (a) On what intervals is f increasing? Decreasing?
   (b) On what intervals is f concave upward? Concave downward?
   (c) Find the x coordinates of the relative extrema and inflection points of f.

9. Use the second derivative test to find the relative maxima and minima of the following functions.
   (a) $f(x) = x^4 - 2x^2 + 3$
   (b) $f(x) = 2x^3 + 3x^2 - 12x - 7$
   (c) $f(x) = (x^2 - 9)^2$
   (d) $f(x) = x^3 + 3x^2 + 1$
   (e) $f(x) = x + 1/x$
   (f) $f(x) = 2x + 1 + 18/x$
   (g) $f(x) = x^2/(x - 2)$

10. At what point does the tangent to the curve $f(x) = 2x^3 - 3x^2 + 6x$ have the smallest slope? What is the slope of the tangent at this point?

11. For what value of x in the interval $-1 \leq x \leq 4$ is the graph of the function $f(x) = 2x^2 - \frac{1}{3}x^3$ steepest? What is the slope of the tangent at this point?

12. Suppose the total cost of manufacturing x items is $x^3 - 30x^2 + 400x + 500$ dollars.
    (a) Find the level of production x at which the total cost function has an inflection point.
    (b) Find the level of production x at which the marginal cost is minimal. (Recall that marginal cost is the derivative of total cost.)
    (c) Is the similarity of the answers in parts (a) and (b) accidental? Explain.
    (d) Sketch the total cost and marginal cost functions.

13. Suppose the marginal cost of manufacturing x items is $6x^2 - 240x + 3{,}000$ dollars per item.
    (a) Find the level of production x at which the marginal cost is minimal.
    (b) Find the level of production at which the *total cost* function has an inflection point.
    (c) Suppose the overhead cost (the cost of producing no items) is \$1,000. Combine this fact with information gained from the marginal cost function to sketch a graph of the *total cost* function.

14. An efficiency study at a certain factory indicates that if an average worker arrives on the job at 8:00 A.M. he will have assembled $-x^3 + 6x^2 + 15x$ transistor radios x hours later (for $0 < x \leq 4$). At what time during the morning does the average worker reach his point of diminishing returns? (That is, at what time is he working most efficiently?)

15. Verify that $2 + 12y - d^2y/dx^2 = 12x^4$ if $y = x^4 + x^2$.

16. Verify that $x(d^2y/dx^2) + 2(dy/dx) = x^3$ if $y = \frac{1}{20}x^4 - 1/x + 2$.

17. Verify that $x(d^2y/dx^2) + 2(dy/dx) = x^3$ if $y = \frac{1}{20}x^4 - A/x + B$, where A and B are arbitrary constants.

*18. Consider a general third-degree polynomial $f(x) = ax^3 + bx^2 + cx + d$ (where $a \neq 0$).
    (a) Find an expression for the x coordinate of the inflection point of f.

(b) What must be true of the coefficients a, b, c, and d if the x coordinate of the inflection point is positive?

(c) What must be true of the coefficients a, b, c, and d if the inflection point is located on the y axis?

(d) Find the equation of a specific third-degree polynomial that is concave downward for $x < 0$, concave upward for $x > 0$, and has an inflection point at $(0, 2)$. Is there more than one such function? Explain.

(e) Prove that the general third-degree polynomial must be concave upward to the right of its inflection point if the coefficient a is positive, and concave downward to the right of its inflection point if the coefficient a is negative.

*19. For each of the following functions, find the second derivative d^2y/dx^2 by implicit differentiation.
(a) $xy = 1$ (b) $x^2 = y^3$ (c) $xy = y^2 + 1$

The nth derivative of a function is defined to be the derivative of the function's $(n-1)$st derivative. Thus, the third derivative is obtained by differentiating the second derivative, the fourth derivative is obtained by differentiating the third derivative, and so on. The nth derivative of the function $y = f(x)$ is denoted by $f^{(n)}(x)$, or by $d^n y/dx^n$.

20. For each of the following functions, compute the third derivative.
(a) $f(x) = x^4 - 2x^3 + 3x + 5$ (b) $y = x^3 + 2x^2 - x + 1$
(c) $f(x) = 8x^2 + 2x + 12$ (d) $y = 1/x$
(e) $f(x) = \sqrt{x}$

*21. Suppose n is a positive integer. For each of the following functions, compute the nth derivative $f^{(n)}(x)$.
(a) $f(x) = x^n$
(b) $f(x) = x^{n-1}$
(c) $f(x) = a_0 + a_1 x + a_2 x^2 + \cdots + a_n x^n$, where $a_0, a_1, a_2, \ldots, a_n$ are constants
(d) $f(x) = x^{n+1}$
(e) $f(x) = 1/x$
(f) $f(x) = \sqrt{x}$

10. RATE OF CHANGE

In Section 2, we introduced the derivative to solve a certain geometric problem: that of finding the slope of the tangent to a curve. From related geometric considerations we were subsequently led to applications of the derivative to the solution of a large class of practical optimization problems.

An alternative approach is to interpret the derivative as the rate at which one quantity changes with respect to another. Viewed in this way, the derivative again has wide practical application. It may denote, for example, the rate at which population grows, a manufacturer's marginal cost, the rate of inflation, the speed of a moving object, or the rate at which natural resources are being depleted.

You may have already sensed the connection between derivatives and rate of change. Indeed, the derivative of a function is the slope of its tangent line, and the slope of any line is the *rate* at which it is rising or falling. The purpose of this section is to make this connection more precise, and to emphasize the important distinction between the *average* and *instantaneous* rates of change. Let us begin with a familiar situation that will serve as a model for the general discussion.

Imagine that a car is moving along a straight road, and let $D(t)$ denote its distance from its starting point after t hours. Suppose we want to determine the speed of the car at a particular time t. Without access to the car's speedometer, we must resort to indirect methods.

Suppose we record the position of the car at time t, and again at some later time $t + \Delta t$. That is, suppose we know $D(t)$ and $D(t + \Delta t)$ (Figure 58). We can then compute the *average speed* of the car between the times t and $t + \Delta t$ as follows:

$$\text{Average speed} = \frac{\text{change in distance}}{\text{change in time}}$$

$$= \frac{D(t + \Delta t) - D(t)}{\Delta t}$$

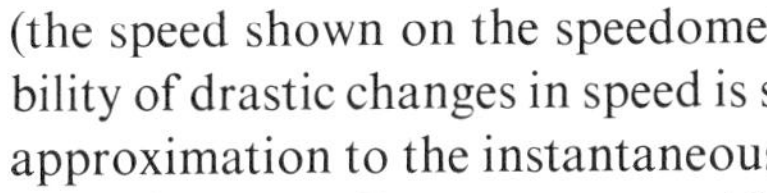

Figure 58

Since the car's speed may fluctuate during the time interval from t to $t + \Delta t$, there is no reason to think that this average speed is equal to the *instantaneous speed* (the speed shown on the speedometer) at time t. However, if Δt is small, the possibility of drastic changes in speed is small, and the average speed may be a fairly good approximation to the instantaneous speed. Indeed, this approximation will improve as Δt decreases. It seems reasonable, therefore, that the instantaneous speed at time t may be computed as the limit of the average speed as Δt approaches zero. That is

$$\text{Instantaneous speed at time } t = \lim_{\Delta t \to 0} \frac{D(t + \Delta t) - D(t)}{\Delta t}$$

Notice that this limit is precisely the derivative of D. Thus the instantaneous speed at time t is just the derivative $D'(t)$ of the distance function D.

These ideas can be extended to more general situations. Suppose that y is a function of x, say $y = f(x)$. Corresponding to a change from x to $x + \Delta x$, the variable y changes by an amount $\Delta y = f(x + \Delta x) - f(x)$. Thus the quotient

$$\frac{\text{Change in } y}{\text{Change in } x} = \frac{\Delta y}{\Delta x} = \frac{f(x + \Delta x) - f(x)}{\Delta x}$$

represents the resulting average rate of change of y with respect to x. As the interval over which we are averaging becomes shorter (that is, as Δx approaches zero), this quotient approaches a quantity that we would intuitively call the instantaneous rate of change of y with respect to x. But this limit is precisely the derivative of f. Thus the instantaneous rate of change of y with respect to x is just the derivative dy/dx.

Let us summarize these important observations.

AVERAGE RATE OF CHANGE

If $y = f(x)$, the *average rate of change of y with respect to x* corresponding to a change from x to $x + \Delta x$ is given by the quotient

$$\frac{\Delta y}{\Delta x} = \frac{f(x + \Delta x) - f(x)}{\Delta x}$$

INSTANTANEOUS RATE OF CHANGE

If $y = f(x)$, the *instantaneous rate of change of y with respect to x* is given by the derivative

$$\frac{dy}{dx} = f'(x)$$

Notice the correlation between these ideas and the geometric concepts associated with our original definition of the derivative. Suppose $y = f(x)$. The average rate of change of y with respect to x corresponding to a change from x to $x + \Delta x$ is precisely the slope of the secant line joining the points $(x, f(x))$ and $(x + \Delta x, f(x + \Delta x))$ on the graph of f (Figure 59a). The instantaneous rate of change for a particular value of x is the slope of the line tangent to the graph at $(x, f(x))$ (Figure 59b).

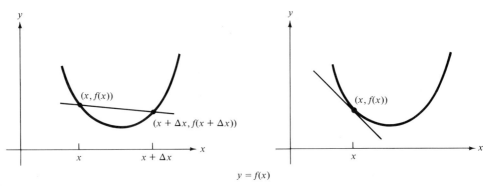

Slope of the secant: average rate of change

a

Slope of the tangent: instantaneous rate of change

b

Figure 59

10. RATE OF CHANGE

When the instantaneous rate of change of y with respect to x is positive, the derivative dy/dx is positive, and so the function is increasing (Figure 60a). Similarly, when the rate of change of y with respect to x is negative, the derivative is negative and the function is decreasing (Figure 60b). On the other hand, when the rate of change is zero, the derivative is zero, and the function levels off for an instant (Figure 61).

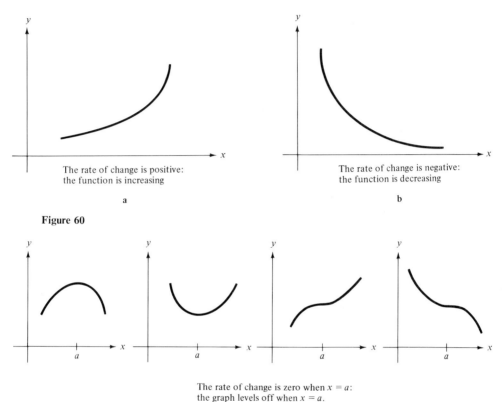

Figure 60

Figure 61

Let us illustrate these ideas by considering a simple practical problem.

Example 10.1

Suppose that t hours past midnight on a certain day the temperature in Miami was
$f(t) = -\frac{1}{4}t^2 + 6t + 50 \quad$ degrees

(a) Derive a formula for the (instantaneous) rate at which the temperature was changing with respect to time.

(b) At what rate was the temperature changing at 1:00 A.M.?

(c) By how much did the temperature actually change between 1:00 A.M. and 2:00 A.M.? Why is this answer different from the answer in part (b)?

(d) At what rate was the temperature changing at 2:00 P.M.? What is the significance of the fact that this rate was negative?

(e) At what rate was the temperature changing at noon?

Solution

(a) To find a formula for the instantaneous rate of change of temperature at time t, we simply differentiate the temperature function $f(t)$. Thus,

$$\text{Rate of change of temperature at time } t = f'(t) = -\tfrac{1}{2}t + 6 \quad \text{degrees per hour}$$

(b) At 1:00 A.M., $t = 1$, and so the temperature was changing at the rate of $f'(1) = 5\tfrac{1}{2}$ degrees per hour.

(c) The actual change in temperature between 1:00 A.M. and 2:00 A.M. was

$$f(2) - f(1) = 61 - 55\tfrac{3}{4} = 5\tfrac{1}{4} \quad \text{degrees}$$

The reason that this answer is different from the hourly rate of change computed in part (b) is that the rate was not constant, but varied during the 1-hour period from 1:00 A.M. to 2:00 A.M. The *instantaneous* rate of change in part (b) can be thought of as the change in temperature that *would* have occurred during the second hour if the rate of change of temperature had remained constant throughout that hour.

(d) The rate of change of temperature at 2:00 P.M. was $f'(14) = -1$ degree per hour. The minus sign indicates that the temperature was *decreasing*.

(e) At noon, $t = 12$, and so the temperature was changing at the rate of $f'(12) = 0$ degrees per hour.

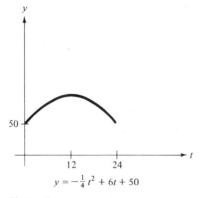

$y = -\tfrac{1}{4}t^2 + 6t + 50$

Figure 62

10. RATE OF CHANGE   173

A graph of the temperature function is sketched in Figure 62. Notice that it has a relative maximum at $t = 12$, the time at which the rate of change of temperature was zero.   □

Let us now consider a typical example from economics, in which the independent variable represents a quantity other than time.

Example 10.2

Suppose the total cost in dollars of manufacturing x units of a certain commodity is $C(x) = 3x^2 + 5x + 10$.

(a) Derive a formula for the rate of change of cost with respect to the number x of units produced.
(b) What is the rate of change of cost with respect to x when 50 units have been produced?
(c) What is the actual cost of manufacturing the 51st unit?
(d) Why are the answers to parts (b) and (c) not identical? Why are these answers approximately the same?

Solution

(a) The rate of change of C with respect to x is the derivative $C'(x) = 6x + 5$ dollars per unit. Economists call this derivative the *marginal cost*.

(b) When 50 units have been produced, $x = 50$ and $C'(50) = 305$ dollars per unit.

(c) The actual cost of manufacturing the 51st unit is the difference between the cost of manufacturing 51 units and the cost of manufacturing 50 units. That is,

Cost of 51st unit $= C(51) - C(50)$

$$= 8{,}068 - 7{,}760 = 308 \quad \text{dollars}$$

(d) In geometric terms, the difference between the answers to parts (b) and (c) is the difference between the slope of a tangent line and the slope of a nearby secant line (Figure 63). The marginal cost $C'(50)$ in part (b) is the slope of the line that is tangent to the cost curve $y = C(x)$ when $x = 50$. The difference between $C(51)$ and $C(50)$ in part (c) is the slope

$$\frac{\text{Change in } y}{\text{Change in } x} = \frac{C(50 + 1) - C(50)}{1}$$

of the secant joining the two points on the cost curve whose x coordinates are 50 and 51.

The answers to parts (b) and (c) are *almost* equal, because the points $(50, C(50))$ and $(51, C(51))$ are close together and happen to lie on a portion of the cost curve that is practically linear. For such points, the slope of the secant line is a good approximation to the slope of the tangent.

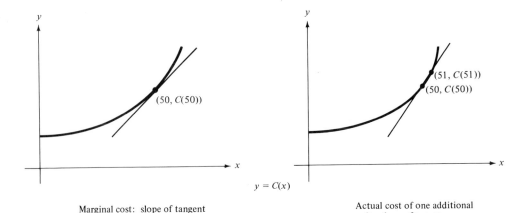

Marginal cost: slope of tangent

Actual cost of one additional unit: slope of secant

Figure 63

Because the similarity of the answers in parts (b) and (c) is fairly typical (and because it is usually easier to compute the marginal cost for one value of x than the total cost for two values of x), economists often use the marginal cost to approximate the actual cost of manufacturing one additional unit. ☐

In many practical problems, the rate at which a quantity changes is not as significant as something called the *percentage rate of change* of that quantity. For example, a yearly rate of change in population of 500 people in a city of 5 million would be negligible, while the same rate of change could have enormous impact in a town of 2,000. The percentage rate of change compares the rate of change of a quantity with the size of that quantity. In particular,

$$\text{Percentage rate of change of } Q = 100 \, \frac{\text{rate of change of } Q}{Q}$$

Thus, a yearly rate of change in population of 500 people in a city of 5 million amounts to a change of only 0.01 percent of the total population per year while the same rate of change in a town of 2,000 is equal to a yearly change of 25 percent of the total population.

PERCENTAGE RATE OF CHANGE

If $y = f(x)$, the *percentage rate of change of y with respect to x* is

$$100 \, \frac{f'(x)}{f(x)} = 100 \, \frac{dy/dx}{y}$$

Consider the following example.

Example 10.3

It is projected that x months from now the population of a certain town will be

$$P(x) = 2x + 4x^{3/2} + 5{,}000$$

people. How fast will the population be changing 9 months from now? What will be the percentage rate of change of the population at that time?

Solution

The rate of change of the population is the derivative

$$P'(x) = 2 + 6\sqrt{x}$$

Since $P'(9) = 20$, we conclude that 9 months from now the population will be growing at a rate of 20 people per month. The corresponding percentage rate of change will be

$$100\,\frac{P'(9)}{P(9)} = 100\,\frac{20}{5{,}126} \approx 0.39$$

percent of the population per month. □

The second derivative of a function may be interpreted as the rate of change of the first derivative. For example, if $D(t)$ denotes the distance from a fixed reference point at time t of an object moving in a straight line, then $D''(t)$ is the rate at which the object's speed changes with respect to time. That is, the second derivative of the distance function is the *acceleration* of the object.

Example 10.4

An object moves along a straight line so that t seconds from now its distance from a fixed reference point will be

$$D(t) = t^3 - 12t^2 + 100t + 12 \quad \text{feet}$$

Find the speed and the acceleration of the object 3 seconds from now.

Solution

We compute the first and second derivatives of D:

$$D'(t) = 3t^2 - 24t + 100$$

and

$$D''(t) = 6t - 24$$

Since $D'(3) = 55$, we conclude that 3 seconds from now the object's speed will be 55 feet per second. Since $D''(3) = -6$, we conclude that the acceleration at that time will be −6 feet per second per second. That is, the speed of the object will be *decreasing* at a rate of 6 feet per second per second. □

The rate of change of a function is *increasing* when its second derivative is positive; that is, when the function is *concave upward*. The rate of change of a function is *decreasing* when its second derivative is negative; that is, when the function is *concave downward*. The various possibilities are illustrated in Figure 64.

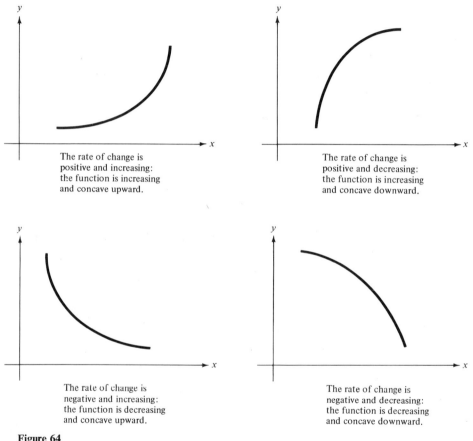

Figure 64

In economics, statements about changes in the rate of inflation are actually statements about a second derivative. Suppose, for example, that $p(x)$ is the price of a certain commodity at time x. The first derivative $p'(x)$ measures the rate at which this price is increasing or decreasing. The second derivative $p''(x)$ tells us how the rate of increase or decrease is changing. For example, if x is measured in months and p in dollars, and if $p'(x) = 0.5$, then the price is increasing at a rate of 50 cents a month at time x. If, in addition, $p''(x) = -0.02$, then the rate of the price increase *decreases* at a rate of 2 cents a month per month. In this situation, we would say that the price is increasing, although the rate of inflation is decreasing. (Which of the pictures in Figure 64 reflects this economic situation?)

Problems

1. For each of the functions sketched below, determine where the rate of change of y with respect to x is positive and where this rate is negative.

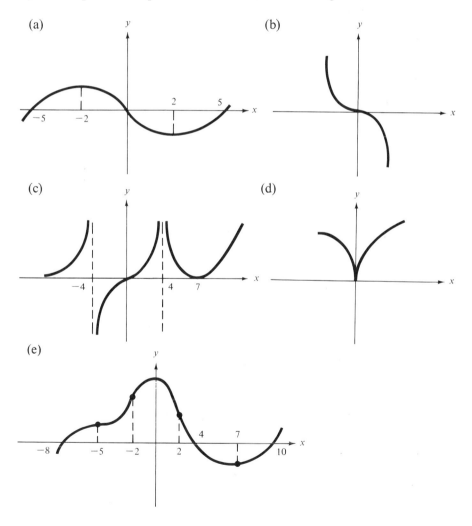

2. For each of the functions sketched in Problem 1, determine where the rate of change of y with respect to x is increasing and where this rate is decreasing.

3. Sketch the graph of a function $y = f(x)$ that has all of the following properties.
   (a) The rate of change of y with respect to x is positive when $x < -1$ and when $x > 5$.
   (b) The rate of change of y with respect to x is negative when $-1 < x < 5$.

4. Translate conditions (a) and (b) in Problem 3 into statements about the derivative of f.

5. Sketch the graph of a function $y = f(x)$ that has all of the following properties.
   (a) The rate of change of y with respect to x is positive when $x < 3$.
   (b) The rate of change of y with respect to x is negative when $x > 3$.
   (c) The rate of change of y with respect to x is increasing when $x < 0$ and when $x > 5$.
   (d) The rate of change of y with respect to x is decreasing when $0 < x < 5$.

6. Translate conditions (a), (b), (c), and (d) in Problem 5 into statements about the first and second derivatives of f.

7. A car is driven on a California freeway for 8 hours so that after t hours it has traveled

   $$D(t) = 48t + \tfrac{5}{2}t^2 - \tfrac{1}{6}t^3 \quad \text{miles}$$

   (a) Derive a formula expressing the speed of the car as a function of time.
   (b) How fast is the car going at the end of 2 hours?
   (c) How far does the car actually travel during the third hour?
   (d) At what times during the 8 hours is the speed of the car increasing?

8. (a) Derive a formula expressing the acceleration of the car in Problem 7 as a function of time.
   (b) At what rate is the speed of the car changing at the end of 6 hours? Is the speed increasing or decreasing at this time?
   (c) By how much does the speed of the car actually change during the seventh hour?
   (d) What is the car's acceleration at the end of 5 hours? Interpret your answer in practical terms.

9. It is estimated that t years from now the population of a certain suburban community will be $20 - 6/(t + 1)$ thousand people.
   (a) Derive a formula expressing the rate at which the population will be changing with respect to time.
   (b) How fast will the community's population be growing at the end of 1 year?
   (c) By how much will the population actually increase during the second year?

10. (a) At what rate will the population of the community in Problem 9 be increasing at the end of 9 years?
    (b) By how much will the population actually increase during the 10th year?
    (c) What will happen to the rate of population growth in the long run (that is, as t approaches plus infinity)?

11. At what percentage rate will the population of the community in Problem 9 be changing at the end of 1 year? At the end of 9 years?

12. An efficiency study at a certain factory indicates that if an average worker arrives on the job at 8:00 A.M. he will have assembled $-x^3 + 6x^2 + 15x$ transistor radios x hours later (for $0 < x \leq 4$).
    (a) Derive a formula for the rate at which the worker is assembling radios after x hours.
    (b) At what rate is he assembling radios at 9:00 A.M.?
    (c) How many radios does he actually assemble between 9:00 A.M. and 10:00 A.M.?
    (d) At what time during the morning is his rate of production the greatest? (Compare this with your answer to Problem 14 on page 167.)

13. Two cars leave an intersection at the same time. One travels east at a constant speed of 50 miles per hour, while the other goes north at a constant speed of 30 miles per hour. Find an expression for the rate at which the distance between the cars is changing. Does this rate depend on time, or is it constant?

14. A truck is 300 miles due east of a car and is traveling west at a constant speed of 30 miles per hour. Meanwhile, the car is going north at a constant speed of 60 miles per hour.
    (a) Derive a formula for the rate at which the distance between the car and the truck is changing with respect to time.
    (b) When is the rate of change in part (a) equal to zero? What can you say about the distance between the car and the truck at this time? (See Problem 14 on page 152.)

15. The gross annual earnings of a certain company were $\sqrt{10t^2 + t + 236}$ thousand dollars, t years after its formation in January 1970.
    (a) At what rate were the gross earnings of the company growing in January 1974?
    (b) At what percentage rate were the gross earnings growing in January 1974?

16. Suppose the total cost of manufacturing q items is

    $C(q) = 3q^2 + q + 500$ dollars

    (a) Use marginal cost to estimate the cost of manufacturing the 41st item.
    (b) Compute the actual cost of manufacturing the 41st item.

*17. At a certain factory, the total cost of manufacturing n items during the daily production run is $0.2n^2 + n + 900$ dollars. From experience, it has been determined that approximately $t^2 + 100t$ items are manufactured during the first t hours of a production run. Find the rate at which the total cost is changing with respect to time 1 hour after production commences.

*18. A retailer has bought several cases of a certain imported wine. As the wine ages, its value initially increases, but eventually the wine passes its prime and its value declines. Suppose that x years from now the value of a case will be changing at a rate of $53 - 10x$ dollars per year. Suppose, in addition, that storage rates will remain fixed at $3 a year per case.
 (a) When should the retailer sell the wine to maximize his profit on the venture?
 (b) Explain in economic terms why it is not necessary to know the purchase price of the wine to answer the question in part (a).

If an object is dropped or thrown vertically, its height (in feet) after t seconds is

$$H(t) = -16t^2 + S_0 t + H_0$$

where S_0 is the initial speed of the object, and H_0 is its initial height. For example, if the object is initially thrown downward with a speed of 45 feet per second from a height of 180 feet, then $S_0 = -45$ and $H_0 = 180$. (In Problem 9 on page 265 you will derive this formula from a simple physical law using integral calculus.) Use this formula in solving the following problems.

19. A stone is dropped (with initial speed zero) from the top of a building 144 feet high.
 (a) When will the stone hit the ground? That is, for what value of t is $H(t)$ equal to zero?
 (b) With what speed will the stone hit the ground?

20. A ball is thrown vertically upward from the ground with an initial speed of 160 feet per second.
 (a) When will the ball hit the ground?
 (b) With what speed will the ball hit the ground?
 (c) When will the ball reach its maximum height?
 (d) How high will the ball rise?

*21. A man standing on the top of a building throws a ball vertically upward. After 2 seconds the ball passes him on the way down, and 2 seconds after that it hits the ground below.
 (a) What was the initial speed of the ball?
 (b) How high is the building?

(c) What is the speed of the ball when it passes the man on the way down?
(d) What is the speed of the ball as it hits the ground?

*22. A ball is thrown vertically upward from the ground with a certain initial speed S_0.
(a) Derive a formula for the time at which the ball hits the ground.
(b) Use the result of part (a) to prove that the ball will be falling at a speed of S_0 feet per second when it hits the ground.

11. RELATED RATES

In many practical problems, a quantity is given as a function of one variable which in turn can be written as a function of a second variable. Using the chain rule, we can calculate the rate of change of the original quantity with respect to the second variable. Consider the following example.

Example 11.1

At a certain factory, the total cost of manufacturing n items during the daily production run is $0.2n^2 + n + 900$ dollars. From experience it has been determined that approximately $t^2 + 100t$ items are manufactured during the first t hours of a production run.

(a) Derive a formula for the rate of change of the total cost with respect to time.

(b) What is this rate of change 1 hour after production commences?

Solution

(a) Let $C = 0.2n^2 + n + 900$. The rate of change of cost with respect to time is dC/dt. By the chain rule,

$$\frac{dC}{dt} = \frac{dC}{dn}\frac{dn}{dt} = (0.4n + 1)(2t + 100)$$

But n, the number of items produced during the first t hours, is just $t^2 + 100t$. Hence, to express dC/dt in terms of t, we substitute $n = t^2 + 100t$ into our equation and get

$$\frac{dC}{dt} = [0.4(t^2 + 100t) + 1](2t + 100)$$

$$= 0.8t^3 + 120t^2 + 4{,}002t + 100$$

(b) After 1 hour, $t = 1$, and so

$$\frac{dC}{dt} = 0.8 + 120 + 4{,}002 + 100 = 4{,}222.8$$

That is, after 1 hour, the total cost is increasing at a rate of $4,222.80 per hour. □

It is possible to compute the numerical rate of change in part (b) of Example 11.1 without actually finding an explicit formula for dC/dt in terms of t. This time-saving technique (which we first encountered in Example 4.6) is illustrated again in the next example.

Example 11.2

For the factory in Example 11.1, compute the rate of change of the total cost with respect to time 1 hour after production commences.

Solution

As in Example 11.1, we let $C = 0.2n^2 + n + 900$ and apply the chain rule to get

$$\frac{dC}{dt} = \frac{dC}{dn}\frac{dn}{dt} = (0.4n + 1)(2t + 100)$$

Since $n = t^2 + 100t$, it follows that $n = 101$ when $t = 1$. Hence, when $t = 1$,

$$\frac{dC}{dt} = [0.4(101) + 1][2(1) + 100] = 41.4(102) = 4{,}222.8 \qquad □$$

In the preceding two examples, the variables C, n, and t were related, and the chain rule $dC/dt = (dC/dn)(dn/dt)$ gave a corresponding relation between their rates of change. For this reason, problems of this sort are frequently called *related rates problems*.

The chain rule has a simple interpretation when viewed in terms of rates of change. Suppose y is a function of u and u is a function of x. The chain rule

$$\frac{dy}{dx} = \frac{dy}{du}\frac{du}{dx}$$

says that the rate at which y is changing with respect to x is the rate at which y is changing with respect to u *times* the rate at which u is changing with respect to x. In Example 11.2, for instance, the total cost was changing at a rate of 41.4 *dollars per item*, and the factory's output was changing at a rate of 102 *items per hour*. The product, 41.4(102), gave the number of *dollars per hour* by which the cost was changing; that is, the rate of change of cost with respect to time.

In the next related rates problem, we do not have explicit formulas relating all the variables. Instead, we are given information about the rate at which one of these variables is changing with respect to another. With the aid of the chain rule (and a little elementary geometry), we can use this information to find the desired rate of change.

Example 11.3

A water tank is in the shape of an inverted cone 20 feet high with a circular base 5 feet in radius. At a certain time, the water begins to run out of the bottom of the tank at a constant rate of 2 cubic feet per minute. How fast is the water level falling when the water is 8 feet deep?

Solution

Let V denote the volume of the water in the tank after t minutes, let h be the corresponding water level, and let r be the radius of the surface of the water (Figure 65).

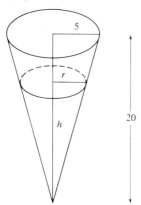

Figure 65

We know that $dV/dt = -2$ (the minus sign indicating that the volume is decreasing). Since our goal is to find dh/dt, we would like to find an equation relating h and V. We begin with the familiar formula for the volume of a cone:

$$V = \tfrac{1}{3}\pi r^2 h$$

Using the ratio

$$\frac{5}{20} = \frac{r}{h}$$

obtained from similar triangles in Figure 65, we can write r in terms of h as

$$r = \frac{h}{4}$$

Hence,

$$V = \tfrac{1}{48}\pi h^3$$

By the chain rule,

$$\frac{dV}{dt} = \frac{dV}{dh}\frac{dh}{dt} = \left(\frac{1}{16}\pi h^2\right)\frac{dh}{dt}$$

We want to find dh/dt when $h = 8$, given that $dV/dt = -2$. Making these substitutions for h and dV/dt, we get

$$-2 = 4\pi\frac{dh}{dt}$$

or

$$\frac{dh}{dt} = -\frac{1}{2\pi}$$

That is, when the water is 8 feet deep, the water level is falling at a rate of $1/(2\pi)$ feet per minute. □

Problems

1. At a certain instant, a ball is rolling toward a furnace at a rate of 80 feet per minute and is at a point where the rate of change of its temperature with respect to its distance from the furnace is 5 degrees per foot. At what rate is the temperature of the ball changing with respect to time?

2. At a certain time, the population of a city is increasing at a rate of 2,000 people per year and has reached a size at which approximately two burglaries are committed each day for each 5,000 people in the population. At what rate is the number of burglaries per day increasing with respect to time?

3. An environmental study of a certain suburban community suggests that the average daily level of carbon monoxide in the air will be $0.5p + 1$ parts per million when the population is p thousand people. It is estimated that t years from now the population of the community will be $10 + t^2/10$ thousand people. At what rate will the daily level of carbon monoxide be increasing with respect to time 4 years from now?

4. At a certain factory, the total cost of manufacturing q units during the daily production run is $0.2q^3 - 0.1q^2 + 0.5q + 600$ dollars. After t hours on a typical workday, $10\sqrt{t^2 + t + 4}$ units have been produced. Compute the rate of change of the total cost with respect to time 3 hours after production commences.

5. When electric blenders are sold for p dollars apiece, local consumers will buy a total of $8{,}000/p$ blenders a month. It is estimated that t months from now the price of the blenders will be $0.04t^{3/2} + 15$ dollars. Compute the rate at which the monthly demand for blenders will be changing with respect to time 25 months from now.

6. A manufacturer has been increasing the total production at his factory by five units each week. His weekly profit is $-x^2/10 + 72x - 140$ dollars when x units are produced during the week.
   (a) Let t denote the number of weeks during which the production increase has been in effect. Find a formula for the rate at which the manufacturer's profit is changing with respect to t if $x = 200$ when $t = 0$.
   (b) At what rate will the profit be changing with respect to time 8 weeks after the start of the production increase? Will the profit be increasing or decreasing at this time?
   (c) When will the rate of change of profit be equal to zero? What is the economic significance of the production level at this time?

7. A water tank is in the shape of an inverted cone 40 feet high with a circular base of radius 20 feet. Water is flowing into the tank at a constant rate of 80 cubic feet per minute. How fast is the water level rising when the water is 12 feet deep?

8. Suppose, in Problem 7, that water is also flowing out of the bottom of the tank. At what rate should the water be allowed to flow out so that the water level will be rising at a rate of only $\frac{1}{2}$ foot per minute when the water is 12 feet deep?

9. A man 6 feet tall is walking away from a street light 20 feet high at a rate of 7 feet per second. How fast is the length of his shadow increasing?

*10. In Problem 9, how fast is the shadow of the man's head moving along the ground?

11. A 20-foot ladder is leaning against a wall. The foot of the ladder is slipping away from the wall at 2 feet per second. At what rate is the top of the ladder moving down the wall when the bottom is 12 feet from the base of the wall?

*12. A ball is dropped from a height of 160 feet. A light is located at the same level, 10 feet away from the initial position of the ball.
   (a) Derive an expression for the rate at which the ball's shadow moves along the ground. (*Hint*: Use the formula for the height of a falling object on page 180.)
   (b) How fast is the shadow moving 1 second after the ball is dropped?

12. REVIEW PROBLEMS

1. Differentiate the following functions.
   (a) $y = 8x^5 - 4x^2 - 3x + 7$
   (b) $f(t) = \sqrt{t - 1/t^3} + \sqrt{2/t + t/3}$
   (c) $g(x) = (x + 2)/(3x - 1)$
   (d) $y = (3u^5 - 6u^3 - 7u^2 + 1)^8$
   (e) $f(x) = 1/\sqrt{x^2 + 2x + 5}$
   (f) $g(t) = (t^2 + 1)^5(3t - 1)^7$
   (g) $y = (x + 1)^3/(x - 1)^2$
   (h) $f(u) = (2u + 1)^2\sqrt{u^3 - 6}$

2. Compute $f'(x)$ for the specified values of x if $f(x) = g(h(x))$.
   (a) $g(u) = 5u^3 - 3u^2 + 2u - 1$, $h(x) = x^3 + 2x^2 + 3$; $x = 1$
   (b) $g(u) = \sqrt{u + 13}$, $h(x) = \sqrt{2x + 1}$; $x = 4$

3. Compute dy/dx for the specified values of x.
   (a) $y = 1/(u + 1)$, $u = x^3 - 2x - 5$; $x = 0$
   (b) $y = 3u^2 - 6u + 2$, $u = 1/x^2$; $x = \frac{1}{3}$

4. For each of the following functions, find the slope and the equation of the tangent line for the specified value of x.
   (a) $y = (2x^3 + 1)^2$; $x = -1$
   (b) $f(x) = (x + 1)/(x - 2)$; $x = 3$

*5. Find all the points (x, y) on the graph of $f(x) = 4x^2$ with the property that the tangent to the graph at (x, y) passes through the point $(2, 0)$.

6. Determine where the functions sketched below are increasing, decreasing, concave upward, and concave downward. Identify the relative extrema and inflection points.

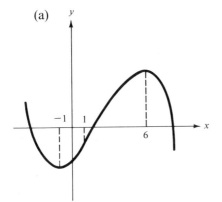
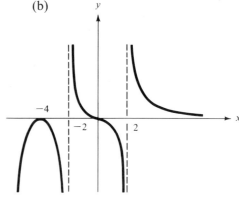

7. The (first) *derivatives* of three functions are sketched below. Determine where the functions themselves are increasing, decreasing, concave upward, and concave downward. Find the x coordinates of their relative extrema and inflection points.

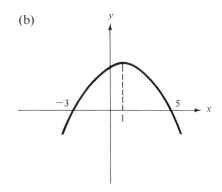

8. Sketch the graph of a function that has all of the following properties.
    (a) $f'(x) > 0$ when $x < 0$ and when $x > 5$
    (b) $f'(x) < 0$ when $0 < x < 5$
    (c) $f''(x) > 0$ when $-6 < x < -3$ and when $x > 2$
    (d) $f''(x) < 0$ when $x < -6$ and when $-3 < x < 2$

9. Determine where each of the following functions is increasing, decreasing, concave upward, and concave downward. Graph the functions, showing their relative extrema and inflection points.
   (a) $f(x) = x^2 - 6x + 1$
   (b) $f(x) = x^3 - 3x^2 + 2$
   (c) $f(x) = x^4 - 4x^3 + 3$
   (d) $f(x) = (x^2 + 3)/(x - 1)$

10. Use the second derivative test to find the relative maxima and minima of the following functions.
    (a) $f(x) = 2x^3 - 3x^2 - 36x + 5$
    (b) $f(x) = (4x^2 + 1)/x$

11. A bus company is willing to charter buses only to groups of 40 or more people. If a group contains exactly 40 people, each is charged $60. However, in larger groups, everybody's fare is reduced by 50 cents for each person in excess of 40. What size group will generate the most revenue for the bus company?

12. An open box is to be made from a piece of cardboard 8 inches long and 5 inches wide by removing a small square from each corner and then folding up the flaps to form the sides. What is the largest possible volume of the box?

13. The consumer demand for a certain commodity is $-200p + 12{,}000$ units per month when the market price is p dollars per unit. (See Problem 19 on page 85.) Express the consumers' total monthly expenditure for the commodity as a function of p, and determine the market price that will result in maximal consumer expenditure.

14. Suppose the consumer demand for a certain commodity is $D(p) = mp + b$ units per month when the market price is p dollars per unit.
    (a) Assume that $m < 0$ and $b > 0$ and sketch this demand function, labeling the points where it intersects the coordinate axes. Explain in economic terms why the assumptions about the signs of m and b are reasonable.
    (b) Express the consumers' total monthly expenditure for the commodity as a function of p, and sketch the graph of this function. Where does the graph cross the p axis?
    (c) Find an expression (in terms of m and b) for the market price at which consumer expenditure will be greatest. Show that this optimal price is the value of p that is midway between the origin and the p intercept of the demand curve.

*15. An efficiency study at a certain factory indicates that if an average worker arrives on the job at 8:00 A.M. he will have assembled $-x^3 + 6x^2 + 15x$ transistor radios x hours later (for $0 \le x \le 4$). A second efficiency study indicates that, after a 15-minute coffee break, the average worker can assemble $-\frac{1}{3}x^3 + x^2 + 23x$ radios in x hours (for $0 \le x \le 4$). Determine the time between 8:00 A.M. and 12:00 noon at which a 15-minute coffee break should be scheduled so that the average worker will assemble the maximum number of radios by lunchtime at 12:15 P.M.

16. Sketch the graph of a function $y = f(x)$ that has all of the following properties.
  (a) The rate of change of y with respect to x is negative when $x < 0$ and when $0 < x < 3$.
  (b) The rate of change of y with respect to x is zero when $x = 0$ and when $x = 3$.
  (c) The rate of change of y with respect to x is positive when $x > 3$.

17. Translate conditions (a), (b), and (c) in Problem 16 into statements about the derivative of f.

18. Sketch the graph of a function $y = f(x)$ that has all of the following properties.
  (a) The rate of change of y with respect to x is positive when $x < -5$ and when $-5 < x < 1$.
  (b) The rate of change of y with respect to x is negative when $x > 1$.
  (c) The rate of change of y with respect to x is increasing when $-5 < x < -2$ and when $x > 3$.
  (d) The rate of change of y with respect to x is decreasing when $x < -5$ and when $-2 < x < 3$.

19. Translate conditions (a), (b), (c), and (d) in Problem 18 into statements about the first and second derivatives of f.

20. The marginal profit (marginal revenue minus marginal cost) of a certain company is $100 - 2x$ dollars per unit when x units are produced. Find the level of production that will generate maximum profit for the company.

21. An object moves along a straight line so that t hours from now its distance from a fixed reference point will be $10t + 5/(t+1)$ miles.
  (a) At what speed will the object be moving 4 hours from now?
  (b) How far will the object actually travel during the fifth hour?

22. It is estimated that t years from now the population of a certain suburban community will be $20 - 6/(t+1)$ thousand people. An environmental study of the community indicates that the average daily level of carbon monoxide in the air will be $0.5\sqrt{p^2 + p + 58}$ parts per million when the population is p thousand people. Find the rate at which the level of carbon monoxide will be increasing with respect to time 2 years from now.

23. Water is flowing into a cylindrical tank 10 feet in diameter at a constant rate of 25 cubic feet per minute. How fast is the water level rising?

24. Use implicit differentiation to find dy/dx if $x^3 + 2xy = 5 + y^2$.

25. (a) Use implicit differentiation to find the slope of the line that is tangent to the curve $(xy + 1)^3 = 27x$ when $x = 1$.
    (b) Check your answer in part (a) by differentiating an explicit formula for y.

3

Exponential and Logarithmic Functions

In this chapter, we pause briefly in our development of calculus to examine two important classes of functions: exponential functions and logarithmic functions. Exponential functions arise naturally in many practical contexts, including the analysis of population growth, product reliability, radioactive decay, and compound interest. Logarithmic functions are closely related to exponential functions and are particularly useful in simplifying computational work.

1. THE NUMBER e

There are many practical problems whose solutions involve the calculation of the following expression:

$$\lim_{n \to +\infty} \left(1 + \frac{1}{n}\right)^n$$

The numerical value of this important limit will be discussed later in this section. First, let us consider a simple practical problem that gives rise to this limit.

Suppose \$1 is deposited in a savings bank that offers an interest rate of 6 percent a year. If the interest is compounded only once a year, the resulting balance at the end of 1 year will be

$1 + 0.06$ dollars

EXPONENTIAL AND LOGARITHMIC FUNCTIONS

When interest is compounded more than once a year, the interest that is added to the account during one period will itself earn interest during the subsequent periods. For example, suppose that the interest is compounded twice a year. This means that every 6 months 3 percent of the existing balance is added to the account. Hence, at the end of 6 months, the balance will be $1 + 0.06/2$ dollars, and at the end of 1 year, the balance will be

$$\left(1 + \frac{0.06}{2}\right) + \left(1 + \frac{0.06}{2}\right)\left(\frac{0.06}{2}\right) = \left(1 + \frac{0.06}{2}\right)^2 \quad \text{dollars}$$

If the interest were compounded three times a year, an amount equal to 2 percent of the existing balance would be added to the account every 4 months. In this case, the balance at the end of 4 months would be

$$1 + \frac{0.06}{3} \quad \text{dollars}$$

the balance at the end of 8 months would be

$$\left(1 + \frac{0.06}{3}\right) + \left(1 + \frac{0.06}{3}\right)\left(\frac{0.06}{3}\right) = \left(1 + \frac{0.06}{3}\right)^2 \quad \text{dollars}$$

and the balance at the end of 1 year would be

$$\left(1 + \frac{0.06}{3}\right)^2 + \left(1 + \frac{0.06}{3}\right)^2\left(\frac{0.06}{3}\right) = \left(1 + \frac{0.06}{3}\right)^3 \quad \text{dollars}$$

Similar computations can be used to show that, if the interest were compounded k times a year, the balance at the end of 1 year would be

$$\left(1 + \frac{0.06}{k}\right)^k \quad \text{dollars}$$

As the frequency with which the interest is compounded increases, the corresponding balance at the end of 1 year also increases. Hence a bank that compounds interest frequently may attract more customers than one that offers the same interest rate, but that compounds the interest less often.

Let us investigate what happens to the balance at the end of 1 year as the frequency with which the interest is compounded increases without bound. That is, let us try to compute the balance if interest is compounded not quarterly, not monthly, not daily, but *continuously*. In mathematical terms, we want to find the limit of the balance $(1 + 0.06/k)^k$ as the number k of interest periods approaches plus infinity. That is, we want to evaluate the limit

$$\lim_{k \to +\infty} \left(1 + \frac{0.06}{k}\right)^k$$

Notice the similarity between this limit and the one at the beginning of this section. The connection will become even more striking if we let $n = k/0.06$. Then,

$$\left(1 + \frac{0.06}{k}\right)^k = \left(1 + \frac{1}{n}\right)^{0.06n} = \left[\left(1 + \frac{1}{n}\right)^n\right]^{0.06}$$

Since n approaches plus infinity as k does, it follows that

$$\lim_{k \to +\infty} \left(1 + \frac{0.06}{k}\right)^k = \left[\lim_{n \to +\infty} \left(1 + \frac{1}{n}\right)^n\right]^{0.06}$$

Hence the answer to our question about the bank balance turns out to depend on the value of the limit

$$\lim_{n \to +\infty} \left(1 + \frac{1}{n}\right)^n$$

What is the numerical value of this limit? Can you guess? Many people would guess that the answer is 1. They reason as follows. As n approaches plus infinity, $1/n$ approaches zero, and so the sum $1 + 1/n$ approaches 1. But $1^n = 1$ for all values of n. Hence $\lim_{n \to +\infty} (1 + 1/n)^n = 1$.

However, an equally good case can be made to support the belief that the limit is equal to plus infinity. Consider the following argument. For any positive value of n, the sum $1 + 1/n$ is greater than 1. If a number b is greater than 1, then b^n increases without bound as n approaches plus infinity. Hence, $\lim_{n \to +\infty} (1 + 1/n)^n = +\infty$.

Actually, the value of our limit is neither 1 nor $+\infty$. The arguments that led to these answers were both fallacious for the following reason. The variable n appears twice in the expression $(1 + 1/n)^n$. In the first argument, we assumed that the n in the sum $1 + 1/n$ approached infinity first. Only after we had found the limit of $1 + 1/n$ did we consider the effect of raising this expression to the nth power. In the second argument, we let the exponent n approach infinity first. To find the actual value of the limit, we must somehow determine what happens to the expression $(1 + 1/n)^n$ as *both* n's approach infinity *simultaneously*. This is quite difficult to do, and a rigorous argument would involve advanced techniques beyond the scope of this text. It turns out that the expression $(1 + 1/n)^n$ approaches a finite number as n approaches plus infinity, and that this number, traditionally denoted by the letter e, is approximately 2.718.

THE NUMBER e

$$e = \lim_{n \to +\infty} \left(1 + \frac{1}{n}\right)^n \approx 2.718$$

If you know a programming language and have access to a computer, you might find it interesting to compute $(1 + 1/n)^n$ for some large values of n.

We shall encounter this limit again in Section 4, when we differentiate a certain logarithmic function. At that time, we shall also need to know that

$$\lim_{n \to -\infty} \left(1 + \frac{1}{n}\right)^n = e$$

This can be proved using the fact that $\lim_{n \to +\infty} (1 + 1/n)^n = e$. The proof is outlined in Problem 8.

Let us now return to our bank balance problem. Recall that we were trying to compute the balance after 1 year if $1 is deposited in a bank offering interest at a rate of 6 percent a year compounded continuously. We found that the balance is given by the expression

$$\left[\lim_{n \to +\infty} \left(1 + \frac{1}{n}\right)^n\right]^{0.06}$$

which we now recognize as simply $e^{0.06}$. Powers of e are listed in a table on page 376. According to the table, $e^{0.06} \approx 1.0618$. Hence we conclude that $1 invested at an interest rate of 6 percent a year compounded continuously will grow to (approximately) $1.0618 after 1 year. In particular, this implies that no matter how often the interest is compounded there is an upper bound of approximately $1.0618 for the possible balance after 1 year if the annual interest rate is 6 percent.

In the preceding discussion, we encountered the expression $e^{0.06}$. We can think of $e^{0.06}$ as the value of the function $f(x) = e^x$ when $x = 0.06$. In the next section, we shall examine functions of the form $f(x) = a^x$ in more detail. As our analysis of the compound interest problem suggests, the particular function $f(x) = e^x$ will prove to be especially useful.

Problems

1. Calculate $(1 + 1/n)^n$ for $n = 1, 2, 3$, and 4.

2. If you know a programming language and have access to a computer, calculate $(1 + 1/n)^n$ for $n = 50, 100, 150, \ldots, 1{,}000$.

3. Suppose $1 is deposited in a savings bank that offers an interest rate of 6 percent a year compounded quarterly (four times a year). Calculate the resulting balance at the end of 1 year and compare your answer with the balance that would be generated if the interest were compounded continuously.

4. Suppose $1 is deposited in a savings bank that offers an interest rate of 8 percent a year.
   (a) Calculate the balance after 1 year if the interest is compounded quarterly.

(b) Find a formula for the balance after 1 year if the interest is compounded k times a year.

(c) Calculate the balance after 1 year if the interest is compounded continuously.

5. How will your answers in Problem 4 change if $100 is initially deposited in the bank?

*6. Suppose $1 is deposited in a savings bank that offers an interest rate of 6 percent a year.
   (a) Find a formula for the balance after 2 years if the interest is compounded k times a year.
   (b) Calculate the balance after 2 years if the interest is compounded continuously.

*7. Suppose P dollars are deposited in a savings bank that offers an interest rate of $100r$ percent a year.
   (a) Find a formula for the balance after t years if the interest is compounded k times a year.
   (b) Find a formula for the balance after t years if the interest is compounded continuously.
   (c) Calculate the balance at the end of 10 years if the initial deposit is $5,000, the interest rate is 6 percent, and the interest is compounded continuously.

*8. Use the fact that $\lim_{n \to +\infty} (1 + 1/n)^n = e$ to prove that $\lim_{n \to -\infty} (1 + 1/n)^n = e$. [*Hint:* Let $m = -n - 1$. Show that $(1 + 1/n)^n = (1 + 1/m)^m (1 + 1/m)$ and take the limit. Notice that m approaches *plus* infinity as n approaches *minus* infinity.]

2. EXPONENTIAL FUNCTIONS

An *exponential function* is a function of the form $f(x) = a^x$, where a is a positive constant. In an exponential function, the independent variable is the *exponent* of a fixed positive number a called the *base* of the function. Notice the difference between an exponential function and a power function $f(x) = x^n$, in which the base is the variable and the exponent is the constant.

You are probably already familiar with the following four conditions that together define a^x for all *rational* values of x.

DEFINITION OF a^x IF x IS RATIONAL (AND $a > 0$)

(1) If $x = n$, where n is a positive integer, then

$$a^x = a^n = a \cdot a \cdots a$$

where n factors, a, appear in the product.

(2) $a^0 = 1$

(3) If $x = n/m$, where n and m are positive integers, then
$$a^x = a^{n/m} = \sqrt[m]{a^n} = (\sqrt[m]{a})^n$$
where $\sqrt[m]{}$ denotes the *positive* mth root.

(4) $a^{-x} = 1/a^x$

For example, $3^4 = 3 \cdot 3 \cdot 3 \cdot 3 = 81$, $3^{-4} = 1/3^4 = \frac{1}{81}$, $4^{1/2} = \sqrt{4} = 2$, $4^{-1/2} = 1/\sqrt{4} = \frac{1}{2}$, $4^{3/2} = (\sqrt{4})^3 = \sqrt{4^3} = 8$, and $4^{-3/2} = 1/(\sqrt{4})^3 = 1/\sqrt{4^3} = \frac{1}{8}$.

Let us try to sketch the function $f(x) = a^x$ (for rational x) when $0 < a < 1$ and when $a > 1$. (What happens when $a = 1$?)

If $0 < a < 1$, then a^x approaches zero as x increases without bound, and a^x approaches plus infinity as x decreases without bound. Moreover, a^x is always positive, and $a^0 = 1$. Hence the graph of a^x resembles the sketch in Figure 1a.

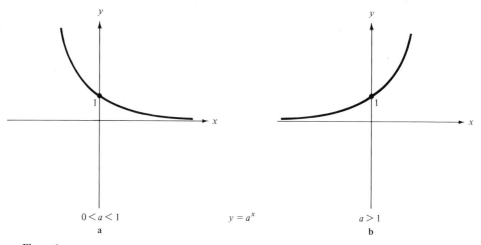

Figure 1

On the other hand, if $a > 1$, then a^x approaches plus infinity as x increases without bound, and it approaches zero as x decreases without bound. In this case, the graph of a^x resembles the sketch in Figure 1b.

Actually, these graphs should not have been drawn as unbroken curves, since a^x has been defined only for *rational* values of x. However, it turns out that because there are "so many" rational numbers there can be only *one* unbroken curve that passes through all the points (x, a^x) for which x is rational. That is, there exists a unique continuous function $f(x)$ that is defined for all real numbers x and that is equal to a^x when x is rational. When x is irrational, we *define* a^x to be the value

$f(x)$ of this unique function. (Don't be discouraged if you do not see how to justify the assertion that there is a *unique* continuous extension of a^x. This is a fairly subtle mathematical fact, and a rigorous proof is beyond the scope of this text. Fortunately, in practical work you will rarely, if ever, have to deal explicitly with a^x for an irrational value of x.)

You may recall that for rational values of the independent variable, exponential functions obey certain *laws of exponents*. It can be shown that these useful laws remain valid when the variable is allowed to assume arbitrary real values.

LAWS OF EXPONENTS

(1) $a^r a^s = a^{r+s}$

(2) $a^r/a^s = a^{r-s}$

(3) $(a^r)^s = a^{rs}$

For the rest of this chapter, we shall focus primarily on the special exponential function $f(x) = e^x$. As we have seen, this function is important because it arises naturally in connection with certain practical problems. Another reason for choosing e as our base will emerge in Section 4, when we differentiate exponential functions. We are not really restricting the scope of our investigation by concentrating on e^x since, as we shall see in Section 3, there is a simple way to get any other exponential function from this one. In particular, if $f(x) = a^x$, there is a constant k such that $f(x) = e^{kx} = (e^x)^k$.

Since $e > 1$, the graph of $f(x) = e^x$ resembles the curve in Figure 2. Notice, in particular, that e^x is always positive, $e^0 = 1$, $\lim_{x \to +\infty} e^x = +\infty$, and $\lim_{x \to -\infty} e^x = 0$.

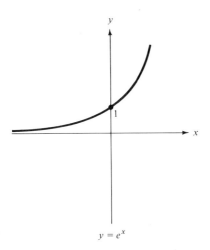

$y = e^x$

Figure 2

Functions involving e^x arise frequently in applications. In Chapter 4, Section 7, we shall use integral calculus to translate certain simple practical relationships into functions of this type. The following example illustrates how we can use the basic properties of e^x to sketch one of these functions.

Example 2.1

Sketch a graph of the function $f(x) = Ae^{-kx}$, where A and k are positive constants.

Solution

Using the laws of exponents, we can rewrite this function as

$$f(x) = A\left(\frac{1}{e^x}\right)^k$$

Clearly, $f(x)$ is always positive, and $f(0) = A$. (Why?) Moreover, since $\lim_{x \to +\infty} e^x = +\infty$, it follows that $\lim_{x \to +\infty} 1/e^x = 0$, and so $\lim_{x \to +\infty} f(x) = 0$. Finally, the fact that $\lim_{x \to -\infty} e^x = 0$ implies that $\lim_{x \to -\infty} f(x)$ is either $+\infty$ or $-\infty$. Since $f(x)$ is always positive, we conclude that $\lim_{x \to -\infty} f(x) = +\infty$. A sketch incorporating all these features is shown in Figure 3.

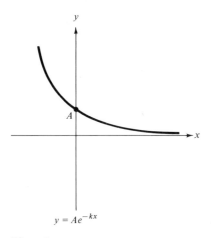

Figure 3

Functions of the form $f(x) = Ae^{-kx}$ are used in actuarial work to describe life expectancy, in physics in connection with radioactive decay, and in industry as reliability functions for certain products. Consider the following typical example.

Example 2.2

A manufacturer of electric toasters has made a statistical study of the reliability of his product. The study indicates that the fraction $f(x)$ of his toasters still in working condition after having been used for x years is approximately

$$f(x) = e^{-0.2x}$$

(a) What fraction of the toasters can be expected to work for at least 3 years?

(b) What fraction of the toasters can be expected to fail during their third year of use?

Solution

(a) Using the table of exponential functions on page 376, we find that

$$f(3) = e^{-0.6} \approx 0.5488$$

That is, about 55 percent of the toasters can be expected to work for at least 3 years.

(b) To find the fraction of the toasters that can be expected to fail during the third year, we subtract the fraction that can be expected to be working after 4 years from the fraction that can be expected to be working after 3 years. We get

$$f(3) - f(4) = e^{-0.6} - e^{-0.8}$$
$$\approx 0.5488 - 0.4493 = 0.0995$$

Hence approximately 10 percent of the toasters can be expected to fail during their third year. ☐

In the next example, we sketch a function of the form $f(x) = C - Ae^{-kx}$. The relationship between a factory worker's productivity and the amount of training he has received can often be expressed (or at least approximated) by such a function. In physics, a function of this form gives the temperature of a cool object x minutes after it has been placed in a warmer environment.

Example 2.3

Sketch a graph of the function $f(x) = C - Ae^{-kx}$, where C, A, and k are positive constants.

Solution

It would not be hard to obtain a sketch using the procedure illustrated in Example 2.1. (Try it.) However, for the sake of reviewing the elementary techniques of translation and reflection, we take a different approach.

The graph of the function $y = -Ae^{-kx}$ (Figure 4a) is just the reflection across the x axis of the graph of $y = Ae^{-kx}$ that we sketched in Example 2.1. To get the graph of $f(x) = C - Ae^{-kx}$, we simply raise the graph in Figure 4a by C units, as shown in Figure 4b. (Notice that $f(0) = C - A$, $\lim_{x \to +\infty} f(x) = C$, $\lim_{x \to -\infty} f(x) = -\infty$, and $f(x) < C$ for all x.)

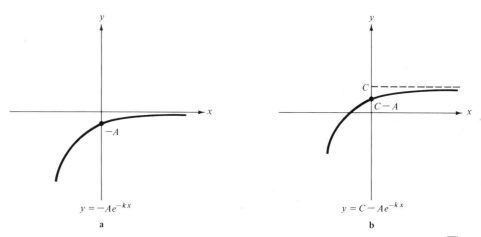

Figure 4

To conclude this section, let us return to the analysis of compound interest that we began in Section 1. We found that if \$1 were deposited in a bank offering interest at an annual rate of 6 percent compounded k times a year, the balance at the end of 1 year would be

$$\left(1 + \frac{0.06}{k}\right)^k \quad \text{dollars}$$

We now generalize this result and derive a formula for the balance after t years if the initial deposit is P dollars and the annual interest rate is $100r$ percent. (The letter P stands for *principal*.) It is easy to see that the balance after 1 year will be

$$P\left(1 + \frac{r}{k}\right)^k \quad \text{dollars}$$

This says that to compute the balance at the end of a year, we simply multiply the expression $(1 + r/k)^k$ by the balance in the account at the beginning of the year.

We now use this observation to compute the balance after 2 years. At the beginning of the second year, the balance is $P(1 + r/k)^k$ dollars. To compute the balance at the end of the second year, we simply multiply this amount by the expression $(1 + r/k)^k$. Hence, the balance after 2 years will be

$$P\left(1 + \frac{r}{k}\right)^k \left(1 + \frac{r}{k}\right)^k = P\left(1 + \frac{r}{k}\right)^{2k} \quad \text{dollars}$$

Similarly, the balance after 3 years will be $P(1 + r/k)^{3k}$ and, in general, the balance $B(t)$ after t years will be

$$B(t) = P\left(1 + \frac{r}{k}\right)^{kt} \quad \text{dollars}$$

If we let k approach plus infinity in this formula, we will obtain the corresponding expression for the balance after t years if the interest is compounded continuously. To simplify the calculation, we let $n = k/r$. Then,

$$P\left(1 + \frac{r}{k}\right)^{kt} = P\left(1 + \frac{1}{n}\right)^{nrt} = P\left[\left(1 + \frac{1}{n}\right)^{n}\right]^{rt}$$

Since n approaches plus infinity as k does, and since $\lim_{n \to +\infty} (1 + 1/n)^n = e$, it follows that when interest is compounded continuously the balance after t years will be

$$B(t) = Pe^{rt} \quad \text{dollars}$$

COMPOUND INTEREST

Suppose P dollars is invested at an interest rate of $100r$ percent a year, and let $B(t)$ denote the balance after t years. Then

$$B(t) = P\left(1 + \frac{r}{k}\right)^{kt}$$

if the interest is compounded k times a year, and

$$B(t) = Pe^{rt}$$

if the interest is compounded continuously.

Not all banks compound interest continuously. However, the formula for the balance in the continuous case is simpler than the corresponding formula in the discrete case, and may be used to approximate the actual balance when interest is compounded frequently. In the next example, we compare a balance generated when interest is compounded continuously with the corresponding balance generated when interest is compounded monthly.

Example 2.4

Suppose $1,000 is deposited in a bank offering an interest rate of 6 percent a year. Compute the balance after 1 year if

(a) the interest is compounded monthly, and

(b) the interest is compounded continuously.

Solution

(a) If the interest is compounded monthly, we use the formula $B(t) = P(1 + r/k)^{kt}$, with $t = 1$, $P = 1,000$, $r = 0.06$, and $k = 12$. We conclude that the balance after 1 year will be

$$B(1) = 1,000\left(1 + \frac{0.06}{12}\right)^{12} \approx \$1,061.68$$

(b) If the interest is compounded continuously, the balance after 1 year will be

$$B(1) = 1,000e^{0.06} \approx 1,000(1.0618) = \$1,061.80 \qquad \square$$

The number 1.0618 that was used to approximate $e^{0.06}$ in part (b) of Example 2.4 came from the table on page 376. A better approximation is 1.061836. Hence, if you had used a different table or an electronic calculator in your solution to part (b), you might have obtained a slightly different answer. For simplicity, all numerical approximations of powers of e in this text will be taken from the table on page 376.

In the next example, we use the compound interest formula to determine the amount of money that should be invested today so that it will be worth a certain desired amount at a specified time in the future.

Example 2.5

A father opens a trust account for his infant son in a bank that pays 7 percent interest compounded continuously. How much should he deposit so that 20 years from now, when his son enters college, the account will contain \$20,000?

Solution

We start with the equation

$$B = Pe^{rt}$$

In this problem, the values of B, r, and t are known, and we want to find P. Solving for P we obtain the equation

$$P = Be^{-rt}$$

into which we substitute $B = 20,000$, $r = 0.07$, and $t = 20$ to get

$$P = 20,000e^{-1.4} \approx 4,932$$

That is, \$4,932 should be deposited now if the account is to contain \$20,000 in 20 years. $\qquad \square$

The amount \$4,932 obtained in Example 2.5 is called the *present value* of \$20,000, payable 20 years from now (relative to an annual interest rate of 7 percent).

In general, the *present value of B dollars payable t years from now* is the amount P that should be invested now so that it will be worth B dollars at the end of t years. As we saw in Example 2.5, there is a simple exponential function that gives the present value of future money when interest is compounded continuously.

PRESENT VALUE OF FUTURE MONEY

If interest is compounded continuously at a rate of $100r$ percent a year, the present value $P(t)$ of B dollars payable t years from now is given by the formula

$P(t) = Be^{-rt}$

Problems

1. Evaluate the following expressions.
    (a) 2^5
    (b) 2^{-5}
    (c) 2^0
    (d) $9^{1/2}$
    (e) $9^{-1/2}$
    (f) $27^{1/3}$
    (g) $27^{-2/3}$
    (h) $(\frac{1}{4})^2$
    (i) $(\frac{1}{4})^{1/2}$
    (j) $(\frac{1}{4})^{-3/2}$
    (k) $2^3(2^4)/(2^2)^3$
    (l) $7^{1/2}(7^{3/2})/7^3$
    (m) $3^{-4}(3^2)/(3^3)^{-2}$
    (n) $2(32^{3/5})/2^5$
    (o) $4^{2/5}(64^{1/5})$
    (p) $[\sqrt{27(3^{5/2})}]^{1/2}$

2. In each of the following equations, determine the numerical value of n.
    (a) $a^3 a^5 = a^n$
    (b) $a^3/a^5 = a^n$
    (c) $a^2/a^2 = a^n$
    (d) $(a^2)^3 = a^n$
    (e) $a^3 a^{-2} = a^n$
    (f) $a^5 a^n = a^2$
    (g) $(a^2)^n = a^6$
    (h) $(a^n)^2 = a^6$
    (i) $\sqrt[n]{a(a^{5/3})} = a^2$
    (j) $a^5(a^{1/3})/a^n = a$
    (k) $a^{2/5} a^n = 1/a$
    (l) $(a^n)^3 = \sqrt{a}$

3. Sketch the curves $y = 2^x$ and $y = 3^x$ on the same set of axes.

4. Sketch the curves $y = (\frac{1}{2})^x$ and $y = (\frac{1}{3})^x$ on the same set of axes.

5. Graph the following functions.
    (a) $f(x) = 2^x$
    (b) $f(x) = (\frac{1}{2})^x$
    (c) $f(x) = 2^{-x}$
    (d) $f(x) = e^{x+1}$
    (e) $f(x) = C + Ae^{-kx}$, where C, A, and k are positive constants.
    (f) $f(x) = C - Ae^{kx}$ where C, A, and k are positive constants.
    (g) $f(x) = e^{x^2}$

6. A manufacturer of light bulbs has made a statistical study of the reliability of his product. The study indicates that the fraction $f(x)$ of his bulbs that burn for at least x hours is approximately $f(x) = e^{-0.02x}$.
   (a) What fraction of the bulbs can be expected to burn for at least 50 hours?
   (b) What fraction of the bulbs can be expected to fail between the 40th and 50th hour of use?
   (c) What fraction of the bulbs can be expected to fail before burning 40 hours?

7. A manufacturer of light bulbs has found that the fraction $f(x)$ of his bulbs that burn for at least x hours is approximately $f(x) = e^{-kx}$, where k is a certain positive constant.
   (a) Sketch the graph of this reliability function.
   (b) Find a function giving the fraction of the bulbs that fail before burning x hours. Sketch this function.

8. Find $f(2)$ if $f(x) = e^{kx}$ and $f(1) = 20$. (*Hint:* You do not have to find the numerical value of k to do this problem.)

9. Find $f(8)$ if $f(x) = Ae^{kx}$, $f(0) = 20$, and $f(2) = 40$.

10. Find $f(4)$ if $f(x) = 50 - Ae^{-kx}$, $f(0) = 20$, and $f(2) = 30$.

11. Biologists have determined that under ideal conditions the number of bacteria in a culture grows exponentially. That is, the number present at time t is given by a function of the form $f(t) = Ae^{kt}$. Suppose that 5,000 bacteria are initially present in a certain culture and that 8,000 are present 10 minutes later. How many bacteria will be present at the end of 30 minutes?

12. A cool drink is removed from a refrigerator on a hot summer day and placed in a room whose temperature is 80 degrees. According to a law of physics, the temperature of the drink x minutes later is given by a function of the form $f(x) = 80 - Ae^{-kx}$. Suppose the temperature of the drink was 45 degrees when it left the refrigerator and was 50 degrees 20 minutes later. How hot will the drink be after 40 minutes?

13. Suppose $1,000 is deposited in a bank offering interest at a rate of 7 percent a year compounded continuously. What will the balance be after 10 years? After 20 years?

14. How much money should be invested today at an annual rate of 6 percent compounded continuously so that 10 years from now it will be worth $10,000?

15. A sum of money is invested at a certain fixed interest rate, and the interest is compounded continuously. After 10 years the money has doubled. How will the balance at the end of 20 years compare with the initial deposit?

*16. A bank offers interest at a rate of 5 percent a year compounded continuously. How much should be deposited today so that withdrawals of $2,000 can be made at the end of each of the next 3 years, after which nothing will be left in the account?

17. When a bank offers an interest rate of 100r percent a year and compounds the interest more than once a year, the total interest earned during a year is greater than 100r percent of the balance at the beginning of that year. The actual percentage by which the balance grows during a year is sometimes called the *effective* interest rate, while the advertised rate of 100r percent is called the *nominal* interest rate. A bank's effective interest rate is the rate that would generate the same total yearly interest if compounded only once a year as the nominal rate would if compounded with the advertised frequency. Find the effective interest rate if the nominal rate is 6 percent and the interest is compounded
    (a) quarterly.
    (b) continuously.

3. THE NATURAL LOGARITHM

In many practical problems, we know a certain number x and need to find the corresponding number y for which $x = e^y$. This number y is called the *natural logarithm* of x. To illustrate how logarithms arise in practice, we begin our discussion with a simple problem about compound interest.

A sum of money is deposited in a bank offering an interest rate of 6 percent a year compounded continuously. How long will it take for the value of this investment to double? Let us try to analyze the problem mathematically.

As we have seen, the balance after t years will be

$$B(t) = Pe^{0.06t} \quad \text{dollars}$$

where P is the initial deposit. We want to find the value of t for which $B(t) = 2P$. Hence we must solve the equation

$$2P = Pe^{0.06t}$$

for t. Dividing by P, we get

$$e^{0.06t} = 2$$

To find t, we need to know the number y for which $2 = e^y$. That is, we need the natural logarithm of 2.

We shall complete the solution to this problem at the end of this section. In the meantime, let us examine logarithms more systematically.

The graph of the exponential function $y = a^x$ (Figure 5) suggests that to each positive number s there corresponds a unique number t such that $s = a^t$. This power t is called the *logarithm of s to the base a*, and is denoted by $\log_a s$. Thus $t = \log_a s$ if and only if $s = a^t$.

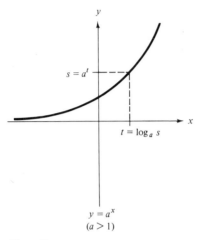

Figure 5

You are probably already familiar with the use of logarithms to the base 10 in the simplification of certain numerical computations. (For numerical work, the number 10 is particularly convenient as the base for logarithms because the standard decimal representation of numbers is based on powers of 10.) Actually, logarithms to *any* base could be used to simplify numerical computations. Because of the importance of the special exponential function e^x, we shall concentrate on the corresponding logarithm, $\log_e$. The logarithm to the base e is called the *natural logarithm* and will be denoted (without explicit reference to the base) by the symbol log. (Some books use the symbol ln to denote the natural logarithm.)

THE NATURAL LOGARITHM

If $x > 0$, we define $\log x$ to be the unique number y such that $x = e^y$. That is, $y = \log x$ if and only if $x = e^y$.

Consider the following examples.

Example 3.1

(a) Find $\log e$.

(b) Find $\log 1$.

Solution

(a) By definition, $\log e$ is the unique number y such that $e = e^y$. Clearly, $y = 1$. Hence $\log e = 1$.

(b) Since $e^0 = 1$, it follows that $\log 1 = 0$. □

In the next example, we derive two important identities that show that logarithmic and exponential functions have a certain "neutralizing" effect on each other.

Example 3.2

Simplify the following expressions.

(a) $e^{\log x}$

(b) $\log e^x$

Solution

(a) According to our definition, $\log x$ is the unique number y for which $x = e^y$. Hence, $e^{\log x} = e^y = x$.

(b) According to the definition of the logarithm, $\log e^x$ is the unique number y for which $e^x = e^y$. Clearly, this number is x itself. Hence, $\log e^x = x$. □

The two identities derived in Example 3.2 show that the composite functions $\log e^x$ and $e^{\log x}$ leave the variable x unchanged. In general, two functions f and g for which $f(g(x)) = x$ and $g(f(x)) = x$ are said to be *inverses* of one another. Thus, the exponential function $y = e^x$ and the logarithmic function $y = \log x$ are inverses of each other.

INVERSE PROPERTIES OF $\log x$ AND e^x

(1) $\log e^x = x$

(2) $e^{\log x} = x$

There is an easy way to obtain the graph of the logarithmic function $y = \log x$ from that of the exponential function $y = e^x$. The method is based on the simple geometric fact, illustrated in Figure 6, that the point (b, a) is the reflection across the line $y = x$ of the point (a, b). It follows that the graph of $y = \log x$ is the reflection across the line $y = x$ of the graph of $y = e^x$. To see this, suppose that (a, b) is a point on the graph of $y = \log x$. Then, $b = \log a$ or, equivalently, $a = e^b$. Hence the reflected point (b, a) can be written as (b, e^b), which is clearly a point on the graph of the function $y = e^x$. Conversely, if (a, b) is on the graph of $y = e^x$, then $b = e^a$, and so $a = \log b$. Hence the reflected point is $(b, a) = (b, \log b)$, which lies on the graph of $y = \log x$.

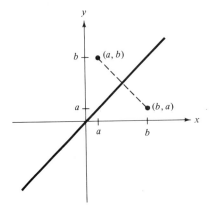

Figure 6

The graph of the exponential function $y = e^x$ is sketched again for reference in Figure 7a. The graph of the logarithmic function $y = \log x$, obtained from the graph in Figure 7a by reflection across the line $y = x$, is shown in Figure 7b.

Notice that $\log x$ is defined only for positive values of x, that $\log 1 = 0$, that $\lim_{x \to +\infty} \log x = +\infty$, and that $\lim_{x \to 0^+} \log x = -\infty$.

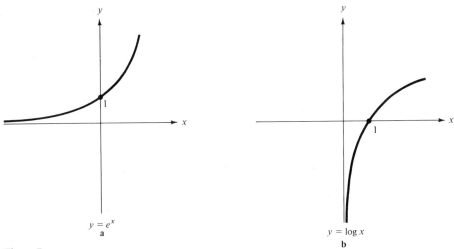

$y = e^x$
a

$y = \log x$
b

Figure 7

As we shall see, the laws of exponents can be rephrased to give the following important properties of logarithms.

PROPERTIES OF LOGARITHMS

(1) $\log uv = \log u + \log v$

(2) $\log u/v = \log u - \log v$

(3) $\log u^v = v \log u$

It is because of these three properties that logarithms are so useful in the simplification of numerical computations. The first property asserts that the logarithm can be used to transform multiplication into the simpler operation of addition. The second property says that the logarithm transforms division into subtraction. And the third shows that logarithms can be used to reduce exponentiation to multiplication. Before deriving these properties from the corresponding laws of exponents, let us consider some typical applications.

Example 3.3

Find the (approximate) value of the expression $\sqrt[5]{0.19(1.3)/2.7}$.

Solution

We let

$$x = \sqrt[5]{0.19(1.3)/2.7}$$

and take the logarithm of both sides of this equation. After simplification based on the properties of logarithms, we get

$\log x = \frac{1}{5}(\log 0.19 + \log 1.3 - \log 2.7)$

Next, we use the table on page 377 to look up the necessary logarithms. We find

$\log 0.19 \approx -1.66073$

$\log 1.3 \approx 0.26236$

and

$\log 2.7 \approx 0.99325$

After performing the required arithmetic we get

$\log x \approx -0.47832$

Remember that our goal is to approximate the value of x. One way to do this is to look through the table of logarithms on page 377 and pick out the number x in the table whose logarithm is closest to -0.47832. If we do this, we find that

$x \approx 0.62$

Another method is to rewrite the equation

$$\log x \approx -0.47832$$

as

$$x \approx e^{-0.47832}$$

and use the table of values for the exponential function on page 376. Since $-0.47832 \approx -0.48$, we have

$$x \approx e^{-0.48} \approx 0.61878 \approx 0.62 \qquad \square$$

The numbers in Example 3.3 were carefully chosen so that we would avoid certain technical difficulties associated with the use of logarithm tables. Since we shall have very little occasion to use such tables in this text, we will not pursue the matter further. A complete discussion of the use of logarithm tables can be found in most algebra texts.

In Section 2, we remarked that any exponential function a^x can be written in terms of the special exponential function e^x. Using the properties of logarithms, we can now prove this assertion.

Example 3.4

Prove that to each positive number a there corresponds a number k such that $a^x = e^{kx}$ for all x.

Solution

We take $k = \log a$. Then,

$$e^{kx} = e^{x \log a} = e^{\log a^x} = a^x \qquad \square$$

In the next example, we prove the first of the three properties of logarithms using the corresponding law of exponents.

Example 3.5

Prove that $\log uv = \log u + \log v$.

Solution

We let

$$r = \log u \qquad s = \log v \qquad \text{and} \qquad t = \log uv$$

Then,

$$uv = e^t \qquad u = e^r \qquad \text{and} \qquad v = e^s$$

It follows that

$$e^t = e^r e^s$$

Using the first law of exponents, we can rewrite this equation as

$$e^t = e^{r+s}$$

which implies that

$$t = r + s$$

or

$$\log uv = \log u + \log v \qquad \square$$

The two remaining properties of logarithms can be proved in a similar way. The details are left as exercises.

To conclude this section, let us return to the compound interest problem with which we began the discussion of logarithms.

Example 3.6

A sum of money is deposited in a bank offering interest at the rate of 6 percent a year compounded continuously. How long will it take for the value of this investment to double?

Solution

The balance after t years will be $B(t) = Pe^{0.06t}$ dollars, where P is the initial deposit. We want to find the value of t for which $B(t) = 2P$. Hence we must solve the equation

$$2P = Pe^{0.06t}$$

Division by P gives

$$e^{0.06t} = 2$$

If we now take the logarithm of both sides of this equation we get

$$0.06t = \log 2$$

or

$$t = \frac{\log 2}{0.06} \approx \frac{0.69315}{0.06} \approx 11.552$$

Hence, we conclude that it takes approximately $11\frac{1}{2}$ years for the money to double.

Notice that the initial deposit P was eliminated from the equation at an early stage, and that the final answer is independent of this quantity. $\square$

Problems

1. Rewrite the following equations using exponential notation (that is, without using the symbol log).
   (a) $x^2 = \log y$
   (b) $-\log F = t/50 + C$
   (c) $\log 5 = 3(\log a - \log b)$
   (d) $\log a = 2 \log b + 3b$

2. Rewrite the following equations using logarithmic notation.
   (a) $y = e^{2x+1}$
   (b) $(1 + r)^t = e^{at}$
   (c) $a/(b - a) = e^c e^t$
   (d) $a^x = 1/b$

3. Evaluate the following expressions.
   (a) $e^{\log 5}$
   (b) $\log e^3$
   (c) $e^{(\log 3)/2}$
   (d) $\log (e^2)^3$
   (e) $e^{3 \log 2 - 2 \log 3}$
   (f) $\log \left(e^3 \sqrt{e}/e^{1/3}\right)$

4. Solve the following equations for x.
   (a) $n = e^{-rx}$
   (b) $-2 \log x = b$
   (c) $5 = 3 \log x - \tfrac{1}{2} \log x$
   (d) $a^{x+1} = b$
   (e) $x^{\log x} = e$
   (f) $a^{-x} = b^{x+1}$

5. Use logarithms to approximate the values of the following expressions.
   (a) $2.9(0.55)^2/1.04$
   (b) $\sqrt[3]{2.8(2.3)}$
   (c) $(\tfrac{1}{2})^e$

6. Graph the following functions.
   (a) $f(x) = -\log x$, for $x > 0$
   (b) $f(x) = \log (-x)$, for $x < 0$
   (c) $f(x) = 4 + \log x$, for $x > 0$
   (d) $f(x) = \log (x - 3)$, for $x > 3$
   (e) $f(x) = 1 - \log (x + 2)$, for $x > -2$
   (f) $f(x) = \log (1/x)$, for $x > 0$

7. How quickly will money double if it is deposited in a bank paying interest at a rate of 7 percent a year if the interest is compounded
   (a) continuously?   (b) annually?   (c) quarterly?

8. How quickly will money triple if it is deposited in a bank offering interest at a rate of 6 percent compounded continuously?

9. Money deposited in a certain bank doubles every 13 years. The bank compounds interest continuously. What interest rate does the bank offer?

10. A certain bank offers an interest rate of 6 percent a year compounded annually. A competing bank compounds its interest continuously. What (nominal) interest rate should the competing bank offer so that the *effective* interest rates of the two banks will be equal? (See Problem 17 on page 205.)

11. A law of physics states that radioactive substances decay exponentially. In particular, if Q_0 grams of a substance are initially present, the amount present after t years is given by a function of the form $Q(t) = Q_0 e^{-kt}$. The *half-life* of a radioactive substance is defined to be the time it takes for 50 percent of a given

amount of the substance to decompose. Suppose that the half-life of a certain substance is 2,000 years. How much of the substance will be left after 1,000 years if 40 grams were originally present?

12. The amount of a radioactive substance present after t years is $Q(t) = Q_0 e^{-kt}$, where Q_0 is the amount of the substance initially present and k is a positive constant. (See Problem 11.) Find a formula relating the constant k to the half-life λ of the substance.

13. Use one of the laws of exponents to prove that $\log(u/v) = \log u - \log v$.

14. Use one of the laws of exponents to prove that $\log u^v = v \log u$.

15. Express $\log_a x$ in terms of the natural logarithm.

16. For each of the following functions f find a function g such that f and g are inverses of one another.
    (a) $f(x) = x^{1/3}$ (b) $f(x) = x^3$ (c) $f(x) = 1/x$
    (d) $f(x) = x + 5$ (e) $f(x) = 2^x$ (f) $f(x) = \log_{10} x$

*17. Prove that if f and g are inverses of one another then the graph of g is the reflection across the line $y = x$ of the graph of f.

*18. (a) Suppose that f and g are functions with the property that $y = g(x)$ if and only if $x = f(y)$. Prove that f and g are inverses of one another.
    (b) Conversely, suppose that f and g are inverses of one another and prove that $y = g(x)$ if and only if $x = f(y)$.

4. DIFFERENTIATION OF LOGARITHMIC AND EXPONENTIAL FUNCTIONS

Both the logarithmic function $y = \log x$ and the exponential function $y = e^x$ turn out to have rather simple derivatives. We begin our discussion of these derivatives by considering the logarithmic function.

THE DERIVATIVE OF $\log x$

$$\frac{d}{dx}(\log x) = \frac{1}{x}$$

To prove this formula, we return to our original definition of the derivative. Recall that in general,

$$\frac{df}{dx} = \lim_{\Delta x \to 0} \frac{f(x + \Delta x) - f(x)}{\Delta x}$$

In this case $f(x) = \log x$, so
$$\frac{f(x + \Delta x) - f(x)}{\Delta x} = \frac{\log(x + \Delta x) - \log x}{\Delta x}$$

Using the properties of logarithms, we can rewrite this quotient as
$$\frac{\log(x + \Delta x) - \log x}{\Delta x} = \log\left(\frac{x + \Delta x}{x}\right)^{1/\Delta x}$$
$$= \log\left(1 + \frac{\Delta x}{x}\right)^{1/\Delta x}$$

This expression will look more familiar if we let $n = x/\Delta x$. Then,
$$\left(1 + \frac{\Delta x}{x}\right)^{1/\Delta x} = \left(1 + \frac{1}{n}\right)^{n/x} = \left[\left(1 + \frac{1}{n}\right)^{n}\right]^{1/x}$$

Hence,
$$\frac{\log(x + \Delta x) - \log x}{\Delta x} = \log\left[\left(1 + \frac{1}{n}\right)^{n}\right]^{1/x}$$
$$= \frac{1}{x}\log\left(1 + \frac{1}{n}\right)^{n}$$

Since $n = x/\Delta x$, it follows that n approaches infinity as Δx approaches zero. (More precisely, n approaches either plus infinity or minus infinity, depending on the sign of Δx.) Since $\lim_{n \to \infty}(1 + 1/n)^n = e$, we conclude that

$$\frac{d}{dx}(\log x) = \lim_{\Delta x \to 0} \frac{\log(x + \Delta x) - \log x}{\Delta x}$$
$$= \lim_{n \to \infty} \frac{1}{x}\log\left(1 + \frac{1}{n}\right)^n$$
$$= \frac{1}{x}\log e$$
$$= \frac{1}{x}$$

Example 4.1

Differentiate the function $f(x) = x \log x$.

Solution

We combine the product rule with the formula for the derivative of $\log x$ to get

$$f'(x) = x\left(\frac{1}{x}\right) + \log x = 1 + \log x \qquad \square$$

Example 4.2

Differentiate the function $f(x) = \log(2x^3 + 1)$.

Solution

In this case, we are dealing with a composite function: $f(x) = g(h(x))$, where $g(u) = \log u$ and $h(x) = 2x^3 + 1$. According to the general form of the chain rule,

$$f'(x) = g'(h(x))h'(x)$$

$$= \frac{1}{h(x)} h'(x)$$

$$= \frac{6x^2}{2x^3 + 1} \qquad \square$$

Example 4.2 illustrates the following general procedure for differentiating the logarithm of a differentiable function of x.

CHAIN RULE FOR LOGARITHMIC FUNCTIONS

$$\frac{d}{dx}[\log h(x)] = \frac{1}{h(x)} h'(x)$$

This says that to differentiate $\log h$ we simply multiply $1/h$ by the derivative of h. Let us try one more example.

Example 4.3

Differentiate the function $f(x) = \log(x^2 + 1)^3$.

Solution

As a preliminary step, we simplify $f(x)$ using one of the properties of logarithms to get

$$f(x) = 3\log(x^2 + 1)$$

Then,

$$f'(x) = 3\left(\frac{1}{x^2 + 1}\right)(2x) = \frac{6x}{x^2 + 1}$$

Convince yourself that the final answer would have been the same had we not made the initial simplification of $f(x)$. $\qquad \square$

216   EXPONENTIAL AND LOGARITHMIC FUNCTIONS

If you were particularly observant, you may have recognized a familiar expression in the statement of the chain rule for logarithms,

$$\frac{d}{dx}[\log h(x)] = \frac{1}{h(x)} h'(x)$$

Except for a factor of 100, the expression $h'(x)/h(x)$ is precisely the formula for the percentage rate of change of h with respect to x that we encountered in Chapter 2, Section 10.

PERCENTAGE RATE OF CHANGE

The percentage rate of change of f with respect to x is

$$100 \frac{f'(x)}{f(x)} = 100 \frac{d}{dx}[\log f(x)]$$

That is, we can find the percentage rate of change of a function by multiplying the derivative of its logarithm by 100. Consider the following example.

Example 4.4

It is projected that x years from now the population of a certain town will be approximately

$$P(x) = 5,000\sqrt{x^2 + 4x + 19} \quad \text{people}$$

At what percentage rate will the population be changing 3 years from now?

Solution

Since

$$\log P(x) = \log 5,000 + \tfrac{1}{2} \log (x^2 + 4x + 19)$$

we have

$$\frac{d}{dx}[\log P(x)] = \frac{x + 2}{x^2 + 4x + 19}$$

Hence, the percentage rate of change 3 years from now will be

$$100 \frac{3 + 2}{9 + 12 + 19} = 12.5 \quad \text{percent per year} \qquad \square$$

Sometimes we can simplify the work involved in differentiating a function if we first take its logarithm. Consider the following example.

Example 4.5

Differentiate the function $f(x) = \sqrt[3]{x+1}/(x-1)$.

Solution

This function can be differentiated by the techniques developed in Chapter 2. However, the resulting computation is somewhat tedious. (Try it.)

A more efficient approach is to work with the logarithm of f,

$$\log f(x) = \tfrac{1}{3} \log(x+1) - \log(x-1)$$

(Notice that by introducing the logarithm we have eliminated the quotient and the cube root.)

We now differentiate both sides of this equation, using the chain rule for logarithms. We get

$$\frac{f'(x)}{f(x)} = \frac{1}{3(x+1)} - \frac{1}{x-1} = \frac{-2(x+2)}{3(x+1)(x-1)}$$

Finally, we solve for $f'(x)$ by multiplying both sides of this equation by $f(x)$.

$$f'(x) = \frac{-2(x+2)}{3(x+1)(x-1)} \frac{\sqrt[3]{x+1}}{x-1}$$

$$= \frac{-2(x+2)}{3(x+1)^{2/3}(x-1)^2} \qquad \square$$

The technique used in Example 4.5 is called *logarithmic differentiation*. Using logarithmic differentiation, we can now derive the formula for the derivative of e^x. Since it is not much harder to get the formula for the derivative of a^x, let us consider this more general case. When we are done, we shall see one more reason why e is the natural choice as the base of an exponential function.

The derivative of a^x is *not* xa^{x-1}. Before you proceed, be sure you see the fallacy in the argument that the rule $d/dx(x^n) = nx^{n-1}$ implies that $d/dx(a^x) = xa^{x-1}$.

Suppose that $y = a^x$. Then,

$$\log y = x \log a$$

We differentiate both sides of this equation with respect to x. Since $\log a$ is a constant, we have

$$\frac{d}{dx}(x \log a) = \log a$$

Since y is a function of x, the chain rule for logarithms gives

$$\frac{d}{dx}(\log y) = \frac{1}{y}\frac{dy}{dx}$$

Equating these two derivatives, we get

$$\frac{1}{y}\frac{dy}{dx} = \log a$$

or

$$\frac{dy}{dx} = y \log a = a^x \log a$$

That is,

$$\frac{d}{dx}(a^x) = a^x \log a$$

Notice that this formula says that the derivative of a^x is a constant multiple of a^x itself. The formula simplifies dramatically when $a = e$. Since $\log e = 1$, it follows that

$$\frac{d}{dx}(e^x) = e^x$$

That is, e^x is its own derivative!

THE DERIVATIVE OF e^x

$$\frac{d}{dx}(e^x) = e^x$$

You should have no trouble combining the formula for the derivative of e^x with the chain rule to get the following formula for the derivative of e^h, where h is a differentiable function of x.

CHAIN RULE FOR EXPONENTIAL FUNCTIONS

$$\frac{d}{dx}[e^{h(x)}] = h'(x)e^{h(x)}$$

Consider the following examples.

Example 4.6

Differentiate the function $f(x) = xe^x$.

Solution

By the product rule,

$$f'(x) = xe^x + e^x = (x + 1)e^x \qquad \square$$

Example 4.7

Differentiate the function $f(x) = e^{x^2+1}$.

Solution

By the chain rule,

$$f'(x) = 2xe^{x^2+1}$$ □

In Chapter 2, Section 3, we stated the rule

$$\frac{d}{dx}(x^n) = nx^{n-1}$$

for differentiating power functions. At that time, we indicated how to prove this rule for integer values of n. Using the formulas for the derivatives of the exponential and logarithmic functions, we can now give a short proof of this rule that is valid for *all* real numbers n.

Example 4.8

Prove that $d/dx(x^n) = nx^{n-1}$.

Solution

Since $x^n = e^{n \log x}$, we have

$$\frac{d}{dx}(x^n) = \frac{d}{dx}(e^{n \log x})$$

$$= \frac{n}{x} e^{n \log x}$$

$$= \frac{n}{x} x^n$$

$$= nx^{n-1}$$ □

In the next example, we use differentiation techniques developed in this section to sketch the function

$$f(x) = \frac{1}{\sigma\sqrt{2\pi}} e^{-(x-\mu)^2/2\sigma^2}$$

where σ and μ are constants and $\sigma > 0$. This complicated looking function is known as the *normal* or *gaussian distribution*, and is one of the most important functions in probability and statistics. For appropriate values of the constants σ and μ, this function gives a fairly accurate description of many natural phenomena, including the distributions of such things as people's heights and IQ scores. As we shall see, the graph of the normal distribution is the famous bell-shaped curve.

The constant μ is called the mean or expected value of the distribution (the average height or IQ score, for example). The constant σ, called the standard deviation, measures the tendency of observed values (of actual heights or IQ scores, for example) to cluster about the expected value μ. The constant $\sqrt{2\pi}$ is included for certain technical reasons and will have no effect on our analysis of the graph.

Example 4.9

Let
$$f(x) = \frac{1}{\sigma\sqrt{2\pi}} e^{-(x-\mu)^2/2\sigma^2}$$

Determine where f is increasing, decreasing, concave upward, and concave downward. Find the relative extrema and inflection points and sketch the graph.

Solution

For simplicity, we take $\mu = 0$. This is not a serious restriction, because once we have sketched the graph for $\mu = 0$, we can obtain the graph for other values of μ by suitable horizontal translations. (Do you see how?)

If
$$f(x) = \frac{1}{\sigma\sqrt{2\pi}} e^{-x^2/2\sigma^2}$$

then
$$f'(x) = \frac{-x}{\sigma^3\sqrt{2\pi}} e^{-x^2/2\sigma^2}$$

Since $e^{-x^2/2\sigma^2}$ is never zero, it follows that $f'(x) = 0$ only when $x = 0$. Moreover, since $e^{-x^2/2\sigma^2}$ is always positive, $f'(x)$ is positive when $x < 0$ and negative when $x > 0$. Hence f is increasing when $x < 0$, decreasing when $x > 0$, and has a relative maximum when $x = 0$ at the point $(0, 1/\sigma\sqrt{2\pi})$.

To determine the concavity of f, we compute the second derivative. Using the product rule we get

$$f''(x) = \frac{x^2}{\sigma^5\sqrt{2\pi}} e^{-x^2/2\sigma^2} - \frac{1}{\sigma^3\sqrt{2\pi}} e^{-x^2/2\sigma^2}$$

$$= \frac{1}{\sigma^3\sqrt{2\pi}} \left(\frac{x^2}{\sigma^2} - 1\right) e^{-x^2/2\sigma^2}$$

It follows that $f''(x) = 0$ if and only if

$$\frac{x^2}{\sigma^2} - 1 = 0$$

or

$$x = \pm\sigma$$

If $-\sigma < x < \sigma$, then $f''(x)$ is negative, and so f is concave downward. If $|x| > \sigma$, $f''(x)$ is positive, and so f is concave upward. The inflection points occur when $x = \pm\sigma$.

The graph of f is sketched in Figure 8a. To get the graph of the general normal distribution $f(x) = 1/(\sigma\sqrt{2\pi})e^{-(x-\mu)^2/2\sigma^2}$, we simply translate the graph in Figure 8a horizontally by μ units. The resulting graph is shown in Figure 8b.

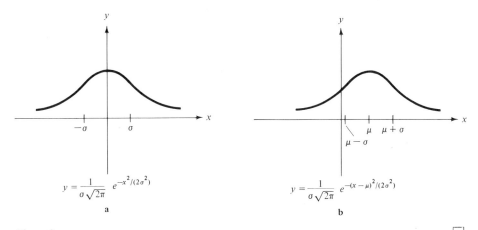

Figure 8

To conclude this section, let us take another look at compound interest.

Example 4.10

A sum of money has been deposited in a bank offering interest at a rate of 7 percent a year compounded continuously. Find an expression for the (instantaneous) rate of change of the balance with respect to time.

Solution

The balance after t years is

$$B(t) = Pe^{0.07t}$$

where P is the initial deposit. The desired rate of change is the derivative

$$\frac{dB}{dt} = 0.07Pe^{0.07t} = 0.07B(t)$$

This says that at any time t the balance is increasing at an annual rate equal to 7 percent of the existing balance. □

Example 4.10 illustrates an important phenomenon. A quantity $Q(t)$ is said to *grow exponentially* if $Q(t) = Ae^{kt}$, where A and k are positive constants. The bank balance in Example 4.10 grew exponentially and its rate of growth at any time was proportional to the existing balance at that time. In general, if a quantity grows exponentially, its rate of growth at any time is proportional to the amount present at that time. (Prove this.) In Chapter 4 we shall see that the converse of this fact is also true. In particular, we shall use integral calculus to show that, if a quantity $Q(t)$ grows at a rate that is proportional to the amount present, then the quantity grows exponentially; that is, there are constants A and k such that $Q(t) = Ae^{kt}$.

Problems

1. Differentiate the following functions.
    (a) $f(x) = e^{5x}$
    (b) $f(x) = 3e^{4x+1}$
    (c) $f(x) = e^{x^2 + 2x - 1}$
    (d) $f(x) = e^{1/x}$
    (e) $f(x) = 30 + 10e^{-0.05x}$
    (f) $f(x) = (1 + e^x)^2$
    (g) $f(x) = x^2 e^x$
    (h) $f(x) = x/e^x$
    (i) $f(x) = (2x^2 + 3x + 5)e^{6x}$
    (j) $f(x) = 3^{2x+1}$
    (k) $f(x) = \log x^3$
    (l) $f(x) = \log(2x^2 + x - 3)$
    (m) $f(x) = \log \sqrt{x^2 + 1}$
    (n) $f(x) = x^2 \log x$
    (o) $f(x) = (1/x) \log x$
    (p) $f(x) = x/\log x$
    (q) $f(x) = \log[(x + 1)/(x - 1)]$
    (r) $f(x) = e^x \log x$
    (s) $f(x) = \log e^{2x}$
    (t) $f(x) = e^{x + \log x}$
    (u) $f(x) = e^{\log x + 2 \log 3x}$

2. Use logarithmic differentiation to find the derivatives of the following functions.
    (a) $f(x) = \sqrt{(x + 1)/(x - 1)}$
    (b) $f(x) = \sqrt{x^2 + 1}/3x$
    (c) $f(x) = (x + 2)^5 (3x + 1)^4/(x - 5)^6$
    (d) $f(x) = a^x$
    (e) $f(x) = x^x$, for $x > 0$

3. Determine where each of the following functions is increasing, decreasing, concave upward, and concave downward. Graph the function, showing the relative extrema and inflection points.
    (a) $f(x) = e^{-2x}$
    (b) $f(x) = xe^x$
    (c) $f(x) = x^2 e^{-x}$
    (d) $f(x) = e^{-x^2}$
    (e) $f(x) = e^x + e^{-x}$
    (f) $f(x) = \log(1/x)$, for $x > 0$
    (g) $f(x) = x - \log x$, for $x > 0$

4. It is projected that t years from now the population of a certain country will be $50e^{0.02t}$ million people.

(a) Show that the rate of population growth at any time will be proportional to the population at that time. What is the constant of proportionality?
(b) What will the percentage rate of change of the population be t years from now? Does the percentage rate of change vary with time?
(c) Compare the constant of proportionality in part (a) and the percentage rate of change in part (b). Is the similarity accidental? Explain.

5. A sum of money has been deposited in a bank offering interest at a rate of 6 percent a year compounded continuously. Find the percentage rate of change of the balance with respect to time.

6. The consumer demand for a certain commodity is $3{,}000 e^{-0.01p}$ units per month when the market price is p dollars per unit. Express the consumers' total monthly expenditure for the commodity as a function of p, and determine the market price that will result in maximal consumer expenditure.

7. Consider the normal distribution
$$f(x) = \frac{1}{\sigma\sqrt{2\pi}} e^{-(x-\mu)^2/2\sigma^2}$$
that was sketched in Example 4.9. Describe how changes in the size of σ will affect the shape of the graph of f. In particular, sketch f when σ is very small and again when σ is large.

8. Derive the formula for the derivative of $\log x$ from the formula for the derivative of e^x. (*Hint:* Rewrite the equation $y = \log x$ as $x = e^y$, and differentiate both sides of this equation with respect to x, keeping in mind that y is a function of x.)

9. Examine the derivation of the formula for the derivative of $\log x$ on page 214. How could you modify this argument to obtain the derivative of $\log_a x$? What is the derivative of $\log_a x$?

*10. Express $\log_a x$ in terms of the natural logarithm $\log x$. Combine this relationship with the formula for the derivative of $\log x$ to obtain a formula for the derivative of $\log_a x$. Compare this formula to your answer in Problem 9.

5. REVIEW PROBLEMS

1. Compute the following limits.
   (a) $\lim\limits_{n \to 0} (1 + n)^{1/n}$ (b) $\lim\limits_{n \to +\infty} (1 + 1/2n)^n$

2. If $\log P = kt + C$, find the constant A for which $P = Ae^{kt}$.

3. Graph the following functions.
   (a) $f(x) = \log (e/x)$, for $x > 0$
   (b) $f(x) = 1 + e^{x-1}$
   (c) $f(x) = C + Ae^{kx}$, where A and k are positive constants and C is a negative constant with $|C| < A$
   (d) $f(x) = 2 - \log(x + 1)$, for $x > -1$

4. Differentiate the following functions.
   (a) $f(x) = 2e^{x^3+1}$
   (b) $f(x) = \log \sqrt{x^2 + 4x + 1}$
   (c) $f(x) = xe^{-x^2}$
   (d) $f(x) = x \log x^2$
   (e) $f(x) = x^{\log x}$
   (f) $f(x) = e^{3 \log (x+1) - \log x}$

5. Use logarithmic differentiation to find the derivatives of the following functions.
   (a) $f(x) = \sqrt{(x^2 + 1)(x^2 + 2)}$
   (b) $f(x) = x^{x^2}$

6. Determine where each of the following functions is increasing, decreasing, concave upward, and concave downward. Graph the function, showing the relative extrema and inflection points.
   (a) $f(x) = e^x - e^{-x}$
   (b) $f(x) = xe^{-kx}$, where k is a positive constant
   (c) $f(x) = x + \log x$, for $x > 0$
   (d) $f(x) = \log (x^2 + 1)$
   (e) $f(x) = 4/(1 + e^{-kx})$, where k is a positive constant

7. A toy manufacturer has found that the fraction of his plastic battery-operated toy oil tankers that sink in less than x days is approximately $f(x) = 1 - e^{-0.03x}$.
   (a) What fraction of the toy tankers can be expected to float for at least 10 days?
   (b) What fraction of the tankers can be expected to sink between the 15th and 20th days?

8. Suppose the population of a certain town doubled between 1960 and 1970. Let $P(t)$ denote the population of the town t years after 1960. How will the population in 1980 compare to the population in 1960 if the population $P(t)$ grows
   (a) linearly [that is, $P(t) = at + b$]
   (b) quadratically, with $P(t) = at^2 + b$
   (c) exponentially [that is, $P(t) = be^{kt}$]

9. The gross national product (GNP) of a certain country was 100 billion dollars in 1960 and 180 billion dollars in 1970. Assuming that the GNP is growing exponentially, what will the GNP be in 1980?

10. How quickly will money triple if it is deposited in a bank offering interest at a rate of 7 percent a year compounded continuously?

11. Money deposited in a certain bank doubles every 13 years. The bank compounds interest continuously. What interest rate does it offer?

12. A certain bank offers a (nominal) interest rate of 8 percent a year compounded quarterly. A competing bank compounds its interest continuously. What (nominal) interest rate should the competing bank offer so that the *effective* interest rates of the two banks will be equal?

13. It is projected that t years from now the population of a certain town will be approximately $P(t) = 4{,}000\sqrt{3t + 5}$ people. At what percentage rate will the population be changing 10 years from now?

Integration

In the first half of this chapter, we study antidifferentiation, *the process of obtaining a function from its derivative. This process arises naturally in the solution of practical problems in which the rate of change of a quantity is known and the goal is to find an expression for the quantity itself. In the second half of the chapter, we examine certain limits of sums that occur frequently in applications. The remarkable fact that these limits, called* definite integrals, *can be computed by antidifferentiation is established in Section 11.*

1. ANTIDERIVATIVES

A sociologist who knows the rate at which the population is growing might wish to use this information to predict future population levels. A physicist who has determined how the speed of a particle varies might want a function expressing the particle's position. A manufacturer who knows the marginal cost might need to compute the total cost of producing a certain number of items.

Although the preceding illustrations are taken from different disciplines, they have an important feature in common. In each case, the derivative of a certain function is known, and the goal is to find the function itself. The process of obtaining a function from its derivative is called *antidifferentiation* or *integration*. Let us begin our discussion of this subject with the following definition.

ANTIDERIVATIVE

An *antiderivative* (or *indefinite integral*) of f is any function F whose derivative equals f.

In our opening illustrations, for example, the sociologist was seeking an antiderivative of the growth rate function, the physicist wanted an antiderivative of speed, and the manufacturer needed an antiderivative of his marginal cost.

To compute antiderivatives, we must apply the rules for differentiation in reverse. Sometimes this is quite easy to do. For example, you should already be able to discover an antiderivative of x^2. And, having succeeded with this function, you will probably see how to integrate any polynomial. (The integration of polynomials will be discussed systematically in Section 2.) In general, however, integration is a more difficult process than differentiation. For example, you should have no trouble differentiating x^2e^x using the product rule. But just try to find a function whose derivative is equal to x^2e^x. There is, unfortunately, no general technique for integrating products. (Actually, this particular product x^2e^x can be integrated rather easily using a method we will develop in Section 4.)

Once you have found what you believe to be an antiderivative of a given function, you can always check your answer by differentiating it. Consider the following examples.

Example 1.1

Verify that $\frac{1}{3}x^3$ is an antiderivative of the function x^2.

Solution

We must show that $d/dx(\frac{1}{3}x^3) = x^2$. This is obvious:

$$\frac{d}{dx}(\tfrac{1}{3}x^3) = \tfrac{1}{3}(3x^2) = x^2 \qquad \square$$

Example 1.2

Verify that $(x^2 - 2x + 2)e^x$ is an antiderivative of the function x^2e^x.

Solution

Using the product rule, we get

$$\frac{d}{dx}[(x^2 - 2x + 2)e^x] = (2x - 2)e^x + (x^2 - 2x + 2)e^x$$

$$= x^2 e^x$$

Hence $(x^2 - 2x + 2)e^x$ is indeed an antiderivative of x^2e^x. $\qquad \square$

A function has more than one antiderivative. For example, $\frac{1}{3}x^3$ is an antiderivative of x^2. But so is $\frac{1}{3}x^3 + 12$, since the derivative of the constant 12 is zero. In general, if F is one antiderivative of f, then the function obtained by adding any constant to F is also an antiderivative of f. In fact, it turns out that all the antiderivatives of f can be obtained by adding constants to a given antiderivative.

THE ANTIDERIVATIVES OF A FUNCTION

If F and G are antiderivatives of f, then there is a constant C such that $F(x) = G(x) + C$.

There is a simple geometric explanation for the fact that any two antiderivatives of the same function must differ by a constant. If F is an antiderivative of f, then, by definition, $dF/dx = f(x)$. For each value of x, then, $f(x)$ is the slope of the tangent to the graph of F. If G is another antiderivative of f, then the slope of its tangent is also $f(x)$. Thus the graphs of F and G are "parallel" curves and must be vertical translations of one another. That is, $F(x) = G(x) + C$ for some constant C.

We illustrate this situation in Figure 1 by sketching several antiderivatives of the function $f(x) = x^2$.

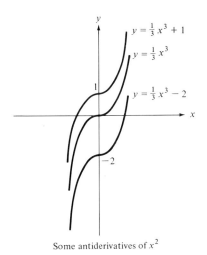

Some antiderivatives of x^2

Figure 1

This geometric interpretation of antiderivatives is further illustrated in the following example.

Example 1.3

Find a function $F(x)$ whose tangent has slope x^2 for each value of x and whose graph passes through the point (3, 5).

Solution

We are looking for a certain antiderivative of x^2. All the antiderivatives of x^2 are of the form $F(x) = \frac{1}{3}x^3 + C$. We want to choose the constant C so that the graph of F will pass through (3, 5). That is, we want

$$5 = F(3) = \tfrac{1}{3}(3^3) + C$$

or

$$C = -4$$

Hence the desired function is $F(x) = \frac{1}{3}x^3 - 4$ (Figure 2).

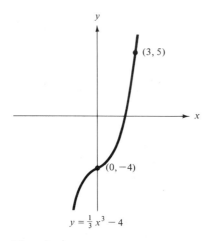

$y = \frac{1}{3}x^3 - 4$

Figure 2

In many practical problems involving integration, specific information is known that allows us to determine the constant C and find the appropriate antiderivative. We conclude this section with two examples illustrating this procedure.

Example 1.4

The rate at which the population of a certain town is growing changes with time. It is estimated that x months from now the rate of population increase will be $2 + 6\sqrt{x}$ people per month. The current population is 5,000. What will the population be 9 months from now?

Solution

Let $P(x)$ denote the population of the town x months from now. Then the derivative of $P(x)$ is the rate at which the population will be changing x months from now. That is,

$$\frac{dP}{dx} = 2 + 6\sqrt{x}$$

It follows that the population function $P(x)$ is an antiderivative of $2 + 6\sqrt{x}$. One such antiderivative is $2x + 4x^{3/2}$. (Check this.) Hence,

$$P(x) = 2x + 4x^{3/2} + C$$

for some constant C. To determine C, we use the information that at present (when $x = 0$) the population is 5,000. That is,

$$5{,}000 = 2(0) + 4(0)^{3/2} + C$$

or

$$C = 5{,}000$$

Hence,

$$P(x) = 2x + 4x^{3/2} + 5{,}000$$

In 9 months, the population will be

$$P(9) = 2(9) + 4(27) + 5{,}000 = 5{,}126 \quad \text{people} \qquad \square$$

Example 1.5

A manufacturer has found that the marginal cost is $x + 3$ dollars per item when x items have been produced. The total cost of producing the first four items is \$520. What is the total cost of producing the first eight items?

Solution

Recall that the marginal cost is the derivative of the total cost function $C(x)$. Hence $C(x)$ must be an antiderivative of $x + 3$. Since one such antiderivative is $\frac{1}{2}x^2 + 3x$, it follows that

$$C(x) = \tfrac{1}{2}x^2 + 3x + k$$

for some constant k. (The letter k is used for the constant to avoid confusion with the cost function C.)

The value of k is determined by the fact that $C(4) = 520$. That is,

$$520 = \tfrac{1}{2}(4)^2 + 3(4) + k$$

or

$$k = 500$$

Hence,

$$C(x) = \tfrac{1}{2}x^2 + 3x + 500$$

and the cost of producing the first eight items is

$$C(8) = \$556 \qquad \square$$

Problems

1. Verify that $F(x)$ is an antiderivative of $f(x)$ if
   (a) $f(x) = x^5 + 1$ and $F(x) = \frac{1}{6}x^6 + x + 8$
   (b) $f(x) = 4 + 5x^{2/3}$ and $F(x) = 4x + 3x^{5/3}$
   (c) $f(x) = a + bx + cx^2$ and $F(x) = ax + \frac{1}{2}bx^2 + \frac{1}{3}cx^3$
   (d) $f(x) = 1/x^2$ and $F(x) = -1/x$
   (e) $f(x) = 1 + 1/(x+1)^2$ and $F(x) = x - 1/(x+1)$
   (f) $f(x) = 2x(x^2 + 1)^9$ and $F(x) = \frac{1}{10}(x^2 + 1)^{10}$
   (g) $f(x) = xe^{x^2}$ and $F(x) = \frac{1}{2}e^{x^2}$
   (h) $f(x) = xe^x$ and $F(x) = (x - 1)e^x$
   (i) $f(x) = ae^{rx}$ and $F(x) = (a/r)e^{rx}$

2. In each of the following examples, decide whether $F(x)$ is an antiderivative of $f(x)$. If it is not, try to find one.
   (a) $f(x) = 2$ and $F(x) = 0$
   (b) $f(x) = x$ and $F(x) = x^2$
   (c) $f(x) = x^5$ and $F(x) = x^6$
   (d) $f(x) = x^{2/3}$ and $F(x) = x^{5/3}$
   (e) $f(x) = 1/x^3$ and $F(x) = 1/x^2$
   (f) $f(x) = 2/x^4$ and $F(x) = -2/x^3$
   (g) $f(x) = e^{2x}$ and $F(x) = e^{2x}$
   (h) $f(x) = x^2 e^{x^3}$ and $F(x) = e^{x^3}$
   (i) $f(x) = \log x$ and $F(x) = x \log x - x$
   (j) $f(x) = (5x + 2)^2$ and $F(x) = \frac{1}{3}(5x + 2)^3$

3. The following functions $F(x)$ are antiderivatives of certain other functions $f(x)$. In each case, find $f(x)$.
   (a) $F(x) = 3x^2 + 6x + 12$ (b) $F(x) = x/(x + 1)$
   (c) $F(x) = \log(3x^2 + 2)$

4. (a) Verify that $\frac{1}{4}x^4$ is an antiderivative of x^3.
   (b) Find an antiderivative of x^4.
   (c) Verify that $\frac{2}{5}x^{5/2}$ is an antiderivative of $x^{3/2}$.
   (d) Find an antiderivative of $x^{4/5}$.
   (e) For $n \neq -1$, find an antiderivative of x^n.
   (f) What happens when $n = -1$?

The antiderivatives needed to solve the following problems can be found in Problem 1.

5. Find a function whose tangent has slope $\frac{1}{3}x^2 + 2x + 5$ for each value of x and whose graph passes through the point $(1, \frac{1}{9})$.

6. Find a function whose tangent has slope e^{5x} and whose graph passes through the point $(0, 2)$.

7. The rate at which the population of a certain town is growing changes with time. Studies indicate that x months from now the rate of population increase will be $4 + 5x^{2/3}$ people per month. The current population is 10,000. What will the population be 8 months from now?

8. In a certain section of the country, the price of large Grade A eggs is currently $1.60 a dozen. Studies indicate that during the next 3 months the price will rise in such a way that, x weeks from now, it will be changing at a rate of $0.2 + 0.003x^2$ cents per week. How much will eggs cost 10 weeks from now?

9. An object is moving so that its speed after t minutes is $3 + 2t + 6t^2$ feet per minute.
   (a) How far does the object travel during the first minute?
   (b) How far does the object travel during the second minute?

10. The resale value of a certain industrial machine decreases over a 10-year period at a rate that changes with time. When the machine is x years old, the rate at which its value is changing is $220(x - 10)$ dollars per year.
    (a) What is the significance of the fact that the rate $220(x - 10)$ is negative for $0 \leq x < 10$?
    (b) If the machine was originally worth $12,000, how much will it be worth after 10 years?

11. A manufacturer has found that his marginal cost is $6x + 1$ dollars per unit when x units have been produced. The total cost of producing the first unit is $130.
    (a) What is the total cost of producing the first 10 units?
    (b) What is the cost of producing the tenth unit?
    (c) What is the overhead (the cost of producing no units)?

12. A tree has been transplanted, and after x years is growing at a rate of $1 + 1/(x + 1)^2$ feet per year. In 2 years it has reached a height of 7 feet. How tall was it when it was transplanted?

2. TECHNIQUES OF INTEGRATION

In computing antiderivatives, we start with the derivative of a function and work backward to find the function itself. This suggests that some general rules for integration can be obtained by stating in reverse the rules for differentiation. Many of the rules that arise in this way are straightforward and quite easy to apply. They will be examined in this section. In Sections 3 and 4 we will consider two important but less elementary methods of integration.

Before proceeding, let us introduce a useful abbreviation for antiderivatives. It is customary to use the symbol

$$\int f(x)\, dx$$

to represent the general form of the antiderivatives of a function $f(x)$. In particular, if $F(x)$ is one antiderivative of $f(x)$, then every antiderivative of $f(x)$ is of the form $F(x) + C$, where C is a constant, and so we write

$$\int f(x)\, dx = F(x) + C$$

For example, we express the fact that every antiderivative of x^2 is of the form $\tfrac{1}{3}x^3 + C$ by writing

$$\int x^2\, dx = \tfrac{1}{3}x^3 + C$$

The symbol $\int$ is called an *integral sign* and tells us that the function following it is to be integrated. The role of the symbol dx is to indicate that x is the variable with respect to which the integration is to be performed. Analogous notation is used if the function is expressed in terms of a variable other than x. For example, $\int t^2\, dt = \tfrac{1}{3}t^3 + C$. Moreover, in the expression $\int px^2\, dx$, the dx tells us that x rather than p is the variable. Thus, $\int px^2\, dx = \tfrac{1}{3}px^3 + C$. (How would you evaluate $\int px^2\, dp$?) In Section 3 we shall see how a formal manipulation of the symbol dx simplifies the integral version of the chain rule.

We begin the discussion of integration techniques by considering power functions. Our goal is to find a function whose derivative is x^n. We know that

$$\frac{d}{dx}(x^{n+1}) = (n+1)x^n$$

This says that x^{n+1} is an antiderivative of $(n+1)x^n$. To get an antiderivative of x^n we merely divide by the constant $n + 1$. In particular, $x^{n+1}/(n+1)$ is an antiderivative of x^n, since

$$\frac{d}{dx}\left(\frac{x^{n+1}}{n+1}\right) = \frac{1}{n+1}\frac{d}{dx}(x^{n+1}) = x^n$$

This conclusion holds for all values of n except, of course, for $n = -1$, in which case $1/(n+1)$ is undefined. (Can you think of an antiderivative of x^{-1}?) Let us exclude this one troublesome case for the moment and summarize our results as follows.

INTEGRATION OF POWER FUNCTIONS

For $n \neq -1$,

$$\int x^n\, dx = \frac{1}{n+1}x^{n+1} + C$$

In other words, to integrate the power function x^n, we increase the power of x by 1 and then divide by the new power.

Example 2.1

Find the following integrals.

(a) $\int x^{3/5} \, dx$

(b) $\int 6 \, dx$

(c) $\int 1/\sqrt{x} \, dx$

Solution

(a) $\int x^{3/5} \, dx = \frac{5}{8} x^{8/5} + C$

(b) Since $6 = 6x^0$, we conclude that $\int 6 \, dx = 6x + C$.

(c) We begin by rewriting $1/\sqrt{x}$ as $x^{-1/2}$. Then,

$$\int 1/\sqrt{x} \, dx = \int x^{-1/2} \, dx = 2x^{1/2} + C = 2\sqrt{x} + C \qquad \square$$

Let us now consider the one remaining power function x^{-1}. Specifically, we want to find a function whose derivative is $1/x$. Since

$$\frac{d}{dx}(\log x) = \frac{1}{x}$$

we might be tempted to conclude that

$$\int \frac{1}{x} \, dx = \log x + C$$

However, this is valid only when x is positive, since $\log x$ is not defined for negative values of x. When x is negative, it turns out that the function $\log |x|$ is an antiderivative of $1/x$. This is easily checked. If x is negative, then $|x| = -x$, and so

$$\frac{d}{dx}(\log |x|) = \frac{d}{dx}(\log - x) = -1\left(\frac{1}{-x}\right) = \frac{1}{x}$$

Since $|x| = x$ when x is positive, we may summarize the situation using a single formula as follows.

THE INTEGRAL OF $1/x$

$$\int \frac{1}{x} dx = \log |x| + C$$

Perhaps you are wondering how to integrate the two functions $\log x$ and e^x. An antiderivative of $\log x$ is not easy to discover using the techniques developed so far, and we postpone this question until Section 4. On the other hand, the integration of e^x is completely trivial.

THE INTEGRAL OF e^x

$$\int e^x \, dx = e^x + C$$

The constant multiple rule and the sum rule for differentiation can easily be rewritten as rules for integration.

CONSTANT MULTIPLE RULE FOR INTEGRALS

$$\int kf(x) \, dx = k \int f(x) \, dx \qquad \text{for any constant } k$$

SUM RULE FOR INTEGRALS

$$\int [f(x) + g(x)] \, dx = \int f(x) \, dx + \int g(x) \, dx$$

To verify the constant multiple rule, let $F(x)$ be an antiderivative of $f(x)$. We want to show that $kF(x)$ is then an antiderivative of $kf(x)$. Using the constant multiple rule for derivatives, we get

$$\frac{d}{dx}[kF(x)] = k \frac{d}{dx}[F(x)] = kf(x)$$

which implies that $kF(x)$ is indeed an antiderivative of $kf(x)$.

The sum rule can be verified just as easily using the sum rule for derivatives. (See Problem 4.) It says that a sum can be integrated term by term.

Example 2.2

Find $\int (3e^x + 2/x - \frac{1}{2}x^2) \, dx$.

Solution

$$\int \left(3e^x + \frac{2}{x} - \tfrac{1}{2}x^2\right) dx = 3\int e^x\,dx + 2\int \frac{1}{x}\,dx - \tfrac{1}{2}\int x^2\,dx$$

$$= 3e^x + 2\log|x| - \tfrac{1}{6}x^3 + C \qquad \square$$

Using these rules we can now integrate any polynomial.

Example 2.3

Find $\int (5x^3 - 4x^2 + 12)\,dx$.

Solution

$$\int (5x^3 - 4x^2 + 12)\,dx = 5\int x^3\,dx - 4\int x^2\,dx + \int 12\,dx$$

$$= 5(\tfrac{1}{4}x^4) - 4(\tfrac{1}{3}x^3) + 12x + C$$

$$= \tfrac{5}{4}x^4 - \tfrac{4}{3}x^3 + 12x + C \qquad \square$$

Sometimes an initial incorrect answer to an integration problem can be modified to produce the correct answer. The constant multiple rule provides the justification for one useful technique for salvaging wrong answers. Consider the following example.

Example 2.4

Find $\int e^{5x}\,dx$.

Solution

Since $\int e^x\,dx = e^x + C$, we might be tempted to think that $\int e^{5x}\,dx = e^{5x} + C$. However, when we check this answer by differentiating, we find that the chain rule introduces an unwanted factor of 5. In particular,

$$\frac{d}{dx}(e^{5x}) = 5e^{5x} \neq e^{5x}$$

and so e^{5x} is not an antiderivative of e^{5x}.

All is not lost, however. Our calculation suggests that the incorrect answer e^{5x} is five times too large, and that if we divide it by 5 we should get a genuine antiderivative of e^{5x}. Let us check this assertion.

$$\frac{d}{dx}(\tfrac{1}{5}e^{5x}) = \frac{1}{5}\frac{d}{dx}(e^{5x}) = \tfrac{1}{5}(5)e^{5x} = e^{5x}$$

It works! Therefore

$$\int e^{5x}\,dx = \tfrac{1}{5}e^{5x} + C$$

Note the crucial role of the constant multiple rule in the preceding example. Multiplication of the initial incorrect answer by $\tfrac{1}{5}$ was successful because the derivative of this product was precisely $\tfrac{1}{5}$ times the derivative of our initial false answer. If the derivative of the initial false answer had been off by anything other than a *constant* factor, this method would not have worked.

Let us consider one more example illustrating this technique.

Example 2.5

Find $\int (3x+5)^6\,dx$.

Solution

Many students will guess that $\tfrac{1}{7}(3x+5)^7$ is an antiderivative of $(3x+5)^6$. However, if we check this answer, we find that

$$\frac{d}{dx}[\tfrac{1}{7}(3x+5)^7] = 3(3x+5)^6$$

which is three times too large. Hence we multiply our wrong answer by $\tfrac{1}{3}$ and conclude that

$$\int (3x+5)^6\,dx = \tfrac{1}{21}(3x+5)^7 + C$$

We will examine this sort of problem more thoroughly in Section 3, when we discuss the integral version of the chain rule.

Consider for a moment the problem of formulating a quotient rule for integrals. The quotient rule for derivatives states that

$$\frac{d}{dx}\left(\frac{f}{g}\right) = \frac{g\dfrac{df}{dx} - f\dfrac{dg}{dx}}{g^2}$$

Hence, the corresponding integration rule would apply only to those quotients that can be put in the special form

$$\frac{g\dfrac{df}{dx} - f\dfrac{dg}{dx}}{g^2}$$

Very few quotients are of this type. Moreover, even if a quotient could theoretically be written in this way, the practical problem of finding the appropriate f and g is usually insurmountable. (If you doubt this, just try to write the quotient $(x^2 - 6x + 15)/(x - 3)^2$ in this special form. You can find the answer on page 109.) Thus, the quotient rule for derivatives cannot be restated as a useful integration rule. In fact, although some particular quotients can be integrated by special means, we will have no general technique for integrating even rational functions.

Occasionally, a quotient can be rewritten in a form in which it can be easily integrated. Here is a simple example.

Example 2.6

Find $\int \dfrac{3x^5 + 2x - 5}{x^3}\, dx$.

Solution

Since

$$\frac{3x^5 + 2x - 5}{x^3} = 3x^2 + \frac{2}{x^2} - \frac{5}{x^3}$$

$$= 3x^2 + 2x^{-2} - 5x^{-3}$$

we have

$$\int \frac{3x^5 + 2x - 5}{x^3}\, dx = \int (3x^2 + 2x^{-2} - 5x^{-3})\, dx$$

$$= x^3 - 2x^{-1} + \tfrac{5}{2}x^{-2} + C \qquad \square$$

There is a way to rephrase the product rule for derivatives to obtain a rule of integration. However, this rule, which will be discussed in Section 4, does not lead to a general method for integrating products. In Section 3, an integral version of the chain rule will be used to integrate products of a certain special type.

Some products, of course, can be expanded and then integrated directly. Consider the following example.

Example 2.7

Find $\int \sqrt{x}(x + 1)^2\, dx$.

Solution

Since

$$\sqrt{x}(x + 1)^2 = x^{1/2}(x^2 + 2x + 1) = x^{5/2} + 2x^{3/2} + x^{1/2}$$

we have

$$\int \sqrt{x}(x+1)^2 \, dx = \int (x^{5/2} + 2x^{3/2} + x^{1/2}) \, dx$$

$$= \tfrac{2}{7}x^{7/2} + \tfrac{4}{5}x^{5/2} + \tfrac{2}{3}x^{3/2} + C \qquad \square$$

Problems

1. Find the following integrals.

    (a) $\int x^{3/4} \, dx$
    (b) $\int 1/t^2 \, dt$
    (c) $\int \sqrt{u} \, du$
    (d) $\int 5 \, dx$
    (e) $\int (3t^2 - 5t + 2) \, dt$
    (f) $\int (2e^u + 6/u - 7u^2) \, du$
    (g) $\int (x^{1/2} - 3x^{2/3} + 6) \, dx$
    (h) $\int (\sqrt[3]{t} + 2/t^5) \, dt$
    (i) $\int (2u^{-1/2} + 3/u^2 - 5\sqrt{u}) \, du$
    (j) $\int (ax^2 + bx + c) \, dx$
    (k) $\int (1/2t - 2/t^2 + 3/\sqrt{t}) \, dt$
    (l) $\int 3e^{5u} \, du$
    (m) $\int e^{-x} \, dx$
    (n) $\int \tfrac{1}{2} e^{-t/2} \, dt$
    (o) $\int \sqrt{u}(u^2 - 1) \, du$
    (p) $\int x^3(2x + 1/x) \, dx$
    (q) $\int t^{-1/2}(t + 1/t)^2 \, dt$
    (r) $\int (u + 1)/u^3 \, du$
    (s) $\int (x^2 + 3x + 5)/2x \, dx$

2. Find the following integrals. (*Hint*: As in Example 2.5, you may want to start with an educated guess which you can later modify.)

    (a) $\int (2x + 3)^5 \, dx$
    (b) $\int \sqrt{6x + 5} \, dx$
    (c) $\int (1 - 3x)^{2/3} \, dx$
    (d) $\int 1/(3x + 1) \, dx$
    (e) $\int 1/\sqrt{5x - 2} \, dx$
    (f) $\int 1/(3 - 2x)^2 \, dx$
    (g) $\int e^{-3x} \, dx$
    (h) $\int e^{2x+5} \, dx$

*3. (a) Find $\int x(x^2 + 2)^5 \, dx$. [Hint: See what happens if you try $\frac{1}{6}(x^2 + 2)^6$.]

(b) Find $\int x\sqrt{x^2 + 5} \, dx$.

(c) Find $\int x^2(x^3 - 2)^{3/4} \, dx$.

(d) Find $\int (2x + 1)(x^2 + x - 5)^9 \, dx$.

(e) Find $\int x^2 e^{x^3 + 2} \, dx$.

(f) Find $\int x^2/\sqrt{x^3 + 1} \, dx$.

4. Prove that $\int [f(x) + g(x)] \, dx = \int f(x) \, dx + \int g(x) \, dx$. (*Hint*: Let F be an antiderivative of f, let G be an antiderivative of g, and use the sum rule for derivatives.)

5. Find a function whose tangent has slope $x^3 - 2/x^2 + 2$ for each value of x and whose graph passes through the point $(1, 3)$.

6. Find a function whose tangent has slope $(2x + 1)^4$ for each value of x and whose graph passes through the point $(-1, 2)$.

*7. Find a function whose graph has a relative minimum when $x = 1$ and a relative maximum when $x = 4$.

8. The resale value of a certain industrial machine decreases at a rate that changes with time. When the machine is t years old, the rate at which its value is changing is $-960e^{-t/5}$ dollars per year. If the machine was bought new for $5,000, how much will it be worth 10 years later?

9. A manufacturer estimates his marginal revenue to be $100q^{-1/2}$ dollars per unit when his output is q units. His corresponding marginal cost has been found to be $0.4q$ dollars per unit. Suppose the manufacturer's profit (revenue minus cost) is $520 when his level of production is 16 units. What is his profit when his level of production is 25 units?

10. The marginal profit (marginal revenue minus marginal cost) of a certain company is $100 - 2x$ dollars per unit when x units are produced. If the company's profit is $700 when 10 units are produced, what is the company's *maximum* possible profit?

11. An object is moving so that its speed after t hours is $10 - 5/(t + 1)^2$ miles per hour.
    (a) How far does the object travel during the first hour?
    (b) How far does it travel during the second hour?

*12. A certain oil well that yields 300 barrels of crude oil a month will run dry in 3 years. The price of crude oil is currently $18 per barrel and is expected to rise at a steady rate of 2 cents per month. How much is the oil well worth? That is, what is the total revenue it will yield during its 3-year life-span? [*Hint*: Let $R(t)$ denote the total revenue from the oil well after t months. Find an expression for the rate at which $R(t)$ increases and then integrate to get a formula for $R(t)$. Assume that the oil is sold as soon as it is extracted from the ground.]

3. INTEGRATION BY SUBSTITUTION

There is no general technique for integrating products. However, products of a certain special type can be integrated fairly easily by applying the chain rule in reverse. As a prelude to developing this integration technique, let us examine a typical application of the chain rule.

If we differentiate the function e^{x^2+3x+5} using the chain rule, we get

$$\frac{d}{dx}(e^{x^2+3x+5}) = (2x+3)e^{x^2+3x+5}$$

Note that the answer comes out in the form of a product, and that one of the factors, $2x+3$, is the derivative of an expression x^2+3x+5 that occurs in the other factor. More precisely, the product is of the form $g(h(x))h'(x)$, where in this case $h(x) = x^2 + 3x + 5$ and g is the exponential function.

In this section we will examine products that can be written in the special form $g(h(x))h'(x)$. Many products of this type can be obtained by differentiation via the chain rule and can be integrated using a technique based on the reversal of this process. Specifically, if G is an antiderivative of g, then

$$\int g(h(x))h'(x)\,dx = G(h(x)) + C$$

since, by the chain rule,

$$\frac{d}{dx}[G(h(x))] = G'(h(x))h'(x)$$
$$= g(h(x))h'(x)$$

For reasons that will soon become clear, this technique is called *integration by substitution*. It says that to integrate a product of the form $g(h(x))h'(x)$ we simply have to find an antiderivative of g and evaluate it at $h(x)$.

INTEGRATION BY SUBSTITUTION

Formula:

$$\int g(h(x))h'(x)\,dx = G(h(x)) + C$$

where G is an antiderivative of g.

Procedure: To compute $\int g(h(x))h'(x)\,dx$, first find $\int g(u)\,du$ and then substitute $u = h(x)$ in your answer.

Consider the following examples.

Example 3.1

Find $\int (2x + 3)e^{x^2 + 3x + 5}\,dx$.

Solution

We notice that $2x + 3$ is the derivative of $x^2 + 3x + 5$, and this suggests that we let $h(x) = x^2 + 3x + 5$. Then, $h'(x) = 2x + 3$ and

$$(2x + 3)e^{x^2 + 3x + 5} = g(h(x))h'(x)$$

where $g(u) = e^u$. Hence we can use the method of substitution to integrate this product.

We first integrate e^u to get

$$\int e^u\,du = e^u + C$$

and then replace u by $h(x)$ in the answer. Thus,

$$\int (2x + 3)e^{x^2 + 3x + 5}\,dx = e^{h(x)} + C$$
$$= e^{x^2 + 3x + 5} + C \qquad \square$$

Example 3.2

Find $\int (6x^2 + 4)(2x^3 + 4x - 6)^5\,dx$.

Solution

Since $6x^2 + 4$ is the derivative of $2x^3 + 4x - 6$, we let $h(x) = 2x^3 + 4x - 6$ and write

$$(6x^2 + 4)(2x^3 + 4x - 6)^5 = g(h(x))h'(x)$$

where $g(u) = u^5$. Then, since

$$\int u^5 \, du = \tfrac{1}{6}u^6 + C$$

it follows that

$$\int (6x^2 + 4)(2x^3 + 4x - 6)^5 \, dx = \tfrac{1}{6}[h(x)]^6 + C$$
$$= \tfrac{1}{6}(2x^3 + 4x - 6)^6 + C \qquad \square$$

The product to be integrated in the next example is not exactly of the form of $g(h(x))h'(x)$. However, it is a constant multiple of such a function and can be integrated by combining the method of substitution with the constant multiple rule.

Example 3.3

Find $\int x^3 \sqrt{x^4 + 2} \, dx$.

Solution

The derivative of $x^4 + 2$ is not x^3, but rather $4x^3$, and so we cannot apply the method of substitution to the product $x^3 \sqrt{x^4 + 2}$ directly. We can, however, apply substitution to the related product $4x^3 \sqrt{x^4 + 2}$ to get

$$\int 4x^3 \sqrt{x^4 + 2} \, dx = \tfrac{2}{3}(x^4 + 2)^{3/2} + C$$

Then, using the constant multiple rule, we can conclude that

$$\int x^3 \sqrt{x^4 + 2} \, dx = \tfrac{1}{4} \int 4x^3 \sqrt{x^4 + 2} \, dx$$
$$= \tfrac{1}{4}(\tfrac{2}{3})(x^4 + 2)^{3/2} + C$$
$$= \tfrac{1}{6}(x^4 + 2)^{3/2} + C \qquad \square$$

The method of substitution may seem more difficult than other computational techniques we have encountered. However, with practice you should eventually be able to integrate products of the form $g(h(x))h'(x)$ mentally, without explicitly writing down the expressions for g, h, and u.

Integration by substitution may be thought of as a technique for simplifying an integral by changing variables. Specifically, we start with $\int g(h(x))h'(x) \, dx$, an integral involving a product of functions of x. By substituting the letter u for the expression $h(x)$ we transform this integral into $\int g(u) \, du$, a simpler integral involving a new variable u.

Observe that the expression $h'(x) \, dx$ in the original integral is replaced in the simplified integral by the symbol du. This correspondence between $h'(x) \, dx$ and du

can be extracted from the equation $u = h(x)$ by the following notational trick. First we differentiate to get

$$\frac{du}{dx} = h'(x)$$

Then we pretend that the symbol du/dx is actually a quotient and solve for du. This gives

$$du = h'(x)\, dx$$

which is precisely the right relationship.

These observations lead to an alternative version of the method of substitution in which we substitute the letter u for an expression $h(x)$ and formally rewrite the integral in terms of u. Let us reconsider the integral in Example 3.3 from this point of view.

Example 3.4

Find $\int x^3 \sqrt{x^4 + 2}\, dx$.

Solution

We let $u = x^4 + 2$. Then, $du/dx = 4x^3$, and we "solve" this equation to get $dx = du/4x^3$. Hence,

$$x^3 \sqrt{x^4 + 2}\, dx = x^3 \sqrt{u}\, \frac{1}{4x^3}\, du$$

$$= \tfrac{1}{4} \sqrt{u}\, du$$

and so

$$\int x^3 \sqrt{x^4 + 2}\, dx = \int \tfrac{1}{4} \sqrt{u}\, du$$

$$= \tfrac{1}{6} u^{3/2} + C$$

Finally, we replace u by $x^4 + 2$ in the answer and conclude that

$$\int x^3 \sqrt{x^4 + 2}\, dx = \tfrac{1}{6}(x^4 + 2)^{3/2} + C \qquad \square$$

Many people find the formal method of substitution more appealing than our original version. They like the fact that it involves straightforward mechanical manipulation of symbols, and they appreciate the convenience of the notation. This formal method is also somewhat more versatile, as we shall see presently. However, for most of the integrals you will encounter, both methods work well and you should feel free to use the one with which you are most comfortable.

At first glance the next example may look similar to Example 3.4. It isn't!

Example 3.5

Find $\int x^2 \sqrt{x^4 + 2}\, dx$.

Solution

Let us try to apply the method of substitution with $u = x^4 + 2$. Then, $dx = du/4x^3$, and

$$x^2\sqrt{x^4 + 2}\, dx = x^2\sqrt{u}\, \frac{1}{4x^3}\, du$$

$$= \frac{1}{4x}\sqrt{u}\, du$$

To eliminate the one remaining x we solve the equation $u = x^4 + 2$ for x and substitute. We find that $x = \sqrt[4]{u - 2}$, and so

$$\int x^2\sqrt{x^4 + 2}\, dx = \int \sqrt{u}/(4\sqrt[4]{u - 2})\, du$$

Unfortunately, this integral is even more complicated than the original one, and our substitution has led nowhere. In fact, there is no easy way to integrate the product $x^2\sqrt{x^4 + 2}$. The crucial difference between this product and the one in Example 3.4 is that the derivative of $x^4 + 2$ is not a constant multiple of the factor x^2. ☐

In Example 2.5 we combined an educated guess with the constant multiple rule to find $\int (3x + 5)^6\, dx$. This integral can also be found by substitution.

Example 3.6

Find $\int (3x + 5)^6\, dx$.

Solution

Let $u = 3x + 5$. Then $du/dx = 3$, so $dx = du/3$. Hence,

$$\int (3x + 5)^6\, dx = \int \tfrac{1}{3}u^6\, du$$

$$= \tfrac{1}{21}u^7 + C = \tfrac{1}{21}(3x + 5)^7 + C \qquad \square$$

Here is a slightly more difficult example.

Example 3.7

Find $\int (\log 2x)/x\, dx$.

Solution

At first glance this integral appears to be different in form from those we have previously encountered. However, the similarity becomes clear once we notice that the derivative of $\log 2x$ is $1/x$. This suggests that we try the substitution $u = \log 2x$. Then $dx = x\, du$, and after substituting we get

$$\int \frac{\log 2x}{x} dx = \int u\, du = \tfrac{1}{2} u^2 + C = \tfrac{1}{2} (\log 2x)^2 + C \qquad \square$$

Our final example is designed to show the versatility of the formal method of substitution. It deals with an integral that does not seem to be of the form $\int g(h(x))h'(x)\, dx$, but that nevertheless can be simplified dramatically by a clever change of variables.

Example 3.8

Find $\int x/(x+1)\, dx$.

Solution

There seems to be no easy way to integrate this quotient as it stands. But watch what happens if we make the substitution $u = x + 1$. Then $du = dx$, and $x = u - 1$. Hence,

$$\int \frac{x}{x+1} dx = \int \frac{u-1}{u} du$$

$$= \int 1\, du - \int \frac{1}{u} du$$

$$= u - \log |u| + C = x + 1 - \log |x+1| + C \qquad \square$$

The integral $\int x/(x+1)\, dx$ in Example 3.8 can actually be put in the form $\int g(h(x))h'(x)\, dx$. Can you identify the functions g and h in this case? And do you see how to apply our original method of substitution to this integral?

Problems

1. Find the following integrals.

(a) $\int (2x + 6)^5\, dx$

(b) $\int \sqrt{4x - 1}\, dx$

(c) $\int (1 - 2x)^{4/5}\, dx$

(d) $\int 1/(3x+5)\,dx$

(e) $\int 1/\sqrt{6x-2}\,dx$

(f) $\int e^{1-x}\,dx$

(g) $\int [(x-1)^5 + 3(x-1)^2 + 6(x-1) + 5]\,dx$

(h) $\int [(1-2x)^5 + 1]/(1-2x)\,dx$

(i) $\int 2xe^{x^2-1}\,dx$

(j) $\int x(x^2+1)^5\,dx$

(k) $\int 3x\sqrt{x^2+8}\,dx$

(l) $\int x^2(x^3+1)^{3/4}\,dx$

(m) $\int xe^{x^2}\,dx$

(n) $\int x^5 e^{1-x^6}\,dx$

(o) $\int 2x^4/(x^5+1)\,dx$

(p) $\int x^2/(x^3+5)^2\,dx$

(q) $\int (x+1)(x^2+2x+5)^{12}\,dx$

(r) $\int (3x^2-1)e^{x^3-x}\,dx$

(s) $\int (x^3+x)\sqrt{x^4+2x^2+5}\,dx$

(t) $\int (3x^4+12x^3+6)/(x^5+5x^4+10x+12)\,dx$

(u) $\int (10x^3-5x)/\sqrt{x^4-x^2+6}\,dx$

(v) $\int x^2(x^5-3)/(x^8-8x^3+1)^3\,dx$

(w) $\int (\log 5x)/x\,dx$

(x) $\int 1/(x\log x)\,dx$

(y) $\int 2x \log(x^2+1)/(x^2+1)\,dx$

(z) $\int e^{\sqrt{x}}/\sqrt{x}\,dx$

2. Use appropriate substitutions to find the following integrals.

   (a) $\int x/(x-1)\,dx$ (b) $\int x\sqrt{x+1}\,dx$

   (c) $\int x/(x-5)^6\,dx$ (d) $\int (x+1)(x-2)^9\,dx$

   (e) $\int (x+3)/(x-4)^2\,dx$ (f) $\int (x+5)/\sqrt{1-x}\,dx$

3. Find a function whose tangent has slope $x\sqrt{x^2+5}$ for each value of x, and whose graph passes through the point (2, 10).

4. Find a function whose tangent has slope $2x/(1-3x^2)$ for each value of x, and whose graph passes through the point (0, 5).

5. It is estimated that x years from now the value of an acre of land near the ghost town of Cherokee, California, will be increasing at a rate of $0.4x^3/\sqrt{0.2x^4+8{,}000}$ dollars per year. If the land is currently worth \$500 per acre, how much will it be worth in 10 years?

4. INTEGRATION BY PARTS

Suppose you want to integrate the product $f(x)g(x)$. If one of the factors, say $g(x)$, can be easily integrated and the other, $f(x)$, can be simplified by differentiation, then *integration by parts* may be the technique to use.

According to this technique, the first step in computing $\int f(x)g(x)\,dx$ is to integrate g and multiply the result by the other factor f. That is, we form the product

$f(x)G(x)$

where G is an antiderivative of g. The second step is to multiply the antiderivative G by the derivative of f and subtract the integral of this product from the result of the first step. At this point we have

$f(x)G(x) - \int f'(x)G(x)\,dx$

This expression is equal to our original integral $\int f(x)g(x)\,dx$. With any luck, the new integral $\int f'(x)G(x)\,dx$ will be easier to find than the original one.

INTEGRATION BY PARTS

$$\int f(x)g(x)\,dx = f(x)G(x) - \int f'(x)G(x)\,dx$$

where G is an antiderivative of g.

The formula for integration by parts is nothing more than a restatement of the product rule for differentiation. Before we verify this, however, let us examine some typical applications of this new technique.

Example 4.1

Find $\int xe^x\,dx$.

Solution

In this case, both factors x and e^x are easy to integrate. Both are also easy to differentiate, but the process of differentiation simplifies x more than e^x. Hence, we try integration by parts, with $g(x) = e^x$ and $f(x) = x$. Since the function $G(x) = e^x$ is an antiderivative of e^x and $d/dx\,(x) = 1$, we have

$$\int xe^x\,dx = xe^x - \int 1e^x\,dx$$
$$= xe^x - e^x + C \qquad \square$$

Example 4.2

Find $\int x\sqrt{x+5}\,dx$.

Solution

Again, both factors can be readily integrated and differentiated. However, the derivative of x is simpler than x, whereas the derivative of $\sqrt{x+5}$ is actually worse than the original. Hence we try integration by parts, with $g(x) = \sqrt{x+5}$ and $f(x) = x$. Since the function $G(x) = \frac{2}{3}(x+5)^{3/2}$ is an antiderivative of $\sqrt{x+5}$ and $d/dx\,(x) = 1$, we have

$$\int x\sqrt{x+5}\,dx = \frac{2}{3}x(x+5)^{3/2} - \int \frac{2}{3}(x+5)^{3/2}\,dx$$
$$= \frac{2}{3}x(x+5)^{3/2} - \frac{4}{15}(x+5)^{5/2} + C \qquad \square$$

Can you think of another way to do Example 4.2?
Using integration by parts, we can finally integrate $\log x$.

Example 4.3

Find $\int \log x \, dx$.

Solution

We don't seem to have a product to work with. However, we can remedy this by writing $\log x$ as $1(\log x)$. Since we do not yet know how to integrate the factor $\log x$, integration by parts is possible only if we let $g(x) = 1$ and $f(x) = \log x$. Then, since the function $G(x) = x$ is an antiderivative of 1 and $d/dx \, (\log x) = 1/x$, we have

$$\int \log x \, dx = x \log x - \int x \frac{1}{x} dx$$

$$= x \log x - \int 1 \, dx$$

$$= x \log x - x + C \qquad \square$$

In the next example we integrate the product $x^2 e^x$ that we encountered in Section 1. Two applications of integration by parts will be needed.

Example 4.4

Find $\int x^2 e^x \, dx$.

Solution

We try integration by parts, with $g(x) = e^x$ and $f(x) = x^2$. Since the function $G(x) = e^x$ is an antiderivative of e^x and $d/dx \, (x^2) = 2x$, we have

$$\int x^2 e^x \, dx = x^2 e^x - \int 2x e^x \, dx$$

$$= x^2 e^x - 2 \int x e^x \, dx$$

As we saw in Example 4.1, the new integral $\int x e^x \, dx$ can be found by applying integration by parts again, this time with $g(x) = e^x$ and $f(x) = x$. Thus,

$$\int x^2 e^x \, dx = x^2 e^x - 2(x e^x - e^x) + C$$

$$= x^2 e^x - 2x e^x + 2e^x + C \qquad \square$$

Our final example is slightly more difficult, because the proper choice of f and g is not immediately obvious.

Example 4.5

Find $\int x^3 \sqrt{x^2+1}\, dx$.

Solution

It seems reasonable to try integration by parts, with $g(x) = x^3$ and $f(x) = \sqrt{x^2+1}$. Then, since the function $G(x) = \frac{1}{4}x^4$ is an antiderivative of x^3 and $d/dx\,(\sqrt{x^2+1}) = x/\sqrt{x^2+1}$, we have

$$\int x^3 \sqrt{x^2+1}\, dx = \tfrac{1}{4}x^4 \sqrt{x^2+1} - \int \tfrac{1}{4}x^4 \frac{x}{\sqrt{x^2+1}}\, dx$$

$$= \tfrac{1}{4}x^4 \sqrt{x^2+1} - \tfrac{1}{4} \int \frac{x^5}{\sqrt{x^2+1}}\, dx$$

Unfortunately, we are now confronted with an integral that is even more complicated than the original one, and our application of integration by parts has not been fruitful.

The situation improves, however, if we think of $x^3 \sqrt{x^2+1}$ as $x^2(x\sqrt{x^2+1})$. Then, the factor $x\sqrt{x^2+1}$ can be integrated easily by substitution, and so integration by parts with $g(x) = x\sqrt{x^2+1}$ and $f(x) = x^2$ looks promising. Let us see what happens.

Since the function $G(x) = \tfrac{1}{3}(x^2+1)^{3/2}$ is an antiderivative of $x\sqrt{x^2+1}$ and $d/dx\,(x^2) = 2x$, we have

$$\int x^3 \sqrt{x^2+1}\, dx = \tfrac{1}{3}x^2(x^2+1)^{3/2} - \tfrac{2}{3} \int x(x^2+1)^{3/2}\, dx$$

$$= \tfrac{1}{3}x^2(x^2+1)^{3/2} - \tfrac{2}{15}(x^2+1)^{5/2} + C \qquad \square$$

Integration by parts is actually nothing more than a restatement of what happens when the product rule is used to differentiate $f(x)G(x)$, where G is an antiderivative of g. In particular,

$$\frac{d}{dx}[f(x)G(x)] = f'(x)G(x) + f(x)G'(x)$$

$$= f'(x)G(x) + f(x)g(x)$$

Expressed in terms of integrals, this says that

$$f(x)G(x) = \int f'(x)G(x)\, dx + \int f(x)g(x)\, dx$$

or

$$\int f(x)g(x)\, dx = f(x)G(x) - \int f'(x)G(x)\, dx$$

which is precisely the formula for integration by parts.

Problems

1. Find the following integrals.

   (a) $\int xe^{-x}\, dx$
   (b) $\int x^2 e^{-x}\, dx$
   (c) $\int xe^{ax}\, dx$
   (d) $\int x^2 e^{ax}\, dx$
   (e) $\int x^3 e^x\, dx$
   (f) $\int (3-2x)e^{-0.06x}\, dx$
   (g) $\int (a+bx)e^{rx}\, dx$
   (h) $\int x \log x\, dx$
   (i) $\int x^2 \log x\, dx$
   (j) $\int x(\log x)^2\, dx$
   (k) $\int x\sqrt{x-6}\, dx$
   (l) $\int x/\sqrt{x+2}\, dx$
   (m) $\int x^3 \sqrt{x^2-1}\, dx$
   (n) $\int x^3/(x^2+3)^4\, dx$
   *(o) $\int x^5 (3x^2-7)^{1/2}\, dx$

2. (a) Find $\int x(x-1)^{3/5}\, dx$ using the methods of Section 3.
   (b) Now find the same integral using integration by parts. Which method is easier?

3. (a) Find $\int (x+1)(x-2)^9\, dx$ using the methods of Section 3.
   (b) Now find this integral using integration by parts. Which method is easier?

4. Use the result of Example 4.3 and the method of substitution to find

   (a) $\int \log(3x+5)\, dx$
   (b) $\int x \log(x^2+1)\, dx$
   (c) $\int (x^2+2x) \log(x^3+3x^2+5)\, dx$

5. Use integration by parts to find $\int x \log x\, dx$. Then find the following related integrals.

   (a) $\int (3x-1) \log(3x-1)\, dx$
   (b) $\int x \log(3x-1)\, dx$
   (c) $\int x^3 \log x^2\, dx$

6. An object is moving so that at the end of t seconds its speed is $te^{-t/2}$ feet per second.
   (a) Express the distance that the object has traveled as a function of time.
   (b) How far has the object gone by the end of 4 seconds?

*7. A certain oil well that yields 300 barrels of crude oil a month will run dry in 3 years. It is estimated that in t months the price of crude oil will be $18 + \tfrac{1}{2} \log(t+1)$ dollars per barrel. What is the total future revenue from the well? (Assume that the oil is sold as soon as it is extracted from the ground.)

8. (a) Use integration by parts to derive the following integration formula.

$$\int x^n e^{ax}\, dx = \frac{1}{a} x^n e^{ax} - \frac{n}{a} \int x^{n-1} e^{ax}\, dx$$

(b) Use this formula to find $\int x^3 e^{5x}\, dx$.

5. TABLES OF INTEGRALS

The integration techniques we have developed in this chapter are sufficient to let us find almost any integral that arises from a practical problem in economics, psychology, or the social sciences. Once in a while, however, an integral turns up that cannot be handled by these methods. For such occasions, it is helpful to know how to use a *table of integrals*.

A table of integrals is a list of integration formulas. Extensive tables listing several hundred formulas can be found in most mathematics handbooks, and condensed versions appear in many calculus texts. Here is a tiny sampling of the formulas that appear in a table of integrals.

$$\int \frac{dx}{p^2 - x^2} = \frac{1}{2p} \log \left| \frac{p + x}{p - x} \right|$$

$$\int \frac{dx}{x(ax + b)} = \frac{1}{b} \log \left| \frac{x}{ax + b} \right|$$

$$\int \frac{dx}{\sqrt{x^2 \pm p^2}} = \log \left| x + \sqrt{x^2 \pm p^2} \right|$$

$$\int x^n e^{ax}\, dx = \frac{1}{a} x^n e^{ax} - \frac{n}{a} \int x^{n-1} e^{ax}\, dx$$

In these formulas, the letters a, b, p, and n denote constants. The term p^2 in the first and third formulas may be any *positive* constant (since any positive number is the square of its square root). The compact fractional notation

$$\frac{dx}{p^2 - x^2}$$

in the first formula is an abbreviation for

$$\frac{1}{p^2 - x^2}\, dx$$

Similar notation occurs in some of the other formulas. Also, for obvious reasons of economy, the constant C is omitted from each of the integrals in the table.

For convenience, most tables of integrals are divided into sections. Integrals containing similar expressions are grouped together in the same section. For example, the first formula in our list might appear in a section entitled "Expressions Containing $p^2 - x^2$." A section called "Expressions Containing $ax + b$" should include our second formula, while our fourth formula would be found in the section "Expressions Containing Exponential and Logarithmic Functions."

In the following examples, we use these four formulas to illustrate the use of a table of integrals.

Example 5.1

Find $\int 1/[x(3x - 6)] \, dx$.

Solution

We apply the second formula, with $a = 3$ and $b = -6$, to get

$$\int \frac{1}{x(3x - 6)} \, dx = -\tfrac{1}{6} \log \left| \frac{x}{3x - 6} \right| + C \qquad \square$$

Example 5.2

Find $\int 1/(6 - 3x^2) \, dx$.

Solution

If the coefficient of x^2 were 1 instead of 3, we could use the first formula. This suggests that we divide numerator and denominator by 3 to get

$$\int \frac{1}{6 - 3x^2} \, dx = \frac{1}{3} \int \frac{1}{2 - x^2} \, dx$$

We now apply the formula, with $p = \sqrt{2}$, to the new integral and conclude that

$$\int \frac{1}{6 - 3x^2} \, dx = \frac{1}{3} \left(\frac{1}{2\sqrt{2}} \right) \log \left| \frac{\sqrt{2} + x}{\sqrt{2} - x} \right| + C$$

$$= \frac{\sqrt{2}}{12} \log \left| \frac{\sqrt{2} + x}{\sqrt{2} - x} \right| + C \qquad \square$$

Example 5.3

Find $\int 1/(3x^2 - 6) \, dx$.

Solution

Since

$$\frac{1}{3x^2 - 6} = \frac{-1}{6 - 3x^2}$$

we can apply the first formula as in Example 5.2 to get

$$\int \frac{1}{3x^2 - 6} dx = -\frac{\sqrt{2}}{12} \log \left| \frac{\sqrt{2} + x}{\sqrt{2} - x} \right| + C \qquad \square$$

Example 5.4

Find $\int 1/(3x^2 + 6) \, dx$.

Solution

We try to match this integral to the one in the first formula by writing

$$\int \frac{1}{3x^2 + 6} dx = -\frac{1}{3} \int \frac{1}{-2 - x^2} dx$$

However, since -2 is negative, it cannot be written as the square, p^2, of any number p, and so our formula does not apply.

There is a formula for integrals of the form $\int 1/(x^2 + p^2) \, dx$ that can be used here. It would be listed in a table of integrals under a heading like "Expressions Containing $x^2 \pm p^2$." However, the antiderivative would be written in terms of what are called inverse trigonometric functions and cannot be expressed in more familiar terms. $\qquad \square$

Example 5.5

Find $\int 1/\sqrt{4x^2 - 9} \, dx$.

Solution

To put this integral into the form of our third formula we divide numerator and denominator by 2. Then,

$$\int \frac{1}{\sqrt{4x^2 - 9}} dx = \frac{1}{2} \int \frac{1}{\sqrt{x^2 - \frac{9}{4}}} dx$$

and we can apply the formula to the new integral, taking $p^2 = \frac{9}{4}$ and using minus signs in place of the symbol $\pm$. Thus,

$$\int \frac{1}{\sqrt{4x^2 - 9}} dx = \tfrac{1}{2} \log \left| x + \sqrt{x^2 - \tfrac{9}{4}} \right| + C \qquad \square$$

Some integral formulas, like the fourth one on our list, express an integral in terms of a simpler integral of the same type. If the formula is subsequently applied to the new integral, further simplification may occur. Successive applications of the formula usually lead to an integral that can be found by elementary means. Formulas of this type are called *recursion formulas*.

Example 5.6

Find $\int x^2 e^{5x}\,dx$.

Solution

If we apply the fourth formula with $n = 2$ and $a = 5$, we get

$$\int x^2 e^{5x}\,dx = \tfrac{1}{5}x^2 e^{5x} - \tfrac{2}{5}\int xe^{5x}\,dx$$

If we now apply the same formula with $n = 1$ and $a = 5$, we get

$$\int xe^{5x}\,dx = \tfrac{1}{5}xe^{5x} - \tfrac{1}{5}\int e^{5x}\,dx$$

$$= \tfrac{1}{5}xe^{5x} - \tfrac{1}{25}e^{5x} + C$$

Combining these results we conclude that

$$\int x^2 e^{5x}\,dx = \tfrac{1}{5}x^2 e^{5x} - \tfrac{2}{25}xe^{5x} + \tfrac{2}{125}e^{5x} + C \qquad \square$$

No special formula was really needed to find the integral in Example 5.6. The problem could have been solved quite easily using integration by parts. Indeed, if the formula had not been so conveniently displayed on page 252, we would have been much better off integrating by parts directly than hunting through a table of integrals for the appropriate formula. Note that the formula we did use in Example 5.6 is nothing more than a special case of the formula for integration by parts.

Try not to succumb to the temptation to rely excessively on tables when computing integrals. Most of the integrals you will encounter can be found quite easily without the aid of formulas. Moreover, before you can use a formula, you must find it, and this can be time-consuming. In general, it is good strategy to use a table of integrals only as a last resort.

Problems

1. Use the integration formulas listed in this section to find the following integrals.

   (a) $\int 3/[4x(x-5)]\,dx$
   (b) $\int 1/\sqrt{9x^2 - 4}\,dx$
   (c) $\int x^3 e^{-x}\,dx$
   (d) $\int 1/(3x^2 - 9)\,dx$
   (e) $\int 1/(3x^2 + 2x)\,dx$
   (f) $\int 1/\sqrt{4 + 9x^2}\,dx$
   *(g) $\int 1/(x^2 - 3x + 2)\,dx$

2. Locate a table of integrals and use it to find the following integrals.

   (a) $\int (x+3)/\sqrt{2x+4}\,dx$ (b) $\int x^3 \sqrt{2x+3}\,dx$

   (c) $\int (\log 2x)^2\,dx$ (d) $\int (x^2+1)^{3/2}\,dx$

   (e) $\int 1/(3x\sqrt{2x+5})\,dx$ (f) $\int 1/\sqrt{3x^2-6x+2}\,dx$

   (g) $\int 1/(2-3e^{-x})\,dx$

*3. One table of integrals lists the formula

$$\int \frac{dx}{\sqrt{x^2 \pm p^2}} = \log \left| \frac{x + \sqrt{x^2 \pm p^2}}{p} \right|$$

while another table lists

$$\int \frac{dx}{\sqrt{x^2 \pm p^2}} = \log \left| x + \sqrt{x^2 \pm p^2} \right|$$

Can you reconcile this apparent contradiction?

4. The following two formulas appear in a table of integrals.

$$\int \frac{dx}{p^2 - x^2} = \frac{1}{2p} \log \left| \frac{p+x}{p-x} \right|$$

$$\int \frac{dx}{a + bx^2} = \frac{1}{2\sqrt{-ab}} \log \left| \frac{a + x\sqrt{-ab}}{a - x\sqrt{-ab}} \right| \quad \text{if} \quad -ab \geq 0$$

   (a) Use the second formula to derive the first.
   (b) Apply both formulas to the integral $\int 1/(9-4x^2)\,dx$. Which do you find easier to use in this problem?

6. ELEMENTARY DIFFERENTIAL EQUATIONS

Any equation that contains a derivative is called a differential equation. For example, the equations $dy/dx = 3x^2 + 6$, $dP/dt = kP$, $d^2H/dt^2 = -32$, and $(dy/dx)^2 + 3(dy/dx) + 2y = e^x$ are all differential equations.

Many practical situations, especially those involving rates, can be described mathematically by differential equations. For example, the assumption that the population grows at a rate proportional to its size can be expressed by the differential equation $dP/dt = kP$, where P denotes the population size, t stands for time, and k is the constant of proportionality. A well-known law of physics states that the acceleration of a falling object due to gravity is (approximately) 32 feet per second per second. This is actually a statement about the second derivative of the object's height H with respect to time t, and the corresponding differential equation is $d^2H/dt^2 = -32$. In economics, statements about marginal cost and marginal revenue can be formulated as differential equations. For instance, if C represents production cost and q the quantity produced, then the differential equation $dC/dq = 2q + 8$ says that the marginal cost when q units are produced is $2q + 8$ dollars per unit.

Example 6.1

Write a differential equation that describes the following situation.

The number of people implicated in a government scandal increases at a rate jointly proportional to both the number of people already implicated and the number of people involved who have not yet been implicated.

Solution

Suppose that N people are involved and let $P(t)$ denote the number who have been implicated by time t. Then,

$N - P(t)$ = number involved but not implicated by time t

and

$$\frac{dP}{dt} = \text{rate at which people are being implicated}$$

The appropriate differential equation is therefore

$$\frac{dP}{dt} = k(N - P)P$$

where k is the constant of proportionality. $\square$

Any function that satisfies a differential equation is said to be a *solution* of that equation. To illustrate this concept, let us consider a simple example.

Example 6.2

Verify that the function $y = e^x + e^{-x} - x$ is a solution of the differential equation $d^2y/dx^2 - y = x$.

Solution

We must compute the second derivative of this function, subtract the function itself, and show that the result is equal to x. Since

$$\frac{dy}{dx} = e^x - e^{-x} - 1$$

it follows that

$$\frac{d^2y}{dx^2} = e^x + e^{-x}$$

and so

$$\frac{d^2y}{dx^2} - y = e^x + e^{-x} - e^x - e^{-x} + x = x$$

Hence $y = e^x + e^{-x} - x$ is indeed a solution of the differential equation. □

You can easily check that the function $y = 3e^x - 2e^{-x} - x$ is also a solution of the differential equation in Example 6.2. In fact, every function of the form $y = C_1 e^x + C_2 e^{-x} - x$, where C_1 and C_2 are constants, is a solution of this equation. Moreover, it turns out that every solution of the equation is of this form. For this reason, the function $y = C_1 e^x + C_2 e^{-x} - x$ is said to be the *general solution* of the differential equation in Example 6.2. The solutions obtained by replacing C_1 and C_2 by specific numbers are sometimes called *particular solutions*.

Example 6.3

Find the particular solution of the differential equation $d^2y/dx^2 - y = x$ satisfying the conditions that $y = 4$ and $dy/dx = 2$ when $x = 0$.

Solution

We use the two given conditions to determine the numerical values of the constants C_1 and C_2 in the general solution $y = C_1 e^x + C_2 e^{-x} - x$. Substituting $y = 4$ and $x = 0$ into the general solution we get

$$4 = C_1 + C_2$$

The derivative of the general solution is $dy/dx = C_1 e^x - C_2 e^{-x} - 1$. Substituting $dy/dx = 2$ and $x = 0$ into this expression we get

$$2 = C_1 - C_2 - 1$$

Solving these two equations for C_1 and C_2 we find that $C_1 = \frac{7}{2}$ and $C_2 = \frac{1}{2}$. Hence the desired particular solution is

$$y = \tfrac{7}{2}e^x + \tfrac{1}{2}e^{-x} - x$$ □

Every time we find an integral, we are actually solving a special type of differential equation. The differential equation in this case is of the form $dy/dx = f(x)$, and its general solution is $y = F(x) + C$, where F is an antiderivative of f.

Differential equations of the form $dy/dx = f(x)$ are particularly easy to solve because the derivative of the quantity in question is given explicitly as a function of the independent variable x. Some practical problems that lead to this type of differential equation will be examined in this section. Notice that our differential equation describing population growth, $dP/dt = kP$, is not of this form, because the derivative of P is expressed in terms of P itself rather than t. This equation can be solved by a method involving two integrations, one with respect to P and the other with respect to t. Details of this procedure will be discussed in Section 7.

The discussion of differential equations in this chapter is intended to be only a brief introduction to a vast subject. Entire courses are devoted to the theory and applications of differential equations. Special techniques for solving many types of differential equations are included in most advanced calculus courses.

Let us now turn to some practical problems that lead to differential equations of the special form $dy/dx = f(x)$.

Example 6.4

The resale value of a certain industrial machine decreases over a 10-year period at a rate that depends on the age of the machine. When the machine is x years old, the rate at which its value is changing is $220(x - 10)$ dollars per year.

(a) Express the value of the machine as a function of its age and initial value.

(b) If the machine was originally worth $12,000, how much will it be worth when it is 10 years old?

Solution

(a) Let $V(x)$ denote the value of the machine when it is x years old. The derivative dV/dx is equal to the rate $220(x - 10)$ at which the value is changing. Hence we are led to the differential equation

$$\frac{dV}{dx} = 220(x - 10) = 220x - 2{,}200$$

To find V we solve this differential equation by integrating:

$$V(x) = \int (220x - 2{,}200)\, dx$$

$$= 110x^2 - 2{,}200x + C$$

Note that C is equal to $V(0)$, the initial value of the machine. A more descriptive symbol than C for this constant is V_0. Using this notation we conclude that

$$V(x) = 110x^2 - 2{,}200x + V_0$$

where V_0 is the initial value of the machine.

(b) If $V_0 = 12{,}000$, then the value after 10 years is

$$V(10) = 11{,}000 - 22{,}000 + 12{,}000 = \$1{,}000$$

Graphs showing the rate of depreciation and the resale value of the machine are sketched in Figure 3. Can you explain why the negative of the rate of change $220(x - 10)$ is used to represent the rate of depreciation?

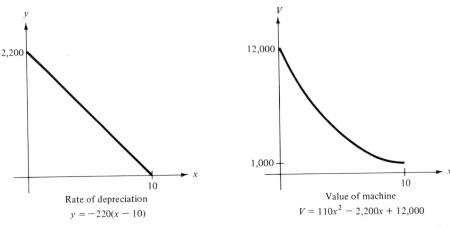

Figure 3

The differential equation in the next example involves the second derivative of a certain quantity. We solve the equation by integrating twice.

Example 6.5

After its brakes are applied, a certain car decelerates at a constant rate of 24 feet per second per second.

(a) Express the distance the car travels in terms of its speed at the moment of braking, and the amount of time that has elapsed since that moment.

(b) Compute the stopping distance if the car was going 60 miles per hour when the brakes were applied.

Solution

(a) Let t be the number of seconds that have elapsed since the brakes were applied and let $D(t)$ be the corresponding distance (measured in feet) that the car has traveled. Since acceleration is the second derivative of distance, we begin with the differential equation

$$\frac{d^2 D}{dt^2} = -24$$

where the minus sign indicates that the speed is *decreasing*.

The first step in solving this equation is to reduce it to a differential equation involving only a first derivative. Since acceleration is the (first) derivative of speed, we let $S(t)$ denote the speed (measured in feet per second) of the car at time t and rewrite the differential equation as

$$\frac{dS}{dt} = -24$$

Integrating we get

$$S(t) = -24t + C_1$$

The constant C_1 is equal to the value of S when $t = 0$. That is, C_1 is the speed of the car at the moment the brakes were first applied. If we denote this initial speed by S_0 instead of C_1, we have

$$S(t) = -24t + S_0$$

Our goal is to find a formula for the distance, not the speed. Recalling that speed is the derivative of distance we replace $S(t)$ by dD/dt to get a new differential equation

$$\frac{dD}{dt} = -24t + S_0$$

This too can be solved easily by integration. We get

$$D(t) = -12t^2 + S_0 t + C_2$$

This time the constant C_2 is equal to $D(0)$ which is zero. Hence,

$$D(t) = -12t^2 + S_0 t$$

where S_0 is the initial speed.

(b) To find the stopping distance, we first determine the time t at which the car stopped moving and then compute $D(t)$ for this value of t.

We take $S_0 = 88$ (since 60 miles per hour is 88 feet per second), set $S(t)$ equal to zero in the formula for the speed, and solve for t:

$$0 = -24t + 88$$

or

$$t = 3.67$$

Then we compute the corresponding distance:

$$D(3.67) = -12(3.67)^2 + 88(3.67) = 161.33$$

That is, if the car is initially going 60 miles per hour, it will stop 161.33 feet beyond the point where the brakes are first applied. □

Read the next example carefully. It is typical of a large class of practical problems to which calculus is applicable only after a continuous mathematical model is introduced to approximate a discrete situation.

Example 6.6

A certain oil well that yields 300 barrels of crude oil a month will run dry in 3 years. It is estimated that t months from now the price of crude oil will be

$P(t) = 18 + 0.3\sqrt{t}$ dollars per barrel

What is the total future revenue from the well?

Solution

This problem cannot be solved! The total future revenue depends on how frequently the oil is sold, and we have not been given this information.

Let us assume that at the end of each month the owner of the oil well sells the entire month's accumulation of oil, for which he receives the market price at that time. With this assumption, we can now solve the revenue problem. The solution requires no calculus whatsoever, just patience.

At the end of the first month, the owner sells 300 barrels of oil at $P(1)$ dollars per barrel. Therefore, his revenue for this month is $300P(1)$ dollars. At the end of the second month he sells 300 barrels for $P(2)$ dollars each. His revenue for the second month is therefore $300P(2)$ dollars. The revenue for the third month is $300P(3)$ dollars, and so on. Hence his total revenue over a 36-month period (3 years) is

$300[P(1) + P(2) + P(3) + \cdots + P(36)]$ dollars

The evaluation of this expression is elementary but tedious. It requires, for example, the computation of 36 square roots, one for each of the 36 values of $P(t) = 18 + 0.3\sqrt{t}$.

The computation would be even more time-consuming if the oil were sold twice a month rather than once. In this case, the total revenue would be

$150[P(0.5) + P(1) + P(1.5) + P(2) + \cdots + P(36)]$ dollars

If the oil were sold at the end of each day, the calculations required to find the total revenue would be prohibitive without the aid of a computer. [The expression for the revenue would contain a sum of $3(365) = 1{,}095$ terms!]

A computer was actually used to calculate the revenue in these three cases. Here are the results. The total future revenue will be \$207,612 if the oil is sold monthly, \$207,489 if the oil is sold twice a month, and \$207,369 if it is sold daily.

All three of these versions of the revenue problem have discontinuous mathematical formulations, because in each case the revenue is accumulated in a finite number of discrete lumps. Let us now adopt a different point of view. We imagine that the oil is sold as soon as it is extracted from the ground. In this case the revenue is accumulated continuously, and the situation can be described by a differential equation. In words,

$$\begin{pmatrix} \text{Number of dollars} \\ \text{coming in per} \\ \text{month at time } t \end{pmatrix} = \begin{pmatrix} \text{number of barrels} \\ \text{produced per} \\ \text{month at time } t \end{pmatrix} \begin{pmatrix} \text{number of dollars} \\ \text{received per} \\ \text{barrel at time } t \end{pmatrix}$$

Thus, if $R(t)$ denotes the total revenue accumulated over the first t months, we have

$$\frac{dR}{dt} = 300P(t) = 5{,}400 + 90\sqrt{t}$$

The solution is

$$R(t) = \int (5{,}400 + 90\sqrt{t})\, dt$$

or

$$R(t) = 5{,}400t + 60t^{3/2} + C$$

Since $R(0) = 0$, it follows that $C = 0$, and so

$$R(t) = 5{,}400t + 60t^{3/2}$$

The total future revenue from the well is then

$$R(36) = 5{,}400(36) + 60(216) = \$207{,}360$$

Notice how closely this answer approximates our previous results, especially the one corresponding to daily sales. Indeed, the continuous version of this problem can be considered to be the *limiting case* of the discrete versions as the frequency of oil sales increases without bound. This idea will be explored more thoroughly in Section 11, where integrals will be characterized as limits of certain sums. □

Problems

1. Verify that the function $y = Ce^{kx}$ is a solution of the differential equation $dy/dx = ky$.

2. Verify that the function $y = \sqrt{2x \log x - 2x + C}$ is a solution of the differential equation $dy/dx = (\log x)/y$.

3. Verify that the function $y = C_1 e^x + C_2 x e^x$ is a solution of the differential equation $d^2y/dx^2 - 2(dy/dx) + y = 0$.

4. Verify that the function $y = \frac{1}{20}x^4 - C_1/x + C_2$ is a solution of the differential equation $x(d^2y/dx^2) + 2(dy/dx) = x^3$.

5. Find the general solutions of the following differential equations.
   (a) $\dfrac{dy}{dx} = 3x^2 + 5x - 6$
   (b) $\dfrac{d^2y}{dx^2} = 3x^2 + 5x - 6$
   (c) $\dfrac{dP}{dt} = \sqrt{t} + e^{-t}$
   (d) $\dfrac{d^2P}{dt^2} = \sqrt{t} + e^{-t}$
   (e) $\dfrac{dV}{dx} = \dfrac{x}{x+1}$
   (f) $\dfrac{d^2V}{dx^2} = \dfrac{x}{x+1}$
   (g) $\dfrac{dA}{dt} = te^t$
   (h) $\dfrac{d^2A}{dt^2} = te^t$
   (i) $\dfrac{dy}{dx} = \log x$
   (j) $\dfrac{d^2y}{dx^2} = \log x$

6. Find the particular solutions of the following differential equations that satisfy the given conditions.
   (a) $\dfrac{dy}{dx} = e^{5x}$; $y = 1$ when $x = 0$
   (b) $\dfrac{d^2y}{dt^2} = 3t^2 + 2t - 1$; $y = 3$ and $\dfrac{dy}{dt} = 0$ when $t = 1$
   (c) $\dfrac{d^2H}{dt^2} = -32$; $H = 160$ and $\dfrac{dH}{dt} = 64$ when $t = 0$
   (d) $\dfrac{d^2A}{dt^2} = e^{-t/2}$; $A = 2$ and $\dfrac{dA}{dt} = 1$ when $t = 0$
   (e) $\dfrac{dP}{dx} = \log x$; $P = \tfrac{1}{2}$ when $x = e$

7. Write differential equations describing the following situations. (Do not try to solve these equations.)
   (a) The number of bacteria in a culture grows at a rate that is proportional to the number of bacteria present.
   (b) The population of a certain town increases at a constant rate.
   (c) The rate of decay of radium is proportional to the amount present.
   (d) The rate at which the temperature of an object changes is proportional to the difference between its temperature and the temperature of the surrounding medium.
   (e) The acceleration of a certain object is proportional to the square root of its speed.
   (f) Money deposited in a certain bank grows so that at any time the rate at which the balance is increasing is equal to 7 percent of the balance at that time.

(g) The rate at which a flu epidemic spreads through a community is jointly proportional to both the number of residents who have already caught the disease and the number who are susceptible to the disease but who have not yet caught it.

8. A manufacturer's marginal cost is $3(q-4)^2$ dollars per unit when his level of output is q units.
   (a) Express the manufacturer's total production cost in terms of his overhead (the cost of producing no units) and the number of units produced.
   (b) What is the cost of producing 14 units if the overhead is $436?

9. A falling object experiences a constant acceleration of 32 feet per second per second due to the force of gravity.
   (a) Express the object's height in terms of its initial height, its initial speed, and time. (You should get the same formula that was given on page 180.)
   (b) With what speed will the object hit the ground if it is dropped from a height of 144 feet?

10. After its brakes are applied, a certain sports car decelerates at a constant rate of 28 feet per second per second.
    (a) Express the distance the car travels in terms of its speed at the moment of braking and the amount of time that has elapsed since that moment.
    (b) Compute the stopping distance if the car was going 60 miles per hour when the brakes were applied.

11. The hero of a popular spy story (who defused the bomb in 5 minutes and survived Problem 15 on page 152) is driving the sports car in Problem 10 at a speed of 60 miles per hour on Highway 1 in the remote republic of San Dimas. Suddenly he sees a camel in the road 199 feet in front of him. After a reaction time of 0.7 second, he steps on the brakes. Will he stop before hitting the camel?

12. Population statistics compiled since 1960 indicate that x years after 1960 a certain county was growing at a rate of approximately $1,500/\sqrt{x}$ people per year. In 1969 the population of the county was 39,000.
    (a) What was the population in 1960?
    (b) If this pattern of population growth continues in the future, how many people will be living in the county in 1985?

13. In a certain section of the country, the price of chicken is currently $1.50 per pound. It is expected that x weeks from now the price will be increasing at a rate of $3\sqrt{x+1}$ cents per week. How much will chicken cost 8 weeks from now?

14. In a certain Los Angeles suburb, a reading of air pollution levels taken at 7:00 A.M. one day in June shows the ozone level to be 0.25 parts per million.

A 12-hour forecast of air conditions indicates that t hours later the ozone level will be changing at a rate of $(0.24 - 0.03t)/\sqrt{36 + 16t - t^2}$ parts per million per hour.
(a) Express the ozone level as a function of t.
(b) At what time will the peak ozone level occur? What will the ozone level be at this time?

15. The resale value of a certain industrial machine decreases at a rate that depends on the age of the machine. When the machine is t years old, the rate at which its value is changing is $-960e^{-t/5}$ dollars per year.
    (a) Express the value of the machine in terms of its age and initial value.
    (b) If the machine was originally worth $5,200, how much will it be worth when it is 10 years old?
    (c) Draw a graph showing the relationship between the machine's resale value and its age.

16. A certain oil well that yields 400 barrels of crude oil a month will run dry in 2 years. The price of crude oil is currently $18 per barrel, and is expected to rise at a steady rate of 3 cents per month. What is the total future revenue from the well? (Assume that the oil is sold as soon as it is extracted from the ground.)

17. A taxi fleet contains 28 cabs, each of which is driven approximately 200 miles a day and averages 14 miles per gallon. The price of gasoline is expected to rise sharply during the next few months, and the best estimates indicate that n days from now the price of gas will be $65 + \sqrt{n}$ cents per gallon. How much can the taxi company expect to spend on gasoline over the next 64 days? Discuss the approximations and assumptions that lie behind your solution.

18. The exclusive manufacturer of the latest fad in casual attire, Niccolò Machiavelli sweatshirts, is currently selling his product at a highly inflated price. As imitations appear on the market, the price of the sweatshirts is expected to drop, so that t weeks from now the price will be $3 + 3e^{-t/10}$ dollars per shirt. A 1-year projection of sales suggests that t weeks from now the manufacturer will be selling $2,000 + 20t - t^2$ shirts a week. What is the total revenue he can expect from the sale of the sweatshirts over the next 50 weeks? Discuss the approximations and assumptions implicit in your solution.

*19. A 2,400-cubic-foot room contains an activated charcoal air filter through which air passes at a rate of 400 cubic feet per minute. The ozone in the air is absorbed by the charcoal as the air flows through the filter, and the purified air is recirculated in the room. Assuming that the remaining ozone is evenly distributed throughout the room at all times, write a differential equation describing the rate at which the ozone content of the air is changing. (Do not try to solve this differential equation.)

7. SEPARABLE DIFFERENTIAL EQUATIONS

Many useful differential equations can be formally rewritten so that all the terms containing the independent variable appear on one side of the equation, and all the terms containing the dependent variable appear on the other. Differential equations with this special property are said to be *separable* and can be solved by a simple procedure involving two integrations.

SEPARABLE DIFFERENTIAL EQUATIONS

A differential equation that can be written in the form

$g(y)\, dy = f(x)\, dx$

is said to be *separable*. The general solution is obtained by integrating both sides of this equation. That is,

$\int g(y)\, dy = \int f(x)\, dx$

This procedure is illustrated in the following example.

Example 7.1

Find the general solution of the differential equation $dy/dx = 2x/y^2$.

Solution

This is a separable differential equation. To separate the variables we pretend that the symbol dy/dx is actually a quotient and write

$y^2\, dy = 2x\, dx$

We now integrate both sides of this equation:

$\int y^2\, dy = \int 2x\, dx$

or

$\tfrac{1}{3} y^3 + C_1 = x^2 + C_2$

Solving for y we get

$y = [3x^2 + 3(C_2 - C_1)]^{1/3}$

Finally, we simplify this solution by using a single letter C to denote the constant $3(C_2 - C_1)$. The general solution of the differential equation is then

$y = (3x^2 + C)^{1/3}$ □

In Section 6, we solved differential equations of the form $dy/dx = f(x)$ by direct integration. The general solution was simply $y = \int f(x)\,dx$. Observe that the differential equation $dy/dx = f(x)$ is separable, because it can be rewritten as $1\,dy = f(x)\,dx$. Let us see what happens if we solve this equation using the new method described in this section. Integrating both sides of the equation, we get

$$\int 1\,dy = \int f(x)\,dx$$

or

$$y = \int f(x)\,dx$$

which is exactly the same solution we obtained previously.

In solving separable differential equations we find it convenient to treat the symbol dy/dx as if it were actually a quotient. Does this disturb you? You certainly have every right to be skeptical of a procedure based on this sort of notational trickery. Fortunately, we can demonstrate the validity of our method without resorting to this questionable practice.

According to our procedure for solving separable equations, the solution of the differential equation

$$g(y)\,dy = f(x)\,dx$$

is

$$G(y) = F(x) + C$$

where G is an antiderivative of g, F an antiderivative of f, and C a constant. To verify this, let us restore the differential equation to its original form. Before the variables were separated by means of the notational trick, the differential equation was

$$\frac{dy}{dx} = \frac{f(x)}{g(y)}$$

or, equivalently,

$$g(y)\frac{dy}{dx} - f(x) = 0$$

We notice a relationship between the terms in this equation and the antiderivatives F and G, namely,

$$\frac{d}{dx}[F(x)] = f(x)$$

and, by the chain rule,

$$\frac{d}{dx}[G(y)] = G'(y)\frac{dy}{dx} = g(y)\frac{dy}{dx}$$

We can therefore rewrite the differential equation as

$$\frac{d}{dx}[G(y) - F(x)] = 0$$

Thus the differential equation asserts that the derivative of the function $G(y) - F(x)$ is zero. But constants are the only functions whose derivatives are identically zero, and so $G(y) - F(x)$ must be a constant. That is, $G(y) - F(x) = C$ or $G(y) = F(x) + C$, and our procedure is verified.

A large number of important practical problems lead to separable differential equations, and we conclude this section with a sampling of such problems.

As our first application, we solve a separable differential equation to obtain the compound interest formula that we encountered in Chapter 3.

Example 7.2

Money deposited in a certain bank grows at a rate that is proportional to the existing balance. Show that the bank compounds interest continuously. What is the relationship between the constant of proportionality and the annual interest rate?

Solution

Let $B(t)$ denote the balance in the account after t years and let k be the constant of proportionality. Since the rate of change of B with respect to t is kB, we begin with the differential equation

$$\frac{dB}{dt} = kB$$

which, after separation of variables, becomes

$$\frac{1}{B}dB = k\,dt$$

Then,

$$\int \frac{1}{B}dB = \int k\,dt$$

and

$$\log |B| = kt + C$$

Since B is positive, we may dispense with the absolute value sign and write

$\log B = kt + C$

Solving this equation for B we get

$B(t) = e^{kt+C} = e^C e^{kt}$

We notice that $B(0) = e^C$. That is, the constant e^C is just the initial deposit or principal which, as in Chapter 3, we denote by P. Hence

$B(t) = Pe^{kt}$

But this is precisely the formula for the balance when interest is compounded continuously at a rate of $100k$ percent a year.

A graph showing the balance as a function of time is sketched in Figure 4.

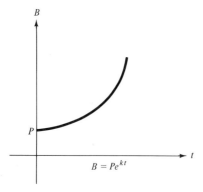

Figure 4

Like the bank balance in the preceding example, many familiar things tend to grow at a rate that is proportional to the amount present at the moment. For example, in the absence of environmental constraints, population increases in this way. This type of growth can be described by the differential equation

$$\frac{dQ}{dt} = kQ$$

where $Q(t)$ denotes the quantity present at time t and k is a positive constant. This is a separable differential equation, and it can be solved in the same way as the one in Example 7.2. The general solution is the exponential function

$Q(t) = Q_0 e^{kt}$

where Q_0 is the amount originally present.

Notice that the preceding calculation shows that if a quantity Q grows at a rate that is proportional to the amount present, then Q grows exponentially. This is the converse of the result we proved at the end of Chapter 3.

Some things *decrease* at a rate that is proportional to the amount present. The corresponding differential equation is $dQ/dt = -kQ$, where k is a positive constant and the minus sign reflects the fact that Q is decreasing. This situation is illustrated in the following example concerning radioactive decay.

Example 7.3

The rate at which a radioactive substance disintegrates is proportional to the amount present. Suppose that one-half of a supply of this substance will decompose in 2,000 years. How much will be left after 1,000 years if 40 grams were originally present?

Solution

Let $Q(t)$ denote the amount of the radioactive substance that remains after t years. Then,

$$\frac{dQ}{dt} = -kQ$$

$$\int \frac{1}{Q} dQ = -\int k \, dt$$

$$\log Q = -kt + C$$

and hence

$$Q(t) = Q_0 e^{-kt}$$

(See Figure 5.)

But $Q_0 = Q(0) = 40$, so we have

$$Q(t) = 40 e^{-kt}$$

We now use the fact that $Q(2,000) = 20$ to find the constant of proportionality k.

$$20 = 40 e^{-2,000k}$$

$$-2,000k = \log \tfrac{1}{2} = -\log 2$$

$$k = \frac{\log 2}{2,000} \approx \frac{0.69315}{2,000} \approx 0.0003465$$

Hence

$$Q(t) \approx 40 e^{-0.0003465 t}$$

Since

$$Q(1,000) \approx 40e^{-0.3465} \approx 28.3$$

we conclude that approximately 28.3 grams of the radioactive substance will be left after 1,000 years.

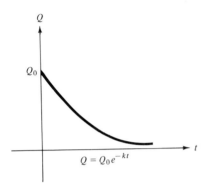

Figure 5

The next example illustrates the use of separable differential equations to solve an important type of *dilution problem*.

Example 7.4

The residents of a certain community have voted to discontinue the fluoridation of their water supply. The local reservoir currently holds 200 million gallons of fluoridated water that contains 1,600 pounds of fluoride. The fluoridated water is flowing out of the reservoir at a rate of 4 million gallons per day and is being replaced at the same rate by unfluoridated water. At all times the remaining fluoride is evenly distributed in the reservoir.

(a) Express the amount of fluoride in the reservoir as a function of time.

(b) When will the fluoride content of the water be reduced to 400 pounds?

Solution

(a) Let $F(t)$ denote the number of pounds of fluoride in the reservoir after t days. Then,

$$\begin{pmatrix} \text{Rate at which} \\ \text{fluoride leaves} \\ \text{the reservoir} \end{pmatrix} = \begin{pmatrix} \text{number of millions} \\ \text{of gallons leaving} \\ \text{the reservoir} \\ \text{each day} \end{pmatrix} \begin{pmatrix} \text{number of pounds} \\ \text{of fluoride in} \\ \text{each million} \\ \text{gallons} \end{pmatrix}$$

The number of pounds of fluoride in a million gallons at time t is $F(t)/200$, the total number of pounds of fluoride in the reservoir at that time divided by the number of million gallons of water in the reservoir. Moreover, there are 4 million gallons leaving the reservoir each day. Hence

$$\frac{dF}{dt} = -4\left(\frac{F}{200}\right) = -\frac{F}{50}$$

where the minus sign shows that the fluoride level is decreasing. We solve this equation by separating the variables:

$$-\int \frac{1}{F}\,dF = \int \frac{1}{50}\,dt$$

$$-\log F = \frac{t}{50} + C$$

or

$$F(t) = F_0 e^{-t/50}$$

Since the initial amount F_0 of fluoride is 1,600 pounds, we can write

$$F(t) = 1{,}600 e^{-t/50}$$

(b) We want to find the value of t for which $F(t) = 400$. That is,

$$400 = 1{,}600 e^{-t/50}$$

or

$$t = 50 \log 4 \approx 69.3$$

Thus, it will take slightly more than 69 days for the fluoride content to drop to 400 pounds. □

In our final example, we analyze a differential equation that describes the spread of an epidemic through a community. A similar equation serves as a model for population growth when environmental factors impose an upper limit on the possible population size.

Example 7.5

Studies have indicated that the rate at which an epidemic spreads through a community is jointly proportional to both the number of residents who have already contracted the disease and the number of residents who are susceptible to the disease but who have not yet caught it. Assuming that this mathematical model for epidemics is correct, derive an expression for the number of residents who will already have caught the disease t months after the outbreak of an epidemic.

Solution

Let S be the total number of residents who are susceptible to the disease, and let $n(t)$ be the number who have caught the disease by time t. Then $S - n(t)$ is the number of susceptible residents who have not yet caught the disease at time t, and we have the separable differential equation

$$\frac{dn}{dt} = k(S-n)n$$

whose solution is

$$\int \frac{1}{(S-n)n} \, dn = \int k \, dt$$

The integral $\int 1/[(S-n)n] \, dn$ is not easy to find. We may either look it up in a table of integrals or use the following algebraic trick. We observe that

$$\frac{1}{(S-n)n} = \frac{1}{S}\left(\frac{1}{n} + \frac{1}{S-n}\right)$$

Hence,

$$\int \frac{1}{(S-n)n} \, dn = \frac{1}{S}\int \frac{1}{n} \, dn + \frac{1}{S}\int \frac{1}{S-n} \, dn$$

$$= \frac{1}{S} \log |n| - \frac{1}{S} \log |S-n| + C_1$$

$$= \frac{1}{S} \log \frac{n}{S-n} + C_1$$

Returning to the equation

$$\int \frac{1}{(S-n)n} \, dn = \int k \, dt$$

we now have

$$\frac{1}{S} \log \frac{n}{S-n} = kt + C$$

Solving this equation for n, we get

$$\log \frac{n}{S-n} = Skt + SC$$

$$\frac{n}{S-n} = e^{SC} e^{Skt}$$

7. SEPARABLE DIFFERENTIAL EQUATIONS

and

$$(1 + e^{SC}e^{Skt})n = Se^{SC}e^{Skt}$$

Hence,

$$n(t) = \frac{Se^{SC}e^{Skt}}{1 + e^{SC}e^{Skt}}$$

We can make this formula more attractive by introducing single letters for the constants Sk and e^{SC}. If we let $A = Sk$ and $B = e^{SC}$, the formula becomes

$$n(t) = \frac{BSe^{At}}{1 + Be^{At}}$$

or, finally,

$$n(t) = \frac{BS}{B + e^{-At}}$$

Observe that this formula contains three unknown constants, A, B, and S. Before $n(t)$ can be computed for a specific epidemic, numerical values must be assigned to these constants on the basis of empirical data.

Even without having specific values for the constants, we can use this formula to obtain some general information about the spread of epidemics. Let us apply some of our curve sketching techniques to the function $n(t)$.

We first observe that

$$\lim_{t \to +\infty} n(t) = S$$

This is not surprising. It says that the number of residents who have had the disease will eventually approach the total number of susceptible residents.

The first derivative of our function is

$$\frac{dn}{dt} = \frac{BSAe^{-At}}{(B + e^{-At})^2}$$

which is always positive (since B, S, and A are positive). This reflects the obvious fact that the number of people who have had the disease increases with time.

After a short calculation we see that the second derivative is

$$\frac{d^2n}{dt^2} = \frac{BSA^2e^{-At}(e^{-At} - B)}{(B + e^{-At})^3}$$

An inflection point occurs when this second derivative is zero, that is, when

$$e^{-At} = B$$

or

$$t = \frac{1}{A} \log \frac{1}{B}$$

If B is greater than 1, this critical value of t will be negative, and so the concavity of the graph will be the same for all positive t. In fact, it is easy to see that the graph will be concave downward in this case. The graph is sketched in Figure 6a.

If B is less than 1, the inflection point occurs for a positive value of t. It is easy to check that d^2n/dt^2 is positive when t is less than $(1/A) \log (1/B)$, and negative when t is greater than this value. Hence the graph is initially concave upward but becomes concave downward for $t > (1/A) \log (1/B)$. This graph is sketched in Figure 6b.

Something interesting emerges if we compute the value of n that corresponds to the inflection point. We get

$$n\left(\frac{1}{A} \log \frac{1}{B}\right) = \frac{BS}{B + e^{-\log(1/B)}} = \frac{BS}{2B} = \frac{S}{2}$$

This means that the concavity of the graph changes when exactly half of the susceptible residents have had the disease. Before this point is reached the disease spreads at an increasing rate; afterward the epidemic begins to taper off. The situation represented in Figure 6a occurs when over half of the susceptible residents contract

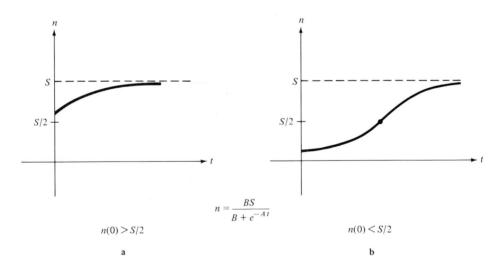

Figure 6

the disease when the epidemic first breaks out. This happens infrequently. The more common situation, represented in Figure 6b, is that fewer than half the susceptible residents initially have the disease. □

Problems

1. Find the general solutions of the following differential equations.

   (a) $\dfrac{dS}{dt} = kS^2$
   (b) $\dfrac{dy}{dx} = e^y$

   (c) $\dfrac{dv}{dt} = e^{v+t}$
   (d) $\dfrac{dy}{dx} = \dfrac{x}{y}$

   (e) $\dfrac{dy}{dx} = \dfrac{y}{x}$
   (f) $\sqrt{u}\,\dfrac{du}{dx} = e^{-x}$

   (g) $\dfrac{dr}{ds} = r^2 s^3$
   (h) $\dfrac{dv}{dt} = \dfrac{t+1}{2vt^3}$

   (i) $\dfrac{dy}{dx} = \dfrac{x\sqrt{x^2+5}}{y^2}$
   (j) $(t-1)\dfrac{dn}{dt} - te^{-n} = 0$

   (k) $2v\dfrac{dv}{dx} - xe^x = 0$
   (l) $3\dfrac{dy}{dx} - \log x = 0$

   (m) $\dfrac{dy}{ds} = \dfrac{(s+1)(y^2+1)}{ys}$

2. Find the particular solutions of the following differential equations that satisfy the given conditions.

   (a) $\dfrac{dy}{dx} = 0.05y;\ y = 500$ when $x = 0$

   (b) $\dfrac{dy}{dx} = 4x^3 y^2;\ y = 2$ when $x = 1$

   (c) $\sqrt{y}\,\dfrac{dy}{dx} = e^{-x};\ y = 4$ when $x = 0$

   (d) $3\dfrac{dy}{dx} + xe^x = 0;\ y = -\tfrac{2}{3}\log 2$ when $x = \log 2$

   (e) $\dfrac{x}{xy+y} = 2\dfrac{dy}{dx};\ y = 4$ when $x = 0$

3. A bank pays interest at an annual rate of 7 percent compounded continuously. If an account is opened for $5,000, what will the balance be at the end of 5 years?

4. A father opens a trust account for his infant son in a bank that pays 6 percent interest compounded continuously. How much should he deposit so that 20 years from now, when his son enters college, the account will contain $30,000?

5. How quickly will money double if it is deposited in a bank paying 7 percent interest compounded continuously?

6. Money deposited in a certain bank doubles every 15 years. The bank compounds interest continuously. What interest rate does it offer?

7. The number of bacteria in a certain culture grows at a rate proportional to the number of bacteria present. If 5,000 bacteria are initially present in the culture and 8,000 are present 10 minutes later, how many bacteria will be present after 1 hour?

8. The gross national product (GNP) of a certain country is increasing at any time at an annual rate of 5 percent of the GNP at that instant. If the GNP was 100 billion dollars in 1960, what will it be in 1980?

9. Radium decays at a rate proportional to the amount present. It takes 1,690 years for 50 percent of a given amount of radium to decay.
   (a) How much of a 50-gram supply of radium will be left after 2,000 years?
   (b) How long will it take for a 50-gram supply of radium to be reduced to 5 grams?

10. A radioactive substance decays at a rate proportional to the amount present. Its *half-life* is defined to be the time it takes for 50 percent of a given amount of the substance to decay. Derive a formula for the amount of the substance remaining in terms of time, the original amount, and the half-life of the substance.

11. The rate at which the temperature of an object changes is proportional to the difference between its own temperature and the temperature of the surrounding medium. A cold drink is removed from a refrigerator on a hot summer day and placed in a room that's temperature is 80 degrees. If the temperature of the drink was 40 degrees when it left the refrigerator and 50 degrees 20 minutes later, how hot will it be after an hour?

12. The rate at which the temperature of an object changes is proportional to the difference between its own temperature and the temperature of the surrounding medium. Let $f(t)$ denote the temperature of the object after it has been in the medium for t minutes.
   (a) Derive a formula for $f(t)$ in terms of the initial temperature of the object and a constant of proportionality if the object is hotter than its surrounding medium. Sketch a graph of $f(t)$ and compute $\lim_{t \to +\infty} f(t)$.
   (b) Repeat part (a) for the case in which the object is colder than its surrounding medium.

13. A tank contains 200 gallons of clear water. Brine containing 2 pounds of salt per gallon flows into the tank at a rate of 5 gallons per minute, and the mixture,

which is stirred so that salt is evenly distributed at all times, runs out of the tank at the same rate.
(a) Express the amount of salt in the tank as a function of time and sketch a graph. What happens to the salt content in the long run?
(b) How much salt is in the tank after 40 minutes?

14. A tank currently holds 200 gallons of brine that contains 3 pounds of salt per gallon. Brine containing 2 pounds of salt per gallon flows into the tank at a rate of 5 gallons per minute, while the mixture, which is kept uniform, runs out of the tank at the same rate.
    (a) Express the amount of salt in the tank as a function of time and sketch a graph. What happens to the salt content in the long run?
    (b) When will the tank contain 500 pounds of salt?

15. A 2,400-cubic-foot room contains an activated charcoal air filter through which air passes at a rate of 400 cubic feet per minute. The ozone in the air is absorbed by the charcoal as the air flows through the filter, and the purified air is recirculated in the room. Assuming that the remaining ozone is evenly distributed throughout the room at all times, determine how long it takes the filter to remove 50 percent of the ozone from the room. Does the answer depend on the amount of ozone initially present in the room?

*16. The number of people implicated in a certain major government scandal increases at a rate jointly proportional to both the number of people already implicated and the number involved who have not yet been implicated. Suppose that 7 people were implicated when a Washington newspaper first made the scandal public, that 9 more were implicated over the next 3 months, and that another 12 were implicated during the following 3 months. Approximately how many people are involved in the scandal?

17. By solving an appropriate differential equation, prove that if a differentiable function is equal to its own derivative then the function must be of the form $y = Ce^x$.

8. THE DEFINITE INTEGRAL

In solving many practical problems, we first find an antiderivative of a certain function, then evaluate this antiderivative for two different choices of the independent variable, and finally subtract one of these values of the antiderivative from the other. The numerical result of such a computation is said to be a *definite integral* of the original function. Here is an elementary problem whose solution is a definite integral.

Example 8.1

If a manufacturer's marginal cost is $6q + 1$ dollars per unit when q units are produced, what is the cost of producing the 10th unit?

Solution

Let $C(q)$ be the total cost of producing q units. Then

$$\frac{dC}{dq} = 6q + 1$$

and $C(q)$ is the antiderivative

$$C(q) = \int (6q + 1)\, dq = 3q^2 + q + C_0$$

where the constant C_0 is the manufacturer's overhead (the cost of producing no units). To find the cost of producing the 10th unit, we take the difference between the cost of producing 10 units and the cost of producing 9 units. That is, the 10th unit costs

$$C(10) - C(9) = (300 + 10 + C_0) - (243 + 9 + C_0) = \$58$$

to produce.

Notice that the answer is independent of the overhead C_0 which appeared in the expressions for both $C(10)$ and $C(9)$ and canceled out. □

Because computations involving definite integrals arise frequently, the following convenient notation is widely used.

THE DEFINITE INTEGRAL

$$\int_a^b f(x)\, dx = F(b) - F(a)$$

where F is an antiderivative of f.

Using this notation, we can express the cost of producing the 10th unit in Example 8.1 as the definite integral, $\int_9^{10} (6q + 1)\, dq$.

The symbol $\int_a^b f(x)\, dx$ is read "the (definite) integral of $f(x)$ from a to b." The numbers a and b are called the *limits of integration*. In computations involving definite integrals, it is often convenient to use the symbol $F(x)|_a^b$ to stand for $F(b) - F(a)$.

Let us consider another elementary practical problem whose solution is a definite integral.

Example 8.2

A study indicates that x months from now the population of a certain town will be increasing at the rate of $2 + 6\sqrt{x}$ people per month. By how much will the population of the town increase during the next 4 months?

Solution

If $P(x)$ denotes the population of the town x months from now, the amount by which the population will increase during the next 4 months is

$$P(4) - P(0) = \int_0^4 (2 + 6\sqrt{x})\, dx$$

$$= (2x + 4x^{3/2} + C)\Big|_0^4 = (40 + C) - (0 + C) = 40 \qquad \square$$

Notice what happens to the constant C in the evaluation of a definite integral. It appears in the expressions for both $F(b)$ and $F(a)$ and is eventually eliminated by the subtraction. We may, therefore, omit the constant C altogether when evaluating definite integrals.

No new techniques are needed for computing definite integrals. We already have techniques for antidifferentiation, and once the antiderivative has been found, the rest is just arithmetic. In the next example, integration by parts is used in the evaluation of a definite integral.

Example 8.3

Evaluate $\int_0^{\log 2} xe^x\, dx$.

Solution

We first use integration by parts to find the indefinite integral.

$$\int xe^x\, dx = xe^x - \int e^x\, dx = xe^x - e^x$$

Then the definite integral is

$$\int_0^{\log 2} xe^x\, dx = (xe^x - e^x)\Big|_0^{\log 2}$$

$$= (2 \log 2 - 2) - (0 - 1)$$

$$= 2 \log 2 - 1$$

Here is a slightly more compact way of writing this solution.

$$\int_0^{\log 2} xe^x \, dx = xe^x \Big|_0^{\log 2} - \int_0^{\log 2} e^x \, dx$$

$$= (xe^x - e^x) \Big|_0^{\log 2}$$

$$= (2 \log 2 - 2) - (0 - 1) = 2 \log 2 - 1 \qquad \square$$

We may summarize the method of integration by parts for definite integrals as follows.

INTEGRATION BY PARTS

$$\int_a^b f(x)g(x) \, dx = f(x)G(x) \Big|_a^b - \int_a^b f'(x)G(x) \, dx$$

where G is an antiderivative of g.

In the next example we use the method of substitution to evaluate a definite integral.

Example 8.4

Evaluate $\int_0^1 x^3(x^4 + 2)^4 \, dx$.

Solution

We let $u = x^4 + 2$. Then $du = 4x^3 \, dx$, and so

$$\int x^3(x^4 + 2)^4 \, dx = \tfrac{1}{4} \int u^4 \, du = \tfrac{1}{20} u^5$$

The limits of integration, 0 and 1, refer to the variable x, not to u. We may therefore proceed in one of two ways. Either we can rewrite the antiderivative in terms of x, or we can find the values of u that correspond to $x = 0$ and $x = 1$.

If we choose the first alternative, we find that

$$\int x^3(x^4 + 2)^4 \, dx = \tfrac{1}{20}(x^4 + 2)^5$$

and so

$$\int_0^1 x^3(x^4 + 2)^4 \, dx = \tfrac{1}{20}(x^4 + 2)^5 \Big|_0^1 = \tfrac{243}{20} - \tfrac{32}{20} = \tfrac{211}{20}$$

If we adopt the second alternative, we use the fact that $u = x^4 + 2$ to conclude that $u = 2$ when $x = 0$, and that $u = 3$ when $x = 1$. Hence,

$$\int_0^1 x^3(x^4 + 2)^4 \, dx = \tfrac{1}{20} u^5 \Big|_2^3 = \tfrac{243}{20} - \tfrac{32}{20} = \tfrac{211}{20}$$

Probably the most efficient approach is to adopt this second alternative and to write the solution compactly as follows:

$$\int_0^1 x^3(x^4 + 2)^4 \, dx = \tfrac{1}{4} \int_2^3 u^4 \, du = \tfrac{1}{20} u^5 \Big|_2^3 = \tfrac{211}{20} \qquad \square$$

We may summarize the method of substitution for definite integrals as follows.

INTEGRATION BY SUBSTITUTION

$$\int_a^b g(u(x))u'(x) \, dx = \int_{u(a)}^{u(b)} g(u) \, du$$

For practice, let us try one more example involving substitution.

Example 8.5

Evaluate $\int_1^e (\log x)/x \, dx$.

Solution

If $u = \log x$, then $du = (1/x) \, dx$, $u(1) = 0$, and $u(e) = 1$. Hence,

$$\int_1^e \frac{\log x}{x} \, dx = \int_0^1 u \, du = \tfrac{1}{2} u^2 \Big|_0^1 = \tfrac{1}{2} \qquad \square$$

Many texts give a different definition of the definite integral. They define it to be a limit of a certain sum and then *prove* that the computation of definite integrals is related to antidifferentiation by the formula $\int_a^b f(x) \, dx = F(b) - F(a)$. This relationship between antiderivatives and limits of sums is often called the *fundamental theorem of calculus*. We, on the other hand, *define* the definite integral by the formula $\int_a^b f(x) \, dx = F(b) - F(a)$. In Section 11 we shall use this definition to obtain the characterization of definite integrals as limits of sums. This characterization is extremely important because it extends significantly the class of practical problems that can be solved by integration. Section 12 will be devoted to some of these new applications. In Sections 9 and 10, as a prelude to this work, we examine a startling connection between definite integrals and the geometric concept of area.

Problems

1. Evaluate the following definite integrals.

   (a) $\int_{-1}^{0} (3x^5 - 3x^2 + 2x - 1)\, dx$

   (b) $\int_{2}^{5} (2 + 2u + 3u^2)\, du$

   (c) $\int_{1}^{9} (\sqrt{t} - 1/\sqrt{t})\, dt$

   (d) $\int_{3}^{3} (1 + 1/x + 1/x^2)\, dx$

   (e) $\int_{\log 1/2}^{\log 2} (e^u - e^{-u})\, du$

   (f) $\int_{-3}^{-1} (t + 1)/t^3\, dt$

   (g) $\int_{-3}^{0} (2x + 6)^4\, dx$

   (h) $\int_{0}^{4} 1/\sqrt{6u + 1}\, du$

   (i) $\int_{-1}^{1} 2te^{t^2 - 1}\, dt$

   (j) $\int_{1}^{2} x^2/(x^3 + 5)^2\, dx$

   (k) $\int_{-1}^{1} (u^3 + u)\sqrt{u^4 + 2u^2 + 6}\, du$

   (l) $\int_{e}^{e^2} 1/(t \log t)\, dt$

   (m) $\int_{2}^{e+1} x/(x - 1)\, dx$

   (n) $\int_{1}^{2} (u + 1)(u - 2)^9\, du$

   (o) $\int_{1}^{e^2} \log t\, dt$

   (p) $\int_{-2}^{2} xe^{-x}\, dx$

   (q) $\int_{0}^{\log 2} u^2 e^{2u}\, du$

2. A study indicates that x months from now the population of a certain town will be increasing at a rate of $5 + 3x^{2/3}$ people per month. By how much will the population of the town increase over the next 8 months?

3. An object is moving so that its speed after t minutes is $5 + 2t + 3t^2$ feet per minute. How far does the object travel during the second minute?

4. The resale value of a certain industrial machine decreases over a 10-year period at a rate that changes with time. When the machine is x years old, the rate at which its value is changing is $220(x - 10)$ dollars per year. By how much does the machine depreciate during the second year?

5. A certain oil well that yields 400 barrels of crude oil a month will run dry in 2 years. The price of crude oil is currently $18 per barrel and is expected to rise at a steady rate of 3 cents per month. What is the total revenue the owner can expect from the well during the 2-year period? (Assume that the oil is sold as soon as it is extracted from the ground.)

6. The promoters of a county fair estimate that t hours after the gates open at 9:00 A.M. visitors will be entering the fair at a rate of $-4(t + 2)^3 + 54(t + 2)^2$ people per hour. How many people will enter the fair between 10:00 A.M. and noon?

7. (a) Prove that $\int_a^b f(x)\,dx + \int_b^c f(x)\,dx = \int_a^c f(x)\,dx$.
   (b) Use the formula in part (a) to evaluate the definite integral $\int_{-1}^1 |x|\,dx$. (*Hint*: Evaluate the two integrals $\int_{-1}^0 |x|\,dx$ and $\int_0^1 |x|\,dx$.)
   (c) Evaluate $\int_0^4 (1 + |x - 3|)^2\,dx$.

*8. A function f is said to be *even* if $f(x) = f(-x)$. For example, $f(x) = x^2$ is an even function.
   (a) Prove that if f is even, then $\int_{-a}^a f(x)\,dx = 2\int_0^a f(x)\,dx$.
   (b) Use part (a) to compute $\int_{-1}^1 |x|\,dx$ and $\int_{-2}^2 x^2\,dx$.

*9. A function f is said to be *odd* if $f(x) = -f(-x)$.
   (a) Give an example of an odd function.
   (b) Complete and prove the following statement. If f is odd, then $\int_{-a}^a f(x)\,dx = \ldots$.
   (c) Compute $\int_{-12}^{12} x^3\,dx$.

9. AREA AND INTEGRATION

Consider the following geometric problem. Suppose $f(x)$ is continuous and non-negative on an interval $a \leq x \leq b$, and let R be the region bounded by the graph of f, the vertical lines $x = a$ and $x = b$, and the x axis (Figure 7). What is the area of the region R?

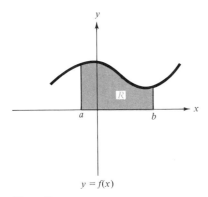

Figure 7

When f is a linear function, the region R is a rectangle, a triangle, or a trapezoid (Figure 8), and you can easily compute the area using a formula from elementary geometry.

But what if f is not linear? In this case you probably have no idea how to begin to find the area and would be astonished to learn that you merely need to compute a definite integral. In fact, the area of R is simply $\int_a^b f(x)\,dx$.

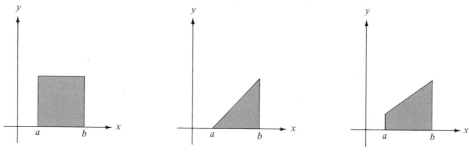

Figure 8

THE AREA UNDER A CURVE

If $f(x)$ is continuous and non-negative on the interval $a \leq x \leq b$, and R is the region bounded by the graph of f, the vertical lines $x = a$ and $x = b$, and the x axis, then

$$\text{Area of } R = \int_a^b f(x)\, dx$$

For convenience, this area is sometimes called the "area under the curve $y = f(x)$ from $x = a$ to $x = b$." Before we verify our integral formula for the area under a curve, it might be interesting to apply the formula to a region whose area we already know.

Example 9.1

Find the area of the region bounded by the lines $y = 2x$, $x = 2$, and the x axis.

Solution

The region in question is the triangle in Figure 9, and its area is clearly 4.

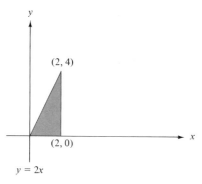

Figure 9

Let us see what happens if we use integration. We apply the integral formula with $f(x) = 2x$. Since the region is bounded on the right by the line $x = 2$, we take $b = 2$. On the left, the boundary consists of the single point (0, 0) which is part of the vertical line $x = 0$. Hence, we take $a = 0$ and get

$$\text{Area} = \int_0^2 2x\, dx = x^2 \Big|_0^2 = 4$$

which is the right answer. □

Let us now verify our integral formula for the area under a curve. Strictly speaking, our argument cannot be called a mathematical *proof*, because we have not actually *defined* the concept of area. Nevertheless, using only a few basic properties of area that should be intuitively clear, we can demonstrate why areas can be computed by finding antiderivatives.

Suppose $f(x)$ is continuous and non-negative on an interval $a \leq x \leq b$. For any value of x in this interval, let $A(x)$ denote the area of the region under the graph of f between a and x (Figure 10).

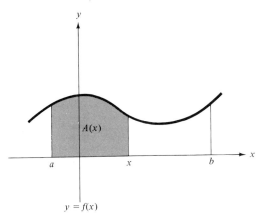

Figure 10

Note that $A(x)$ is a function of the variable x and that $A(b)$, the value of this function when $x = b$, is precisely the area we are attempting to compute. Our goal, then, is to show that $A(b) = \int_a^b f(x)\, dx$. The key step will be to establish that the derivative of the area function $A(x)$ is equal to $f(x)$. To do this, we consider the quotient

$$\frac{A(x + \Delta x) - A(x)}{\Delta x}$$

Since $A(x + \Delta x)$ represents the area under the curve from a to $x + \Delta x$, and $A(x)$ is the area from a to x, the difference $A(x + \Delta x) - A(x)$ is just the area under the

288 INTEGRATION

curve between x and $x + \Delta x$ (Figure 11). If Δx is small, this area is approximately the same as the area of the rectangle in Figure 12, whose height is $f(x)$ and width is Δx. That is,

$$A(x + \Delta x) - A(x) \approx f(x)\,\Delta x$$

or

$$\frac{A(x + \Delta x) - A(x)}{\Delta x} \approx f(x)$$

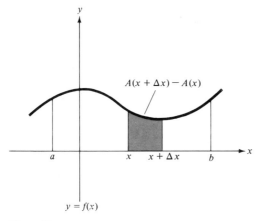

Figure 11

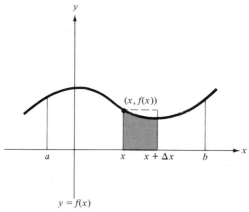

Figure 12

We can make the difference between $f(x)$ and the quotient $[A(x + \Delta x) - A(x)]/\Delta x$ as small as we like by taking Δx sufficiently small. This means that

$$\lim_{\Delta x \to 0} \frac{A(x + \Delta x) - A(x)}{\Delta x} = f(x)$$

But this limit is precisely the derivative of the area function A, and so we conclude that

$$\frac{dA}{dx} = f(x)$$

This implies that A is an antiderivative of f, and hence

$$\int_a^b f(x)\,dx = A(b) - A(a)$$

But $A(a)$ is the area under the curve between $x = a$ and $x = a$, which is clearly zero. Hence,

$$\int_a^b f(x)\,dx = A(b)$$

and the formula is verified.

We conclude this section with two more examples illustrating the use of this procedure for computing areas.

Example 9.2

Find the area of the region bounded by the curve $y = x^2$, the x axis, and the vertical line $x = 1$.

Solution

From the graph of the function $y = x^2$ in Figure 13 we see that the region in question is "bounded" on the left by the line $x = 0$. Hence,

$$\text{Area} = \int_0^1 x^2\,dx = \tfrac{1}{3}x^3 \Big|_0^1 = \tfrac{1}{3}$$

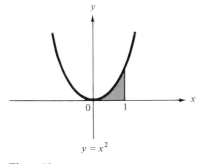

Figure 13

Example 9.3

Find the area of the region bounded by the curve $y = -x^2 + 4x - 3$ and the x axis.

Solution

From the graph of the function

$$y = -x^2 + 4x - 3 = -(x - 3)(x - 1)$$

in Figure 14 we see that the region in question is "bounded" on the left by $x = 1$ and on the right by $x = 3$. Hence,

$$\text{Area} = \int_1^3 (-x^2 + 4x - 3)\, dx$$

$$= \left(-\tfrac{1}{3}x^3 + 2x^2 - 3x \right) \Big|_1^3$$

$$= 0 - (-\tfrac{4}{3}) = \tfrac{4}{3}$$

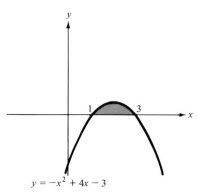

$y = -x^2 + 4x - 3$

Figure 14

In the next section we shall look at some refinements of the area problem. We shall consider, for example, what happens when the function $f(x)$ is negative over part of the interval. Since this material has relatively few practical applications in the social sciences, you may want to skip Section 10 and turn directly to Section 11. There, the basic relationship between area and integration will be used to give a geometric explanation of why definite integrals can be viewed as limits of sums.

Problems

1. Use calculus to find the area of the triangle bounded by the line $y = 4 - 3x$ and the coordinate axes.

2. Use calculus to find the area of the triangle with vertices $(-4, 0)$, $(2, 0)$, and $(2, 6)$.

3. Use calculus to find the area of the rectangle with vertices $(1, 0)$, $(-2, 0)$, $(-2, 5)$, and $(1, 5)$.

4. Use calculus to find the area of the trapezoid bounded by the lines $y = x + 6$ and $x = 2$ and the coordinate axes.

5. Find the area of the region bounded by the curve $y = \sqrt{x}$, the lines $x = 4$ and $x = 9$, and the x axis.

6. Find the area of the region bounded by the curve $y = 4x^3$, the x axis, and the line $x = 2$.

7. Find the area of the region bounded by the curve $y = 1 - x^2$ and the x axis.

8. Find the area of the region bounded by the curve $y = -x^2 - 6x - 5$ and the x axis.

9. Find the area of the region bounded by the curve $y = e^x$, the lines $x = 0$ and $x = \log \frac{1}{2}$, and the x axis.

10. Find the area of the region bounded by the curve $y = \log x$, the x axis, and the line $x = e$.

*11. Find the area of the region bounded by the curve $y = x^2 - 2x$ and the x axis. (*Hint*: Reflect the region across the x axis and integrate the corresponding function.)

*12. Find the area of the region bounded by the curves $y = x^3$ and $y = x^2$.

*13. Suppose $f(x)$ is non-negative and continuous for $x \geq a$. Then $\int_a^\infty f(x)\,dx$ is called an *improper integral* of f and is defined by

$$\int_a^\infty f(x)\,dx = \lim_{N \to +\infty} \int_a^N f(x)\,dx$$

(a) Give a geometric interpretation of this improper integral.
(b) Evaluate $\int_1^\infty 1/x\,dx$.
(c) Evaluate $\int_1^\infty 1/x^2\,dx$.
(d) Find the area of the region to the right of the y axis that lies between the curve $y = e^{-x}$ and the x axis.

10. COMPUTATION OF AREAS

In this section, we consider some variations on the basic area problem discussed in Section 9. We begin with the case in which $f(x)$ is continuous and *negative* on an interval $a \leq x \leq b$, and R is the region bounded by the graph of f, the vertical lines

292 INTEGRATION

$x = a$ and $x = b$, and the x axis (Figure 15). The area of R is the same as the area of the region R_0 obtained by reflecting R across the x axis (Figure 16). But R_0 is just the region under the graph of the non-negative function $y = -f(x)$ between $x = a$ and $x = b$, and so we can compute its area using the method of Section 9. Hence,

$$\text{Area of } R = \text{area of } R_0 = \int_a^b -f(x)\, dx = -\int_a^b f(x)\, dx$$

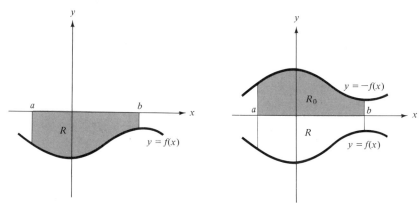

Figure 15   **Figure 16**

Thus, if the region between the graph of a function and the x axis lies below the axis, its area is the *negative* of the definite integral of the function over the appropriate interval.

THE AREA ABOVE A NEGATIVE CURVE

If $f(x)$ is continuous and negative on the interval $a \leq x \leq b$, and R is the region bounded by the graph of f, the vertical lines $x = a$ and $x = b$, and the x axis, then

$$\text{Area of } R = -\int_a^b f(x)\, dx$$

Consider the following example.

Example 10.1

Find the area of the region bounded by the curve $y = x^2 - 4x + 3$ and the x axis.

Solution

From the graph of the function

$$y = x^2 - 4x + 3 = (x - 3)(x - 1)$$

in Figure 17, we see that the region in question lies below the x axis and is bounded on the left by $x = 1$ and on the right by $x = 3$. Hence,

$$\text{Area} = -\int_1^3 (x^2 - 4x + 3)\, dx = \tfrac{4}{3}$$

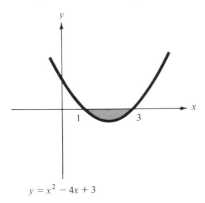

$y = x^2 - 4x + 3$

Figure 17

Compare this answer with the related area computed in Example 9.3. □

If the sign of $f(x)$ changes on the interval $a \le x \le b$, we first divide this interval into subintervals on which $f(x)$ does not change sign, then compute the areas of the corresponding regions, and finally add these areas together to obtain the total area between the x axis and the graph of f from $x = a$ to $x = b$. For example, if R is the shaded region in Figure 18 between the x axis and the graph of a function $f(x)$, then

Area of R = area of R_1 + area of R_2 + area of R_3

$$= \int_a^c f(x)\, dx - \int_c^d f(x)\, dx + \int_d^b f(x)\, dx$$

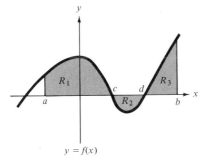

$y = f(x)$

Figure 18

Example 10.2

Find the area of the region bounded by the curve $y = x^2 - x$, the lines $x = -1$ and $x = 2$, and the x axis.

Solution

We graph the function $y = x^2 - x = x(x - 1)$ in Figure 19 and observe that the region in question can be divided into three pieces, R_1, R_2, and R_3, each lying entirely on one side of the x axis. Hence,

Total area = area of R_1 + area of R_2 + area of R_3

$$= \int_{-1}^{0} (x^2 - x) \, dx - \int_{0}^{1} (x^2 - x) \, dx + \int_{1}^{2} (x^2 - x) \, dx$$

$$= \tfrac{5}{6} + \tfrac{1}{6} + \tfrac{5}{6} = \tfrac{11}{6}$$

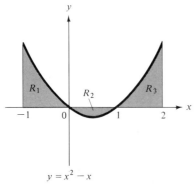

Figure 19

We conclude this section with the following variation on the area problem. Suppose f and g are continuous functions with $f(x) \geq g(x)$ on the interval $a \leq x \leq b$. Find the area of the region R bounded by the graphs of f and g and the vertical lines $x = a$ and $x = b$ (Figure 20). That is, find the area of the region *between* the two curves and the vertical lines $x = a$ and $x = b$.

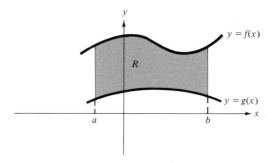

Figure 20

If, as in Figure 20, both $f(x)$ and $g(x)$ are non-negative on the interval $a \le x \le b$, we simply subtract the area under the lower curve $y = g(x)$ from the area under the upper curve $y = f(x)$ to get the area between the curves. Thus,

$$\text{Area of } R = \int_a^b f(x)\,dx - \int_a^b g(x)\,dx = \int_a^b [f(x) - g(x)]\,dx$$

This formula remains valid even when f or g is negative over part of the interval, as in Figure 21a. Here is the reason why. By translating this region vertically by some amount K, we can obtain a region R_0 having the same area as R but lying entirely above the x axis (Figure 21b). Since R_0 lies between the two *non-negative* curves $y = f(x) + K$ and $y = g(x) + K$, we can use the integral formula to compute the area of R_0. Hence,

$$\text{Area of } R = \text{area of } R_0$$
$$= \int_a^b \{[f(x) + K] - [g(x) + K]\}\,dx$$
$$= \int_a^b [f(x) - g(x)]\,dx$$

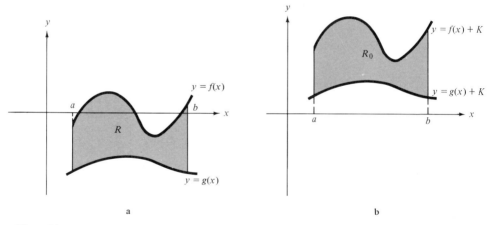

Figure 21

THE AREA BETWEEN TWO CURVES

Suppose $f(x)$ and $g(x)$ are continuous with $f(x) \ge g(x)$ on the interval $a \le x \le b$, and R is the region bounded by the graphs of f and g and the vertical lines $x = a$ and $x = b$. Then

$$\text{Area of } R = \int_a^b [f(x) - g(x)]\,dx$$

Example 10.3

Find the area of the region bounded by the curves $y = x^3$ and $y = x^2$.

Solution

We graph the curves (Figure 22) and solve their equations simultaneously to find the points of intersection, (0, 0) and (1, 1). We see that the region in question is bounded above by the curve $y = x^2$ and below by the curve $y = x^3$ and extends from $x = 0$ to $x = 1$. Hence,

$$\text{Area} = \int_0^1 (x^2 - x^3)\, dx = (\tfrac{1}{3}x^3 - \tfrac{1}{4}x^4)\Big|_0^1 = \tfrac{1}{12}$$

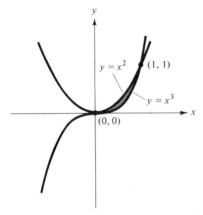

Figure 22 ☐

Problems

1. Find the area of the region bounded by the curves $y = x^2 + 5$ and $y = -x^2$, the line $x = 3$, and the y axis.

2. Find the area of the region bounded by the curve $y = e^x$ and the lines $y = 1$ and $x = 1$.

3. Find the area of the region bounded by the curve $y = x^2$ and the line $y = x$.

4. Find the area of the region bounded by the curve $y = x^2$ and the line $y = 4$.

5. Find the area of the region bounded by the curves $y = \sqrt{x}$ and $y = x^2$.

6. (a) Find the area of the region to the right of the y axis bounded by the curves $y = x$ and $y = x^3$.
   (b) Find the total area of the region bounded by the curves $y = x$ and $y = x^3$.

7. (a) Find the area of the region to the right of the y axis that is bounded above by the curve $y = 4 - x^2$ and below by the line $y = 3$.

*(b) Find the area of the region to the right of the y axis that lies below the line $y = 3$ and is bounded by the curve $y = 4 - x^2$, the line $y = 3$, and the coordinate axes.

8. Find the area of the region bounded by the curves $y = x^3 - 6x^2$ and $y = -x^2$.

*9. Find the area of the region that lies below the curve $y = x^2 + 4$ and is bounded by this curve, the line $y = -x + 10$, and the coordinate axes.

*10. Find the area of the region that lies below the line $y = x$ and is bounded by this line, the x axis, and the curve $y = 12 - x^2$.

11. THE DEFINITE INTEGRAL AS A LIMIT OF A SUM

In Example 6.6 we solved a practical problem by using an integral to approximate a special type of sum. We saw that, in a certain sense, the integral was a limit of sums of this type. Using the geometric interpretation of definite integrals as area, we can now make the connection between integrals and sums more precise.

Suppose that $f(x)$ is non-negative and continuous on the interval $a \leq x \leq b$. Then, as we have seen, the area under the graph of f between $x = a$ and $x = b$ is given by the definite integral $\int_a^b f(x)\,dx$. We now approximate this area using rectangles as follows. First, we divide the interval $a \leq x \leq b$ into n equal subintervals of width $\Delta x = (b - a)/n$ and label the endpoints of these intervals consecutively as x_1, $x_2, \ldots, x_{n+1}$ (with $x_1 = a$ and $x_{n+1} = b$). Next we draw n rectangles (Figure 23) such that for each $j = 1, 2, \ldots, n$ the base of the jth rectangle is the jth subinterval and the height of the jth rectangle is $f(x_j)$. The area of this jth rectangle is $f(x_j)\,\Delta x$ and is an approximation to the area under the curve from $x = x_j$ to $x = x_{j+1}$. The sum of the areas of all n rectangles is

$$f(x_1)\,\Delta x + f(x_2)\,\Delta x + \cdots + f(x_n)\,\Delta x$$

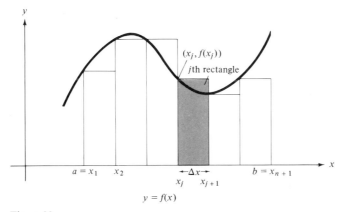

Figure 23

INTEGRATION

This is an approximation to the total area under the curve from $x = a$ to $x = b$ and hence an approximation to the corresponding definite integral. That is,

$$f(x_1)\,\Delta x + f(x_2)\,\Delta x + \cdots + f(x_n)\,\Delta x \approx \int_a^b f(x)\,dx$$

We can obtain a better approximation of the area under the curve by using more subintervals with smaller widths in our partition of the interval $a \le x \le b$. In fact, as Figure 24 suggests, we can approximate the area under the curve as closely as we like by taking n sufficiently large. This says that the area under the curve is the limit of the sum of the areas of the approximating rectangles as the number, n, of rectangles approaches $+\infty$. That is,

$$\lim_{n \to +\infty} [f(x_1)\,\Delta x + f(x_2)\,\Delta x + \cdots + f(x_n)\,\Delta x] = \int_a^b f(x)\,dx$$

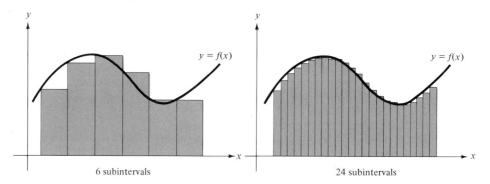

Figure 24

where $x_1 = a$, $x_{n+1} = b$, and $x_1, x_2, \ldots, x_{n+1}$ divide the interval $a \le x \le b$ into n equal subintervals.

This is the crucial relationship between sums and integrals that we were seeking. Although we have established it only for non-negative functions, it actually holds for any function that is continuous on the interval $a \le x \le b$. For example, if $f(x)$ is negative for $a \le x \le b$, we can apply the result for non-negative functions to $-f(x)$ which is positive on the interval. This gives the equation

$$\lim_{n \to +\infty} [-f(x_1)\,\Delta x - f(x_2)\,\Delta x - \cdots - f(x_n)\,\Delta x] = \int_a^b -f(x)\,dx$$

from which we can get the desired equality by multiplying both sides by -1. If $f(x)$ changes sign on the interval $a \le x \le b$, we first break this interval into subintervals on which $f(x)$ does not change sign, then write the equations relating the limit of sums and the definite integral for each of these subintervals separately, and, finally, add these equations together to obtain the equality that holds for the entire interval $a \leq x \leq b$.

The characterization of definite integrals as limits of sums can be used to solve many practical problems. Before considering some of these applications, however, we pause to introduce a convenient, compact way of writing sums.

To describe the sum

$$f(x_1) \Delta x + f(x_2) \Delta x + \cdots + f(x_n) \Delta x$$

it suffices to specify the general term $f(x_j) \Delta x$ and to indicate that n terms of this form are to be added together, starting with the term in which $j = 1$ and ending with the term in which $j = n$. It is customary to use the symbol

$$\sum_{j=1}^{n} f(x_j) \Delta x$$

to convey this information. Thus,

$$\sum_{j=1}^{n} f(x_j) \Delta x = f(x_1) \Delta x + f(x_2) \Delta x + \cdots + f(x_n) \Delta x$$

Similar *summation notation* can be used to represent any sum whose general term is known.

SUMMATION NOTATION

$$\sum_{j=1}^{n} a_j = a_1 + a_2 + \cdots + a_n$$

Let us consider some simple examples illustrating the use of this notation.

Example 11.1

Use summation notation to express the following sums.

(a) $1 + 4 + 9 + 16 + 25 + 36 + 49 + 64$

(b) $g(1) + g(2) + g(3) + \cdots + g(25)$

(c) $(1 - x_1)^2 \Delta x + (1 - x_2)^2 \Delta x + \cdots + (1 - x_{15})^2 \Delta x$

Solution

(a) This is a sum of eight terms of the form j^2, starting with $j = 1$ and ending with $j = 8$. Hence,

$$1 + 4 + 9 + 16 + 25 + 36 + 49 + 64 = \sum_{j=1}^{8} j^2$$

(b) The jth term in this sum is $g(j)$. Hence
$$g(1) + g(2) + g(3) + \cdots + g(25) = \sum_{j=1}^{25} g(j)$$

(c) This time, the jth term is $(1 - x_j)^2 \Delta x$. Hence,
$$(1 - x_1)^2 \Delta x + (1 - x_2)^2 \Delta x + \cdots + (1 - x_{15})^2 \Delta x = \sum_{j=1}^{15} (1 - x_j)^2 \Delta x \qquad \square$$

We can use summation notation to express the characterization of the definite integral as a limit of a sum as follows.

THE DEFINITE INTEGRAL AS A LIMIT OF A SUM

$$\int_a^b f(x)\, dx = \lim_{n \to +\infty} \sum_{j=1}^n f(x_j)\, \Delta x$$

where $x_1 = a$, $x_{n+1} = b$, and $x_1, x_2, \ldots, x_{n+1}$ divide the interval $a \le x \le b$ into n equal subintervals.

Actually, this is a somewhat restricted version of a more general characterization of definite integrals. The equality between the definite integral and the limit of the sum would still hold if $f(x_j)$ in the jth term of the sum were replaced by $f(x_j')$, where x_j' is any point whatsoever in the jth subinterval. Moreover, the n subintervals need not have equal width, as long as the width of the largest eventually approaches zero as n increases. For most applications, however, our restricted characterization is quite sufficient, and we will have no use for the more general result.

The characterization of the definite integral as a limit of a sum is useful because many practical problems lead naturally to expressions that can be put in the form $\lim_{n \to +\infty} \sum_{j=1}^n f(x_j)\, \Delta x$. Whenever such an expression arises, it can be replaced by the corresponding definite integral, and calculus can then be used to solve the problem. As our first application, we consider a simple problem that can be solved quite easily without the use of our new characterization of integrals.

Example 11.2

A car enters the Pennsylvania Turnpike at noon and is driven so that x hours later its speed is $f(x)$ miles per hour. How far does the car go between 1:00 P.M. and 3:00 P.M.?

Solution

Using techniques we have already encountered, we see immediately that the desired distance is given by the definite integral $\int_1^3 f(x)\, dx$.

Let us see what happens if we apply our new characterization of integrals to this problem. We begin by dividing the time interval from $x = 1$ to $x = 3$ into n equal subintervals (Figure 25) whose endpoints are $x_1, x_2, \ldots, x_{n+1}$, where $x_1 = 1$ and $x_{n+1} = 3$. We let Δx denote the length of each of these intervals. During the jth interval (from $x = x_j$ to $x = x_{j+1}$), the speed of the car may vary. However, we may approximate the actual speed by assuming that throughout the entire interval the car maintains the speed $f(x_j)$ at which it was moving at the start of the interval. If Δx is small, this assumption is almost correct. The distance covered by a car going $f(x_j)$ miles per hour for Δx hours is $f(x_j) \Delta x$ miles. Hence we can use this expression to approximate the distance actually covered by our car during the jth time interval. The sum

$$f(x_1) \Delta x + f(x_2) \Delta x + \cdots + f(x_n) \Delta x = \sum_{j=1}^{n} f(x_j) \Delta x$$

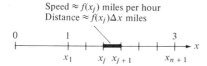

Figure 25

is therefore an approximation to the total distance the car travels between $x = 1$ and $x = 3$. That is,

$$\text{Total distance} \approx \sum_{j=1}^{n} f(x_j) \Delta x$$

This approximation will improve if the number, n, of subintervals is increased, since this will make Δx smaller and hence reduce the error introduced when we assume that the car maintains a constant speed throughout the jth interval. In fact, we can make the approximate distance as close as we like to the actual distance by taking n sufficiently large. Thus,

$$\text{Total distance} = \lim_{n \to +\infty} \sum_{j=1}^{n} f(x_j) \Delta x$$

According to our characterization of definite integrals, this limit is precisely the definite integral of $f(x)$ from $x = x_1$ to $x = x_{n+1}$. Since $x_1 = 1$ and $x_{n+1} = 3$, we conclude that

$$\text{Total distance} = \int_{1}^{3} f(x) \, dx$$

which is the same expression we obtained previously using other methods. □

If you are thinking that our new method resulted in a relatively inefficient solution of the preceding problem, you are, of course, quite right. Many problems that can be solved by the substitution of a definite integral for a limit of a sum can be handled more efficiently if viewed in terms of differential equations. Nevertheless, there are many problems that we could not easily solve without using the characterization of definite integrals as limits of sums, and we will examine some of these in Section 12.

It might be interesting at this point to return to the oil well problem (Example 6.6) in which we first encountered the connection between integrals and sums. This time, let us solve the problem using our new technique.

Example 11.3

A certain oil well that yields 300 barrels of crude oil a month will run dry in 3 years. It is estimated that in t months the price of crude oil will be $P(t) = 18 + 0.3\sqrt{t}$ dollars per barrel. If the oil is sold as soon as it is extracted from the ground, what is the total future revenue from the well?

Solution

To approximate the revenue, we divide the time interval from $t = 0$ to $t = 36$ (months) into n equal subintervals whose endpoints are $t_1, t_2, \ldots, t_{n+1}$, where $t_1 = 0$ and $t_{n+1} = 36$. We let Δt denote the length of each of these intervals.

During any one of these time intervals, the oil well will yield exactly 300 Δt barrels of crude oil. The price of crude oil varies during an interval, and therefore we cannot directly compute the revenue obtained during that period of time. So we approximate. We assume that throughout the jth interval the price remains fixed at $P(t_j)$, the price in effect at the start of the interval. Then the revenue obtained during the jth interval is $P(t_j)(300 \, \Delta t)$, and the revenue accumulated during all n intervals is given by the sum $\sum_{j=1}^{n} 300 P(t_j) \, \Delta t$. This approximation of the actual total revenue from the well improves as n increases and, in fact,

$$\text{Total revenue} = \lim_{n \to +\infty} \sum_{j=1}^{n} 300 P(t_j) \, \Delta t$$

According to our characterization of definite integrals, this limit is just the definite integral of the function $300P(t)$ from $t = t_1$ to $t = t_{n+1}$. Since $t_1 = 0$ and $t_{n+1} = 36$, we have

$$\text{Total revenue} = \int_0^{36} 300 P(t) \, dt = \int_0^{36} (5{,}400 + 90\sqrt{t}) \, dt = 207{,}360 \quad \text{dollars}$$

which is precisely the same answer we obtained in Example 6.6.

Perhaps this solution would look less formidable if we omitted some of the explanation. Here is a more compact way of writing the solution.

We divide the interval $0 \le t \le 36$ into n equal subintervals of length Δt and let t_j denote the beginning of the jth interval. Then,

Price during jth interval $\approx P(t_j)$ dollars per barrel

and

Yield during jth interval $= 300 \, \Delta t$ barrels

Hence,

Revenue from jth interval $\approx 300 P(t_j) \, \Delta t$ dollars

and

Total revenue $\approx \sum_{j=1}^{n} 300 P(t_j) \, \Delta t$ dollars

In fact,

Total revenue $= \lim_{n \to +\infty} \sum_{j=1}^{n} 300 P(t_j) \, \Delta t$ dollars

and so

Total revenue $= \int_{0}^{36} 300 P(t) \, dt$ dollars $\qquad \square$

In the solution of the two preceding examples, we used finite sums to approximate quantities that were actually equal to definite integrals. In the next example, we reverse our point of view and use a definite integral to approximate a finite sum.

Example 11.4

The author of a popular new book of eggplant recipes has just received his first royalty check for $2,000. He plans to deposit it, and all subsequent royalty checks, in a bank offering 8 percent interest compounded continuously. If he receives his royalty checks quarterly and sales of his book continue at the same rate, how much will the author have in the bank account 3 years from now (immediately before he deposits his 13th check)?

Solution

Let t_j denote the time (the number of years from now) at which the author receives his jth royalty check. (For example, $t_1 = 0$, $t_2 = \frac{1}{4}$, $t_3 = \frac{1}{2}$, and $t_{13} = 3$.) Recall that, if P dollars is deposited in a bank offering 8 percent interest compounded continuously, then the balance at the end of t years will be $Pe^{0.08t}$ dollars. At the end of the 3-year period, the $2,000 that was deposited at time t_j will have been in the bank $3 - t_j$ years and hence will have grown to $2,000 e^{(0.08)(3 - t_j)}$ dollars. The total amount in the bank at the end of 3 years will therefore be equal to the sum

$$\sum_{j=1}^{12} 2,000 e^{(0.08)(3 - t_j)}$$

Evaluation of this sum would be cumbersome. We can avoid this tedious calculation if we approximate the sum by a definite integral. As it stands, this sum does not appear to be of the same form as those in the characterization of definite integrals. However, if we let Δt denote the length of time between successive deposits, then $\Delta t = \frac{1}{4}$ and we can write $2{,}000 = 8{,}000\,\Delta t$. Thus the sum giving the balance in the account after 3 years can be rewritten as

$$\sum_{j=1}^{12} 8{,}000 e^{(0.08)(3 - t_j)} \Delta t$$

This is a sum of the form $\sum_{j=1}^{n} f(t_j)\,\Delta t$, with $f(t) = 8{,}000 e^{(0.08)(3-t)}$, $t_1 = 0$, and $t_{n+1} = 3$. Hence it can be approximated by the definite integral

$$\int_0^3 8{,}000 e^{(0.08)(3 - t)}\, dt$$

Thus,

Balance after 3 years $\approx \displaystyle\int_0^3 8{,}000 e^{(0.08)(3 - t)}\, dt$

$$= 8{,}000 e^{0.24} \int_0^3 e^{-0.08 t}\, dt$$

$$= -100{,}000 e^{0.24} e^{-0.08 t}\Big|_0^3$$

$$\approx -100{,}000 + 127{,}120$$

$$= \$27{,}120 \qquad \square$$

The replacement of a sum by a definite integral as in Example 11.4 can simplify the solution of a practical problem by eliminating the need for cumbersome calculations. The price we pay for this convenience, however, is that our answer is only approximately correct. Let us consider this approximation in geometric terms. For simplicity, suppose that $f(x)$ is non-negative on the interval $a \leq x \leq b$. When we replace the sum $\sum_{j=1}^{n} f(x_j)\,\Delta x$ (with $x_1 = a$ and $x_{n+1} = b$) by the definite integral $\int_a^b f(x)\, dx$, we are simply using the area under the curve $y = f(x)$ from $x = a$ to $x = b$ (Figure 26a) to approximate the area of the n rectangles in Figure 26b. The closeness of the approximation clearly depends on the width Δx of the rectangles and the shape of the curve $y = f(x)$. Unfortunately, without some further analysis (which would be beyond the scope of this text), we are usually unable to determine just how good our approximation really is.

11. THE DEFINITE INTEGRAL AS A LIMIT OF A SUM

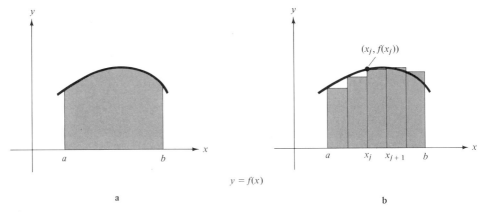

Figure 26

Problems

1. Use summation notation to express the following sums.
   (a) $1^3 + 2^3 + 3^3 + \cdots + 12^3$
   (b) $1 + \frac{1}{2} + \frac{1}{3} + \frac{1}{4} + \frac{1}{5} + \cdots + \frac{1}{20}$
   (c) $3 + 6 + 9 + 12 + 15 + 18 + 21 + 24 + 27 + 30$
   (d) $f(1) + f(2) + f(3) + \cdots + f(n)$
   (e) $f(x_1) + f(x_2) + f(x_3) + \cdots + f(x_n)$
   (f) $(3x_1 - 1) + (3x_2 - 1) + \cdots + (3x_9 - 1)$
   (g) $(300 \log t_1) \Delta t + (300 \log t_2) \Delta t + \cdots + (300 \log t_n) \Delta t$
   (h) $(x_1^2 - x_1 + 2) \Delta x + (x_2^2 - x_2 + 2) \Delta x + \cdots + (x_n^2 - x_n + 2) \Delta x$

2. Evaluate the following sums.
   (a) $\sum_{j=1}^{5} j^2$
   (b) $\sum_{j=1}^{4} (3j + 1)$
   (c) $\sum_{j=1}^{6} 2^j$
   (d) $\sum_{j=1}^{3} 2/(j+1)$
   (e) $\sum_{j=1}^{10} (-1)^j$
   (f) $\sum_{j=1}^{4} (-1)^{j+1}(j-1)^2$

   Solve Problems 3 through 9 twice: first using the characterization of the definite integral as a limit of a sum, and then by solving an appropriate differential equation.

3. An object is moving so that its speed after t minutes is $f(t) = 1 + 4t + 3t^2$ feet per minute. How far does the object travel during the third minute?

4. It is estimated that t years from now the value of a certain parcel of land will be increasing at a rate of $r(t)$ dollars per year. Find an expression for the amount by which the value of the land will increase during the next 5 years.

5. A tree has been transplanted and after x years is growing at a rate of $f(x) = 1 + 1/(x + 1)^2$ feet per year. By how much does the tree grow during the second year?

6. The promoters of a county fair estimate that t hours after the gates open at 9:00 A.M. visitors will be entering the fair at a rate of $r(t)$ people per hour. Find an expression for the number of people who will enter the fair between 11:00 A.M. and 1:00 P.M.

7. A bicycle manufacturer expects that n months from now he will be selling 5,000 bicycles a month at a price of $p(n) = 80 + 3\sqrt{n}$ dollars per bicycle. What is the total revenue he can expect from the sale of the bicycles over the next 16 months? Discuss the approximations and assumptions that lie behind your solution.

8. A bicycle manufacturer expects that n months from now he will be selling $f(n) = 5,000 + 50\sqrt{n}$ bicycles per month at a price of $p(n) = 80 + 3\sqrt{n}$ dollars per bicycle. What is the total revenue he can expect from the sale of the bicycles over the next 16 months? Discuss the approximations and assumptions that lie behind your solution.

9. Suppose that t months from now an oil well will be producing crude oil at a rate of $r(t)$ barrels per month and that the price of crude oil will be $p(t)$ dollars per barrel. Assuming that the oil is sold as soon as it is extracted from the ground, find an expression for the total revenue from the oil well over the next 2 years.

*10. After t years, a certain investment will generate income continuously at a rate of $g(t)$ dollars per year. Suppose that as the income is received it is deposited in a bank where it earns interest at an annual rate of 7 percent compounded continuously. Find an expression for the amount of money that will be in the bank account at the end of 5 years.

11. Suppose that the average automatic dishwasher consumes water at a rate of q gallons per hour and that t hours after 8:00 A.M. on a certain day $N(t)$ automatic dishwashers are being used in New York City. What quantity is represented by the integral $\int_4^7 qN(t)\, dt$?

12. A study of employees at a plant producing transistor radios indicates that the average new worker with no previous experience will take $15 + 15e^{-0.01n}$ minutes to assemble his nth radio.
    (a) Write an expression for the total time required for a new worker to assemble his first 100 radios. (Do not try to evaluate this expression.)
    (b) Approximate the total time in part (a) by computing an appropriate definite integral.

(c) If you know a programming language and have access to a computer, calculate the total time in part (a). Compare this value with your approximation in part (b).

13. The owner of a chain of health food restaurants has just received a shipment of 12,000 pounds of soybeans which he immediately places in storage. Every Friday he will remove 300 pounds of the beans from the warehouse for use in his restaurants during the following week. The cost of storing the soybeans is 0.2 cent per pound for each week (or fraction of a week).
    (a) Find (but do not evaluate) an expression for the total storage cost the owner will have paid by the time the supply of soybeans is exhausted. (Assume that the shipment arrived on a Monday.)
    (b) Approximate this total storage cost by computing an appropriate definite integral. (*Hint*: Assume that the supply of soybeans decreases linearly with time.)
    (c) Is your approximation in part (b) greater or less than the actual total storage cost? Explain.
    (d) If you know a programming language and have access to a computer, calculate the actual total storage cost.

12. FURTHER APPLICATIONS OF THE DEFINITE INTEGRAL

In this section we consider some practical problems whose solutions are based on the characterization of the definite integral as a limit of a sum. We begin with the problem of finding the average value of a continuous function over an interval.

Suppose that t hours past midnight on July fourth the temperature in Chicago was $f(t)$ degrees. What was the *average* temperature in the city between 9:00 A.M. and noon? In other words, what is the average value of the function $f(t)$ over the interval $9 \le t \le 12$? Before we can compute the answer, we must decide what we mean by the average value of a function over an interval.

One way to approach this problem is to divide the time interval from $t = 9$ to $t = 12$ into n equal subintervals of length Δt and then, if t_j denotes the beginning of the jth interval, compute the average of the n temperature readings $f(t_1), f(t_2), \ldots, f(t_n)$. In particular, we add these n values together and divide by n to get

$$\frac{f(t_1) + f(t_2) + \cdots + f(t_n)}{n}$$

This seems like a reasonable approximation to what we would intuitively think of as the average temperature between 9:00 A.M. and noon. Indeed, if the temperature remained unchanged throughout each of the subintervals, this expression would be precisely the average we are seeking. We obtain an average that is more sensitive to rapid temperature fluctuations if we decrease the size of the subintervals by increasing the value of n. This suggests the following definition of average value.

AVERAGE VALUE OF A FUNCTION OVER AN INTERVAL

The average value of the function $f(t)$ over the interval $a \leq t \leq b$ is the limit

$$\lim_{n \to +\infty} \frac{f(t_1) + f(t_2) + \cdots + f(t_n)}{n}$$

where $t_1, t_2, \ldots, t_{n+1}$ divide the interval into n equal subintervals.

In a moment we shall give a geometric interpretation of this quantity that will reinforce our belief that this is the "correct" definition of average value. But first, let us consider how we might calculate such a limit.

Observe that if the interval $a \leq t \leq b$ is divided into n equal subintervals, then the length Δt of each is $(b-a)/n$. Hence we can write

$$\frac{f(t_1) + f(t_2) + \cdots + f(t_n)}{n} = \frac{1}{b-a}\left[f(t_1)\frac{b-a}{n} + f(t_2)\frac{b-a}{n} + \cdots + f(t_n)\frac{b-a}{n}\right]$$

$$= \frac{1}{b-a} \sum_{j=1}^{n} f(t_j) \, \Delta t$$

But the limit of this sum is just the definite integral of f from a to b, and so

$$\lim_{n \to +\infty} \frac{f(t_1) + f(t_2) + \cdots + f(t_n)}{n} = \frac{1}{b-a} \lim_{n \to +\infty} \sum_{j=1}^{n} f(t_j) \, \Delta t$$

$$= \frac{1}{b-a} \int_a^b f(t) \, dt$$

COMPUTATION OF AVERAGES

The average value of a continuous function $f(t)$ over the interval $a \leq t \leq b$ is given by the formula

$$\text{Average value} = \frac{1}{b-a} \int_a^b f(t) \, dt$$

Note that if we multiply both sides of this equation by $b - a$, we get

$$(b-a)(\text{average value}) = \int_a^b f(t) \, dt$$

If $f(t)$ is non-negative, the integral on the right-hand side of this equation is just the area under the graph of f from $t = a$ to $t = b$. The product on the left-hand side is the area of a rectangle whose width is $b - a$ and whose height is the average value of f over the interval. In other words, the average value of $f(t)$ over the interval $a \leq t \leq b$ is equal to the height of the rectangle whose base is this interval and whose area is the same as the area under the graph of $f(t)$ from $t = a$ to $t = b$ (Figure 27).

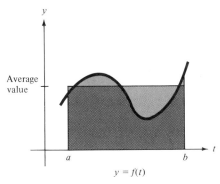

Figure 27

Let us now return to the temperature problem with which we began our discussion of average value.

Example 12.1

Suppose that t hours past midnight on July fourth the temperature in Chicago was $-\frac{1}{4}t^2 + 6t + 45$ degrees. What was the average temperature in the city between 9:00 A.M. and noon?

Solution

We want the average value of the function $f(t) = -\frac{1}{4}t^2 + 6t + 45$ over the interval $9 \leq t \leq 12$. Using our formula for average value, we get

$$\text{Average temperature} = \tfrac{1}{3} \int_9^{12} (-\tfrac{1}{4}t^2 + 6t + 45)\, dt$$

$$= \tfrac{1}{3}(-\tfrac{1}{12}t^3 + 3t^2 + 45t)\Big|_9^{12}$$

$$= 80\tfrac{1}{4} \quad \text{degrees} \qquad \square$$

The next application is related to the problem we solved in Chapter 3, Section 2, of finding the *present value* of future money. Recall that the problem was to determine the amount of money that would have to be invested today at an annual interest rate r so that it will be worth B dollars t years from now. We found that if the interest is compounded continuously this amount turns out to be Be^{-rt} dollars. We shall now examine a generalization of this problem.

PRESENT VALUE

The *present value* of a business venture or investment scheme that will generate income continuously at a certain rate (or discretely, through a certain sequence of payments), over a specified period of time, is the amount of money that must be deposited today at the prevailing interest rate to generate the same income stream (or sequence of payments).

Clearly, the present value of a future income stream is less than the total future income from that stream. (Why?) We can think of the present value of future income as the maximum price that one should be willing to pay today to obtain this income. It would be foolish to pay more for the income, because any amount of money that exceeds this present value would yield a better return just sitting in a bank earning interest.

In Example 12.3 we will use integration to compute the present value of a continuous income stream. First, however, we illustrate the concept of present value in the simpler discrete case.

Example 12.2

Suppose banks offer interest at a rate of 5 percent compounded continuously. What is the present value of an investment plan that guarantees annual payments of $2,000 for each of the next 3 years?

Solution

The first payment of $2,000 will arrive after 1 year. Its present value is therefore $2,000e^{-0.05}$ dollars. The second payment of $2,000 will arrive after 2 years, so its present value is $2,000e^{-(0.05)(2)} = 2,000e^{-0.10}$ dollars. The present value of the third payment is $2,000e^{-0.15}$ dollars. Hence the present value of the investment plan that guarantees these three payments is

$$2,000e^{-0.05} + 2,000e^{-0.10} + 2,000e^{-0.15} \approx \$5,433.56$$

That is, if $5,433.56 were deposited in a bank today, withdrawals of $2,000 could be made at the end of each of the next 3 years, after which nothing would be left in the account. □

Let us now consider the case of a continuous income stream.

Example 12.3

The management of a national chain of ice cream parlors is selling a 5-year franchise to operate its newest outlet in Madison, Wisconsin. Past experience in similar localities suggests that t years from now the profit flow from the ice cream parlor will be $g(t) = 14,000 + 700t$ dollars per year. What is the present value of the franchise if banks offer interest at a rate of 7 percent compounded continuously?

Solution

Our strategy will be to approximate the continuous profit flow from the franchise by discrete sequences of payments. The present values of these sequences can be computed as in the preceding example and are sums of a certain form. To get the present value of the franchise itself, we take a limit of these approximating sums. This limit turns out to be a definite integral.

To approximate the present value of the franchise, we divide the 5-year time interval into n equal subintervals of length Δt and let t_j denote the beginning of the jth interval. We assume that throughout the jth interval profit is generated at the constant rate $g(t_j)$ dollars per year and that the resulting profit of $g(t_j)\,\Delta t$ dollars is received in one lump sum at the beginning of the interval (when $t = t_j$). The present value of the jth payment is

$$e^{-0.07t_j}g(t_j)\,\Delta t$$

and hence the present value of the entire sequence of n payments is

$$\sum_{j=1}^{n} e^{-0.07t_j}g(t_j)\,\Delta t$$

This approximation to the present value of the franchise clearly improves as n increases and the length Δt of the subintervals approaches zero. In fact, we can make the difference between this sum and the present value of the franchise as small as we like by taking n sufficiently large. Hence,

$$\text{Present value of franchise} = \lim_{n \to +\infty} \sum_{j=1}^{n} e^{-0.07t_j}g(t_j)\,\Delta t$$

But this limit is precisely the definite integral from $t = 0$ to $t = 5$ of the function $f(t) = e^{-0.07t}g(t)$. Thus

$$\text{Present value of franchise} = \int_0^5 e^{-0.07t}g(t)\,dt$$

$$= \int_0^5 (14{,}000 e^{-0.07t} + 700 t e^{-0.07t})\,dt$$

We evaluate this integral, using integration by parts on the term $700te^{-0.07t}$, and get

$$\text{Present value of franchise} = \left. -342{,}857.14 e^{-0.07t} - 10{,}000 t e^{-0.07t} \right|_0^5$$

$$\approx \$66{,}015 \qquad \square$$

In our final example, we are given a function $f(t)$ representing the fraction of the membership of a certain group that can be expected to remain in the group for at least t months. This function is assumed to apply to individuals newly entering the group, as well as to the original members. The rate at which new members arrive is known, and the problem is to predict the size of the group at some future time. Problems of this general form arise in many fields, including sociology, demography, and ecology.

Example 12.4

A new county mental health clinic has just opened. Statistics compiled at similar facilities suggest that the fraction of patients who will still be receiving treatment at the clinic t months after their initial visit is given by the function $f(t) = e^{-t/20}$. The clinic initially accepts 300 people for treatment, and plans to accept new patients at a rate of 10 per month.

(a) Approximately how many people will be receiving treatment at the clinic 15 months from now?

(b) Approximately how many patients will be receiving treatment x months from now?

(c) In the long run, approximately how many people will be receiving treatment at the clinic?

Solution

So that we may apply calculus to this problem, let us assume that the number of patients is a continuous rather than discrete variable. Let $N(t)$ denote the number of patients receiving treatment t months from now.

(a) Since $f(15)$ is the fraction of patients whose treatment continues at least 15 months, it follows that of the current 300 patients only $300f(15)$ will still be receiving treatment in 15 months.

To approximate the number of *new* patients who will be receiving treatment 15 months from now, we divide the 15-month time interval into n equal subintervals of length Δt and let t_j denote the beginning of the jth interval. Since new patients are accepted at a rate of 10 per month, the number of new patients accepted during the jth time interval is $10 \, \Delta t$. Fifteen months from now, approximately $15 - t_j$ months will have elapsed since these $10 \, \Delta t$ new patients had their initial visits, and so approximately $f(15 - t_j)(10 \, \Delta t)$ of these will still be receiving treatment at that time. It follows that the total number of new patients still receiving treatment 15 months from now can be approximated by the sum

$$\sum_{j=1}^{n} 10 f(15 - t_j) \, \Delta t$$

Adding the number of current patients who will still be receiving treatment in 15 months, we get

$$N(15) \approx 300 f(15) + \sum_{j=1}^{n} 10 f(15 - t_j) \, \Delta t$$

We can make this approximation as close as we like to the actual value $N(15)$ by taking n sufficiently large. Thus,

$$N(15) = 300f(15) + \lim_{n \to +\infty} \sum_{j=1}^{n} 10f(15 - t_j) \Delta t$$

$$= 300f(15) + \int_0^{15} 10f(15 - t)\, dt$$

$$= 300e^{-3/4} + 10e^{-3/4} \int_0^{15} e^{t/20}\, dt$$

$$= 200 + 100e^{-3/4}$$

$$\approx 247$$

That is, 15 months from now, the clinic will be treating approximately 247 patients.

(b) To derive a formula for the number of patients being treated x months from now, we simply repeat the preceding argument with 15 replaced by x. We get

$$N(x) = 300f(x) + \int_0^x 10f(x - t)\, dt$$

$$= 300e^{-x/20} + \int_0^x 10e^{-(x-t)/20}\, dt$$

$$= 200 + 100e^{-x/20}$$

(c) Since

$$\lim_{x \to +\infty} N(x) = \lim_{x \to +\infty} (200 + 100e^{-x/20}) = 200$$

we conclude that in the long run the clinic will be treating approximately 200 people. A graph of the function $N(x)$ is sketched in Figure 28.

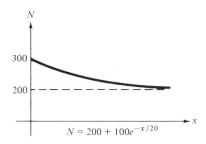

Figure 28

Problems

1. For each of the following non-negative functions, find the average value over the specified interval. In each case, graph the function and draw the rectangle whose base is the given interval and whose area is equal to the area under the graph and above the interval.

   (a) $f(x) = x; 0 \leq x \leq 4$ (b) $f(t) = 2t - t^2; 0 \leq t \leq 2$
   (c) $f(x) = x^3; 0 \leq x \leq 2$ (d) $f(t) = (t+2)^2; -4 \leq t \leq 0$

2. For each of the following functions, find the average value over the specified interval.

   (a) $f(x) = x^3; -5 \leq x \leq 5$ (b) $f(t) = t^2 - t - 2; -1 \leq t \leq 2$
   (c) $f(x) = 1/x; 1 \leq x \leq e$ (d) $f(t) = \log t; 1 \leq t \leq e$

3. A car is driven for 2 hours so that after t hours its speed is $48 + 4t - t^2$ miles per hour.
   (a) What is the average speed of the car during the first hour?
   (b) What is the average speed of the car during the second hour?
   (c) What is the average speed of the car during the first 2 hours?
   (d) How is your answer to part (c) related to your answers to parts (a) and (b)? Can you explain this relationship?

4. A car is driven so that after t hours its speed is $f(t)$ miles per hour. Prove that the car's average speed during the first 2 hours is the same as the average of its average speeds during the first and second hours.

5. A car is driven so that after t hours its speed is $f(t)$ miles per hour.
   (a) Find an expression (involving a definite integral) for the average speed of the car during the first N hours.
   (b) Find an expression (involving a definite integral) for the total distance the car travels during the first N hours.
   (c) Discuss the relationship between your answers to parts (a) and (b).

6. Having been left partially disabled after a head-on collision with a camel (see Problem 11 on page 265), the hero of a popular spy story has been retired from the Secret Service. As compensation for his many long years of dedicated public service, the government has offered him a choice between a 10-year pension of 5,000 pounds (sterling) per year or a flat sum of 35,000 pounds to be paid immediately. Assuming that an interest rate of 10 percent compounded continuously will be available at banks throughout this period, decide which offer he should accept. (*Hint*: Compare the flat sum of 35,000 pounds with the present value of the pension. Assume that the pension is paid continuously.)

7. Suppose an investment scheme will generate income at a fixed rate of $1,000 per year. Assume that the $1,000 is dispensed uniformly throughout the year and that an interest rate of 100r percent compounded continuously is available at banks.
  (a) Derive a formula for the present value of the investment scheme if the income is to be paid for N years.
  (b) What is the present value of the investment scheme if the income is guaranteed forever? [*Hint*: Let N approach $+\infty$ in part (a).]
  (c) Explain in economic terms why it is possible to have a *finite* answer in part (b).

8. A special act of Parliament has directed the government to increase the retirement benefits paid to the spy in Problem 6. Pursuant to this directive, the chancellor of the exchequer offers the spy a choice between a pension that pays 5,000 pounds a year forever (to the spy or designated heirs) or a flat sum of 51,000 pounds to be paid immediately. Assuming that an interest rate of 10 percent will always be available at banks, decide which offer the spy should accept.

*9. Consider an investment scheme that, after t years, will generate income continuously at a rate of $g(t)$ dollars per year. Suppose that as the income is received it is deposited in a bank where it earns interest at an annual rate of 7 percent compounded continuously. The total amount of money that will be accumulated in this way during the next N years is called the *future value* of the investment scheme after N years.
  (a) Express the future value of this investment scheme after N years as a definite integral.
  (b) Find an expression for the amount of money that would have to be deposited in the bank today so that its value after N years would equal the future value obtained in part (a). Show that this amount is the same as the *present value* of the investment scheme. (In technical jargon, this says that the present value of an income stream is the same as the present value of the future value of that income stream.)

10. Let $f(t)$ denote the fraction of the membership of a certain type of group that will remain in the group for at least t years. Suppose that a group of this type with P members has just been formed and that t years from now new members will be added to the group at the rate of $r(t)$ individuals per year.
  (a) Find an expression for the size of the group 10 years from now.
  (b) Find an expression for the size of the group x years from now.

11. A national consumers' association has compiled statistics suggesting that the fraction of its members who are still active t months after joining the group

is given by the function $f(t) = \frac{1}{10} + \frac{9}{10}e^{-t/8}$ (for $t \leq 60$). A new local chapter has 150 charter members and expects to attract new members at the rate of 10 per month.
   (a) Express the expected number of active chapter members as a function of the age of the chapter.
   (b) How many active members can this chapter expect to have at the end of 8 months? At the end of 2 years?
   (c) When will the number of active chapter members be smallest?
   (d) Why would it be unreasonable to suppose that the fraction of members active for more than t months might be represented by the given function $f(t)$ for all values of t?

*12. Does the number of patients who will be receiving treatment in the long run at the clinic in Example 12.4 depend on the number of people initially accepted for treatment? Justify your answer mathematically and explain in practical terms why your answer is reasonable.

*13. At what constant rate should the mental health clinic in Example 12.4 accept new patients so that, in the long run, only about 160 patients will be receiving treatment at the clinic?

*14. A certain atomic power plant produces radioactive waste in the form of strontium 90 at a constant rate of 500 pounds per year. The half-life of strontium 90 is 28 years. How much of this radioactive waste from the atomic power plant will be present after 140 years? (*Hint*: See Problem 10 on page 278.)

*15. A manufacturer receives a shipment of a certain raw material which is initially placed in storage. Subsequently, the raw material is withdrawn from storage and used until the supply is exhausted 1 year later.
   (a) Suppose that the storage rate is p dollars per unit of the raw material per year. Show that the manufacturer's total yearly storage cost is pA where A is the average number of units of the raw material in storage during the year.
   (b) Suppose that the raw material is withdrawn from storage at a *constant* rate, so that the supply decreases linearly and is exhausted at the end of 1 year. Show that the average number of units of the raw material in storage during the year is $N/2$ where N is the number of units in the original shipment.
   (c) Generalize your answers to parts (a) and (b) to the case in which the manufacturer receives more than one shipment of the raw material during the year. In generalizing part (b), assume that the shipments are of equal size, that the rate at which the raw material is used remains constant

throughout the year, and that the last shipment is used up exactly 1 year after the arrival of the first shipment.

(d) Use your results in part (c) to justify the assertion about storage cost that was stated without proof in Chapter 2, Example 8.4.

13. REVIEW PROBLEMS

1. Find the following integrals.

   (a) $\int (x^2 - 5 + 2x^{3/2} + 1/x^2)\, dx$ (b) $\int \sqrt{3x + 1}\, dx$

   (c) $\int (3x + 1)\sqrt{3x^2 + 2x + 5}\, dx$ (d) $\int x\sqrt{x - 1}\, dx$

   (e) $\int 5e^{3x}\, dx$ (f) $\int 5xe^{3x}\, dx$

   (g) $\int (2x + 1)e^x e^{x^2}\, dx$ (h) $\int (x^2 - 1)/(x^3 - 3x + 1)\, dx$

   (i) $\int x/\sqrt{1 + 3x}\, dx$ (j) $\int (x - 5)^{12}\, dx$

   (k) $\int x(x - 5)^{12}\, dx$ (l) $\int x \log 3x\, dx$

   (m) $\int (\log 3x)/x\, dx$ (n) $\int \log 2x\, dx$

2. Evaluate the following definite integrals.

   (a) $\int_0^1 (x^4 - 3x^3 + 1)\, dx$ (b) $\int_1^4 (\sqrt{x} + x^{-3/2})\, dx$

   (c) $\int_1^2 (2x - 4)^5\, dx$ (d) $\int_{-1}^0 [(x + 1)^4 + 2(x + 1)^3 - (x + 1)^2 + 6]\, dx$

   (e) $\int_0^1 2xe^{x^2 - 1}\, dx$ (f) $\int_{\log 1/2}^0 e^{-x}\, dx$

   (g) $\int_{-1}^1 xe^x\, dx$ (h) $\int_1^e \log x\, dx$

3. Find the general solutions of the following differential equations.

   (a) $\dfrac{dy}{dx} = x^3 - 3x^2 + 5$ (b) $\dfrac{d^2y}{dx^2} = x^3 - 3x^2 + 5$ (c) $\dfrac{dy}{dx} = kx$

   (d) $\dfrac{d^2y}{dx^2} = kx$ (e) $\dfrac{dy}{dx} = ky$ (f) $\dfrac{dy}{dx} = k(80 - y)$

   (g) $\dfrac{dy}{dx} = y^3 x^2$ (h) $\dfrac{dy}{dx} = \dfrac{x+1}{2y}$ *(i) $\dfrac{dy}{dx} = y(1 - y)$

   (j) $y \dfrac{dy}{dx} - xe^x = 1$

4. Find the particular solutions of the following differential equations that satisfy the given conditions.

   (a) $\dfrac{dy}{dx} = 5x^4 - 3x^2 - 2$; $y = 4$ when $x = 1$

   (b) $\dfrac{d^2y}{dx^2} = e^{x/2}$; $y = 4e$ and $\dfrac{dy}{dx} = 3e$ when $x = 2$

   (c) $\dfrac{dy}{dx} = 0.06y$; $y = 100$ when $x = 0$

   (d) $\dfrac{dy}{dx} = \dfrac{x}{y^2}$; $y = 3$ when $x = 2$

5. Find the area of the region bounded by the curve $y = 1/x^2$, the x axis, and the vertical lines $x = 1$ and $x = 4$.

6. Find the area of the region bounded by the curve $y = x^2 - 6x + 5$ and the x axis.

7. Find the area of the region bounded by the curve $y = x^3 - 4x$ and the x axis.

8. Find the area of the region bounded by the curve $y = x^4$ and the line $y = x$.

9. Find a function whose tangent has slope $1/\sqrt{x} + 3x^2$ for each value of x and whose graph passes through the point $(4, 10)$.

10. A car leaves Omaha, Nebraska, at noon and is driven for 6 hours so that x hours later its speed is $60 + 10x - 3x^2$ miles per hour.
    (a) How far does the car go between noon and 5:00 P.M.?
    (b) What is the car's average speed between noon and 5:00 P.M.?

11. The resale value of a certain industrial machine decreases over a 10-year period at a rate that depends on the age of the machine. When the machine is t years old, the rate at which its value is changing is $100t - 1{,}000$ dollars per year.
    (a) Express the value of the machine in terms of its age and initial value.
    (b) If the machine was originally worth $5,500, how much will it be worth when it is 10 years old?
    (c) If the machine was originally worth $5,500, what is the average value of the machine over the 10-year period?

12. A manufacturer estimates his marginal revenue to be $200 - 4q$ dollars per unit (for $q < 40$) when q units are produced. His corresponding marginal cost has been found to be $0.6q$ dollars per unit. Suppose the manufacturer's profit is $1,070 when 10 units are produced. What is his profit when 20 units are produced?

13. The rate of growth of bacteria in a certain culture is proportional to the number of bacteria present. If 6,000 bacteria are initially present in the culture and 8,000 are present 10 minutes later, how long will it take for the number of bacteria to double?

14. A bank pays interest at an annual rate of 8 percent compounded continuously. If an account is opened for $10,000, what will the balance be at the end of 5 years?

15. The rate at which the temperature of an object changes is proportional to the difference between its own temperature and that of the surrounding medium. A 90-degree object is placed in a 60-degree room. Twenty minutes later, the temperature of the object is 80 degrees.
    (a) When will the temperature of the object be 70 degrees?
    (b) How hot will the object be at the end of 1 hour?

16. A tank currently holds 200 gallons of brine that contains 3 pounds of salt per gallon. Clear water flows into the tank at a rate of 4 gallons per minute, while the mixture (which is kept uniform) runs out of the tank at a rate of 5 gallons per minute. How much salt is in the tank at the end of 100 minutes?

17. A manufacturer expects that x months from now he will be selling $n(x)$ lamps per month at a price of $p(x)$ dollars per lamp. Find an expression for the total revenue he can expect from the sale of the lamps over the next 12 months.

18. The publisher of a new ecology newsletter estimates that the fraction of subscribers who will retain their subscriptions for at least t months is given by the function $f(t) = e^{-t/20}$. There are 2,000 charter subscribers to the newsletter, and the publisher expects that t months from now his circulation will be increasing at a rate of $r(t) = 4t$ *new* subscriptions per month.
    (a) How many subscriptions can he expect to have x months from now?
    (b) At approximately what rate will the total number of subscriptions be increasing in the long run?

Functions of Several Variables

In this chapter, we study differential calculus of functions that have more than one independent variable. Although somewhat more complicated, the theory will be similar to that developed in Chapter 2 for functions of a single variable. Indeed, many of the results in this chapter will be derived from analogous results in Chapter 2. As in the one-variable case, the tools we develop can be applied to practical problems involving optimization and rate of change.

1. FUNCTIONS AND GRAPHS

In many practical situations, the value of one quantity depends on the values of several others. For example, the number of copies of a calculus text that will be sold may depend on the number of salesmen employed by the publisher, the amount of money spent on advertising (including the distribution of complimentary copies to instructors), and the amount spent on development (including reviewing, designing, and editing). Relationships of this sort can often be represented mathematically by functions having more than one independent variable. Let us begin our discussion of such functions with the following simple example.

Example 1.1

A service station sells three grades of gasoline, regular, low-lead, and premium, at prices of 70, 73, and 75 cents a gallon, respectively. Express the station's total revenue from the sale of the gasoline as a function of the number of gallons of each grade sold.

Solution

We introduce three variables x_1, x_2, and x_3 to denote the number of gallons sold of regular, low-lead, and premium, respectively. Then,

Total revenue = $70x_1 + 73x_2 + 75x_3$ cents   □

The convenient functional notation that we have been using for functions of a single variable can be extended to describe functions of several variables. In Example 1.1 for instance, we could let f denote the total revenue and write

$$f(x_1, x_2, x_3) = 70x_1 + 73x_2 + 75x_3$$

In this case, f is a function of three independent variables x_1, x_2, and x_3. Its domain consists of all ordered triples (x_1, x_2, x_3) of real numbers. (Of course, only if the numbers x_1, x_2, and x_3 are non-negative does this function have a practical economic interpretation here.) Similar notation can be used to describe functions having any number of independent variables. The use of this functional notation is illustrated in the following examples.

Example 1.2

If $f(x_1, x_2) = x_1 e^{x_2} + 3x_1^2$, determine the domain of f and compute $f(-2, 0)$.

Solution

The domain of f consists of all ordered pairs (x_1, x_2) of real numbers. To compute $f(-2, 0)$, we simply substitute $x_1 = -2$ and $x_2 = 0$ into the formula for f. We get

$$f(-2, 0) = -2e^0 + 3(-2)^2 = 10 \qquad \square$$

Example 1.3

If $f(r, s, t) = (3r^2 + 5s)/(r - t)$, determine the domain of f and compute $f(2, 3, -1)$.

Solution

The function f is defined whenever its denominator $r - t$ is not zero. Thus the domain of f consists of all ordered triples (r, s, t) for which $r \neq t$. Triples such as $(r, s, t) = (1, 3, 1)$ are not in the domain of f. On the other hand, $(r, s, t) = (2, 3, -1)$ is in this domain, and we find that

$$f(2, 3, -1) = \frac{3(2)^2 + 5(3)}{2 - (-1)} = 9 \qquad \square$$

In the next example, a certain quantity is given as a function of two variables, each of which can be written in terms of a third variable. By forming an appropriate *composite function*, we can express the original quantity as a function of the third variable.

Example 1.4

A certain liquor store in Minneapolis carries two brands of inexpensive white table wine, one from California and the other from a vineyard in New York State. The owner of the liquor store has found that the demand for the California wine depends not only on its own price but on the price of the competing New York wine as well. In particular, if the California wine sells for x dollars a bottle and the

New York wine for y dollars a bottle, the demand for the California wine is approximately

$$300 - 20x^2 + 30y \quad \text{bottles per month}$$

It is expected that t months from now the price of a bottle of the California wine will be

$$x = 2 + 0.05t \quad \text{dollars}$$

and the price of a bottle of New York wine will be

$$y = 2 + 0.1\sqrt{t} \quad \text{dollars}$$

Express the demand for the California wine as a function of time.

Solution

Let $Q(t)$ denote the monthly demand for the wine t months from now. To find a formula for $Q(t)$ we simply replace x by $2 + 0.05t$ and y by $2 + 0.1\sqrt{t}$ in the expression $300 - 20x^2 + 30y$. Thus

$$Q(t) = 300 - 20(2 + 0.05t)^2 + 30(2 + 0.1\sqrt{t})$$
$$= 280 - 4t - 0.05t^2 + 3\sqrt{t} \qquad \square$$

The function $Q(t)$ in Example 1.4 can be thought of as the composite function $f(g(t), h(t))$, where $f(x, y) = 300 - 20x^2 + 30y$, $g(t) = 2 + 0.05t$, and $h(t) = 2 + 0.1\sqrt{t}$. Note that in this case the composite function turns out to be a function of a *single* variable.

In the next example, we start with a function of a single variable and then write that variable as a function of two other variables. The result is a composite function of *two* variables.

Example 1.5

Let $g(x) = \sqrt{x + 2}$ and $h(r, s) = 1 + r + 2s$. Find the composite function $f(r, s) = g(h(r, s))$ and specify its domain.

Solution

We replace x by the expression $1 + r + 2s$ in the formula for $g(x)$ to get

$$f(r, s) = g(h(r, s)) = \sqrt{3 + r + 2s}$$

The domain of f consists of all ordered pairs (r, s) of real numbers for which $3 + r + 2s \geq 0$. $\qquad \square$

In our investigation of functions of a single variable, we made extensive use of the geometric representation of functions as graphs. Using a three-dimensional coordinate system, we can represent functions of *two* variables graphically as surfaces. Unfortunately, we have no analogous way of visualizing functions of more than two variables.

When graphing a function f of two variables, it is customary to use the letters x and y to denote the two independent variables, and the letter z for the corresponding dependent variable. That is, we let $z = f(x, y)$.

To construct our three-dimensional coordinate system, we adjoin a third axis (the z axis) to the familiar xy coordinate plane as shown in Figure 1. Notice that the xy plane is taken as horizontal. The z axis is perpendicular to the xy plane and the upward direction is chosen to be the positive z direction. (For simplicity, only the positive coordinate axes are drawn in Figure 1.)

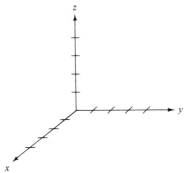

Figure 1

We can describe the location of a point in three-dimensional space by specifying three coordinates. For example, the point that is 4 units above the xy plane and that lies directly over the point $(x, y) = (1, 2)$ is represented by the coordinates $x = 1$, $y = 2$, and $z = 4$. (See Figure 2.) For convenience, we express the coordinates of a point compactly as an ordered triple (x, y, z). Thus the point in Figure 2 is represented by the ordered triple $(1, 2, 4)$. Similarly, $(2, -1, -2)$ represents a point that is 2 units *below* the xy plane and that lies directly under the point $(x, y) = (2, -1)$.

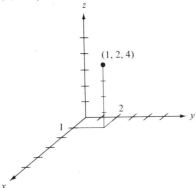

Figure 2

Suppose that $z = f(x, y)$. We can identify the ordered pairs (x, y) in the domain of f with points in the xy plane, and we can think of the function f as assigning a height z to each such point. Thus if $f(1, 2) = 4$, we would express this fact geometrically by plotting the point $(1, 2, 4)$ in three-dimensional space. The *graph* of f consists of all points (x, y, z) for which $z = f(x, y)$. The function may assign different heights to different points in its domain and, in general, its graph will be some sort of sheet or *surface* lying above the xy plane (Figure 3). (Of course, the surface might lie *below* the xy plane or be partially above and partially below this plane. For simplicity, let us agree that the word "above" will include all these cases.)

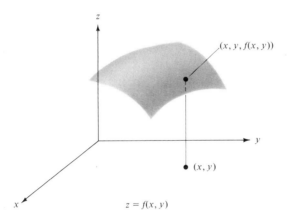

Figure 3

In practical work, you will have very little occasion to graph functions of two variables. Therefore, we shall not take time to develop techniques for sketching surfaces. However, since you may be interested in seeing some surfaces that arise as graphs of actual functions, portions of two such surfaces are sketched in Figure 4.

Although we shall not have to graph any other functions of two variables in this chapter, the fact that such functions can be represented geometrically as surfaces will be helpful in Section 5 when we study relative maxima and minima.

The next section of this chapter deals with the differentiation of functions of several variables. If we wanted to give a mathematically rigorous treatment of this subject, we would have to define the notion of continuity for functions of several variables. We can characterize the continuous functions of two variables informally as those functions whose graphs are "unbroken" surfaces. However, when more than two variables are involved, we can give no such geometric characterization. Precise definitions of continuity for functions of several variables are rather technical and can be found in more advanced texts. In this chapter, we shall assume without further discussion that all our functions are continuous and have continuous

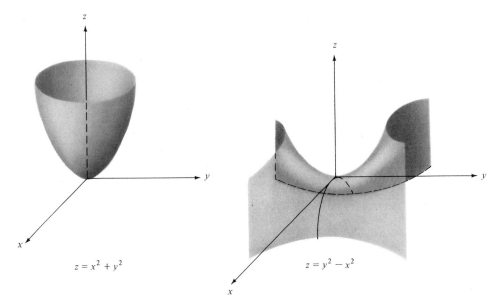

$z = x^2 + y^2$ $z = y^2 - x^2$

Figure 4

derivatives at all points that are of interest. Virtually all the functions you are likely to encounter as you apply calculus to practical problems in the social sciences will have this property.

Problems

1. For each of the following functions, specify the domain and compute the values indicated.
   (a) $f(x_1, x_2, x_3) = x_1(x_2 - 1)^2 + 3x_1 x_3$; $f(1, 0, -2)$, $f(0, 8, 4)$
   (b) $g(x, y) = (x + y)/(x - y)$; $g(2, -2)$, $g(-1, 3)$
   (c) $h(r, s) = \sqrt{s^2 - r^2}$; $h(0, 1)$, $h(-1, 3)$
   (d) $f(x, y, z) = x + y^2 + \sqrt{1 - z}$; $f(1, 2, 0)$, $f(-1, 3, -3)$
   (e) $g(x_1, x_2) = (x_1 - 1)/\sqrt{x_2^2 + 4}$; $g(1, 5)$, $g(2, 0)$
   (f) $h(r, s, t) = te^{r+s} + \log(t + 2)$; $h(\log 2, 0, -1)$, $h(1, 1, 1)$

2. In each of the following cases, find the indicated composite function and specify its domain.
   (a) $f(g(x), h(x))$, where $f(s, t) = s + t(s + 1)$, $g(x) = x^2$, $h(x) = x + 2$
   (b) $g(h(x_1, x_2))$, where $g(z) = \sqrt{z}$, $h(x_1, x_2) = 2x_1 - 3x_2$
   (c) $f(g(x_1, x_2), h(x_1, x_2))$, where $f(y, z) = y^2 + 2z$, $g(x_1, x_2) = x_1 - x_2$, $h(x_1, x_2) = 2x_1 x_2$

(d) $h(f(t), g(t))$, where $h(x, y) = (x + y)/y$, $f(t) = 1/t$, $g(t) = t + 1$
(e) $f(g(x, y))$, where $f(t) = \log t$, $g(x, y) = ye^x$
(f) $f(g(x, y), h(x, y))$, where $f(u, v) = ve^u$, $g(x, y) = x + y$, $h(x, y) = x^2/y$

3. A manufacturer can produce electric typewriters at a cost of $80 apiece, and manual typewriters at a cost of $20 apiece.
   (a) Express the manufacturer's total monthly production cost as a function of the number of electric typewriters and the number of manual typewriters produced each month.
   (b) Compute the total monthly cost if the manufacturer is currently producing 500 electric and 800 manual typewriters a month.
   (c) Suppose the manufacturer wants to increase his production of electric typewriters by 100 a month from the level in part (b). What corresponding change should he make in his monthly output of manual typewriters so that his total monthly cost will not change?

*4. Suppose the manufacturer in Problem 3 wishes to keep his production costs fixed at $56,000 a month.
   (a) Write an equation relating the monthly output of electric typewriters and the monthly output of manual typewriters that must be satisfied if this goal is to be achieved.
   (b) On a two-dimensional coordinate system, draw the graph of your equation in part (a). (This graph is said to be a *level curve* of the total cost function.)
   (c) Determine the maximum monthly output of electric typewriters compatible with the manufacturer's goal of holding his monthly costs fixed at $56,000. Explain how this information can be obtained from your level curve in part (b).

5. A publisher has found that in a certain city each of his salesmen will sell approximately

$$\frac{r^2}{2{,}000p} + \frac{s^2}{100} - s \quad \text{sets of encyclopedias a month}$$

where s denotes the total number of salesmen he has working in the city, p the price of a set of encyclopedias, and r the amount of money spent each month on local advertising.
   (a) Express the total number of sets of encyclopedias that will be sold each month in the city as a function of s, r, and p.
   (b) How many sets will be sold in the city each month if the publisher employs 10 salesmen, spends $6,000 a month on local advertising, and sells the books for $800 a set?
   (c) Express the publisher's total monthly *revenue* from the local sale of encyclopedias as a function of s, r, and p.

6. Let c denote the cost of producing each set of the encyclopedias in Problem 5 and let w denote the monthly salary of each salesman. Express the publisher's total monthly *profit* from the local sale of encyclopedias as a function of s, r, p, c, and w.

7. Suppose that in addition to his salary each salesman in Problem 6 receives a commission equal to 8 percent of the price of each set of encyclopedias he sells. Find the publisher's monthly profit function in this case.

8. A contractor employs 10 skilled workers and 25 unskilled workers. A new union contract calls for wages to increase so that t months from now skilled workers will earn $5 + 0.1t$ dollars an hour and unskilled workers will earn $2 + 0.02t^2$ dollars an hour.
   (a) Express the total hourly wages paid by the contractor as a function of t.
   (b) Find functions $f(x, y)$, $g(t)$, and $h(t)$ so that your answer to part (a) can be written as the composite function $f(g(t), h(t))$. Give economic interpretations of the functions f, g, and h.

9. Using x trained and y untrained workers, a manufacturer can produce $10x^2y$ items a day. Currently the manufacturer has 20 trained and 40 untrained workers on the job.
   (a) How many items are currently being produced each day?
   (b) By how much will the daily production level change if one more trained worker is added to the current work force?
   (c) By how much will the daily production level change if one more untrained worker is added to the current work force?
   (d) By how much will the daily production level change if one more trained worker *and* one more untrained worker are added to the current work force?

*10. Suppose f is a function of two variables.
   (a) Is it necessarily true that
   $$f(x + 1, y + 1) - f(x, y) = [f(x + 1, y) - f(x, y)] + [f(x, y + 1) - f(x, y)]$$
   (*Hint*: See Problem 9.)
   (b) Find a function $f(x, y)$ for which the equality in part (a) is satisfied for all points (x, y).

11. A liquor store carries two brands of inexpensive white table wine, one from California and the other from a vineyard in New York. If the California wine sells for x dollars a bottle and the New York wine for y dollars a bottle, the monthly demand for the California wine is approximately $300 - 20x^2 + 30y$ bottles and the monthly demand for the New York wine is approximately $240 - 12y^2 + 20x$ bottles.

(a) What happens to the demand for the New York wine as its price increases? As its price decreases?

(b) What happens to the demand for the California wine as the price of the New York wine increases? As the price of the New York wine decreases?

(c) Express the total monthly demand for the two wines as a function of x and y.

(d) Suppose both the California and New York wines currently sell for $2 a bottle. By how much will the total monthly demand for the two wines change if the price of the California wine is increased by $1 a bottle and the price of the New York wine by 50 cents a bottle?

*12. At a certain factory, the daily output of a certain commodity is approximately $3K^{1/2}L^{1/3}$ units, where K denotes the firm's capital investment measured in thousands of dollars and L the size of the labor force measured in man-hours. Suppose the current capital investment is $400,000 and that 1,331 man-hours of labor are used each day. Use marginal analysis to *estimate* the effect that an additional capital investment of $1,000 would have on the daily output if the size of the labor force is not changed. (*Hint*: Substitute $L = 1,331$ into the function representing daily output and then differentiate the resulting function of K.)

2. PARTIAL DERIVATIVES

In many practical problems involving functions of several variables, we are interested in finding the rate of change of a function with respect to one of its variables when all the others are kept fixed. That is, we want to *differentiate* the function with respect to the particular variable in question, keeping all the other variables fixed. This process is known as *partial differentiation*, and the resulting derivative is said to be a *partial derivative* of the function. To see how partial derivatives might be used in a practical problem, consider the following example.

Example 2.1

A manufacturer estimates that the number of transistor radios he can produce each week is given by the function

$$f(x, y) = 1{,}200x + 500y + x^2 y - x^3 - y^2$$

where x is the number of skilled workers and y is the number of unskilled workers employed in his plant. Currently his work force consists of 40 skilled workers and 60 unskilled workers. Estimate the change in his weekly output of radios that would result from the addition of one more skilled worker if the number of unskilled workers is kept fixed.

Solution

The precise change in productivity due to the addition of one skilled worker can be found by computing the difference $f(41, 60) - f(40, 60)$. However, the arithmetic involved in this computation is somewhat tedious and we can simplify the situation by using calculus.

Suppose we substitute the fixed value $y = 60$ into our expression for f. The resulting equation,

$$f = 1{,}200x + 60x^2 - x^3 + 26{,}400$$

tells us how f is related to x when y is held fixed at 60. If we now differentiate with respect to the variable x, we get

$$\frac{df}{dx} = 1{,}200 + 120x - 3x^2$$

which is the rate of change of the output f with respect to the number x of skilled workers when 60 unskilled workers are employed. When $x = 40$, the value of this derivative is

$$1{,}200 + 120(40) - 3(40)^2 = 1{,}200$$

and should be an *approximation* to the number of additional radios that will be produced each week due to the hiring of the 41st skilled worker. (Incidentally, the actual increase in productivity turns out to be 1,139 radios per week.) □

In Example 2.1 we started with a function $f(x, y)$ into which we substituted a numerical value for y and then differentiated with respect to x. The result was an expression giving the rate of change of f with respect to x for the *particular* fixed value of y. A more versatile approach would be to differentiate f with respect to x, *pretending* that y is a constant but without actually specifying a numerical value for y. This would give us a general formula in terms of x and y for the rate of change of f with respect to x when y is held fixed. To illustrate this approach, let us reconsider the function in Example 2.1.

Example 2.2

Suppose $f(x, y) = 1{,}200x + 500y + x^2 y - x^3 - y^2$.

(a) Find a function of x and y expressing the rate of change of f with respect to x when y is held fixed.

(b) Find the rate of change of f with respect to x when $x = 40$ and $y = 60$.

Solution

(a) We differentiate f term by term with respect to x, treating y as if it were a constant. If we let f_x denote the resulting function, we find that

$$f_x(x, y) = 1{,}200 + 0 + 2xy - 3x^2 + 0$$
$$= 1{,}200 + 2xy - 3x^2$$

(b) To compute the rate of change of f with respect to x when $x = 40$ and $y = 60$, we simply evaluate the function $f_x(x, y)$ when $(x, y) = (40, 60)$. We get

$$f_x(40, 60) = 1{,}200 + 2(40)(60) - 3(40)^2 = 1{,}200 \qquad \square$$

The function f_x that we obtained in part (a) of the preceding example is called the *partial derivative of f with respect to x*. We now define partial derivatives in general for functions of n variables.

PARTIAL DERIVATIVES

Suppose $f(x_1, x_2, \ldots, x_n)$ is a function of n variables. The *partial derivative of f with respect to its jth variable x_j* is denoted by f_{x_j} and is defined to be the function obtained by differentiating f with respect to x_j, treating all the other variables as constants.

No new rules are needed for the computation of partial derivatives. To compute f_{x_j} we simply differentiate f with respect to the single variable x_j, pretending that all the other variables are constants. Consider the following examples.

Example 2.3

Find the partial derivatives f_x, f_y, and f_z if $f(x, y, z) = x^2 + 2xy + yz^3$.

Solution

To compute f_x, we think of f as a function of x and differentiate, treating y and z as constants. We get

$$f_x(x, y, z) = 2x + 2y$$

Similarly, to compute f_y we treat x and z as constants. Hence

$$f_y(x, y, z) = 2x + z^3$$

Finally,

$$f_z(x, y, z) = 3yz^2 \qquad \square$$

Example 2.4

Find the partial derivatives f_x and f_y if $f(x, y) = (2x + y)/(x^2 y)$.

Solution

Using the quotient rule in each case we find that

$$f_x(x, y) = \frac{(x^2y)(2) - (2x + y)(2xy)}{(x^2y)^2} = \frac{-2x - 2y}{x^3y}$$

and

$$f_y(x, y) = \frac{(x^2y)(1) - (2x + y)(x^2)}{(x^2y)^2} = -\frac{2}{xy^2} \qquad \square$$

Example 2.5

Find the partial derivatives g_r and g_s if $g(r, s) = (2rs^2 + r^3)^5$.

Solution

Using the chain rule we get

$$g_r(r, s) = 5(2rs^2 + r^3)^4(2s^2 + 3r^2)$$

and

$$g_s(r, s) = 5(2rs^2 + r^3)^4(4rs) = 20rs(2rs^2 + r^3)^4 \qquad \square$$

Some texts give a more formal definition of partial derivatives. We include it here so that it will look familiar if you happen to encounter it somewhere else. For simplicity, we state the definition for a function of only two variables.

FORMAL DEFINITION OF PARTIAL DERIVATIVES

$$f_x(x, y) = \lim_{\Delta x \to 0} \frac{f(x + \Delta x, y) - f(x, y)}{\Delta x}$$

Notice that this definition of f_x resembles our original definition of the derivative of a function of one variable in Chapter 2. In fact, except for the presence of the variable y that is held constant throughout the limit process, this is precisely the definition of the derivative of f with respect to x. Hence this formal definition agrees with our original definition of f_x. It says that the partial derivative f_x is simply the derivative of f with respect to x, computed when y is held fixed.

An analogous definition can be given for the other partial derivative f_y. See if you can write down the appropriate limit.

When a dependent variable such as z is used to denote a function $f(x_1, x_2, \ldots, x_n)$, the symbol $\partial z/\partial x_j$ is frequently used instead of f_{x_j} to denote the partial derivative of f with respect to x_j. (Sometimes the two notations are combined and we write

$\partial f/\partial x_j$.) This new notation, which is reminiscent of the notation dy/dx for the ordinary derivative of a function of one variable, has certain advantages. For example, as we shall see in Section 3, it leads to a particularly nice formulation of the chain rule for partial derivatives. However, our original notation is more convenient when the partial derivative is to be evaluated numerically. For example, using our original notation, we write $f_x(3, 2)$ to denote the value of the partial derivative f_x when $x = 3$ and $y = 2$. The standard analogous expression when the symbol $\partial f/\partial x$ is used is

$$\left.\frac{\partial f}{\partial x}\right|_{(3,2)}$$

which is clearly more cumbersome.

In the next example, we use partial derivatives to analyze an economic situation in which two commodities that are substitutes for one another compete in the same market.

Example 2.6

A liquor store carries two brands of inexpensive white table wine, one from California and the other from a vineyard in New York. If the California wine sells for x dollars a bottle and the New York wine for y dollars a bottle, the monthly demand for the California wine is given by the function $f(x, y) = 300 - 20x^2 + 30y$ and the monthly demand for the New York wine by the function $g(x, y) = 240 - 12y^2 + 20x$. Analyze the effects of price changes on the demand for the wines.

Solution

The partial derivative $f_x = -40x$ is the rate at which the demand for the California wine changes with respect to its price. Since f_x is *negative* (for all positive values of x), we conclude that the demand for the California wine *decreases* as its price increases.

The partial derivative $f_y = 30$ is the rate at which the demand for the California wine changes with respect to the price of the competing New York wine. The fact that f_y is always *positive* tells us that the demand for the California wine *increases* as the price of its competitor increases.

A similar analysis of the partial derivatives of g leads to analogous conclusions about the demand for the New York wine. □

Partial derivatives can themselves be differentiated. The resulting functions are called *second-order partial derivatives*. Let us introduce notation for the four second-order partial derivatives of a function of two variables.

SECOND-ORDER PARTIAL DERIVATIVES

Suppose $z = f(x, y)$.

The partial derivative of f_x with respect to x is denoted by f_{xx} (or by $\partial^2 z/\partial x^2$).
The partial derivative of f_x with respect to y is denoted by f_{xy} (or by $\partial^2 z/\partial y \, \partial x$).
The partial derivative of f_y with respect to x is denoted by f_{yx} (or by $\partial^2 z/\partial x \, \partial y$).
The partial derivative of f_y with respect to y is denoted by f_{yy} (or by $\partial^2 z/\partial y^2$).

Consider the following example.

Example 2.7

Compute the four second-order partial derivatives of the function $f(x, y) = xy^3 + 5xy^2 + 2x + 1$.

Solution

Since

$$f_x = y^3 + 5y^2 + 2$$

it follows that

$$f_{xx} = 0 \quad \text{and} \quad f_{xy} = 3y^2 + 10y$$

Since

$$f_y = 3xy^2 + 10xy$$

it follows that

$$f_{yx} = 3y^2 + 10y \quad \text{and} \quad f_{yy} = 6xy + 10x \qquad \square$$

The two derivatives f_{xy} and f_{yx} are sometimes called the *mixed* second-order partial derivatives of f. Notice that the mixed partials in Example 2.7 are equal. This is not an accident. It turns out that for virtually all the functions you will encounter in practical work the mixed partial derivatives will be equal. That is, you will get the same answer if you first differentiate f with respect to x and then differentiate the resulting function with respect to y as you would if you perform the differentiation in the opposite order. Here is a precise statement of the situation.

EQUALITY OF THE MIXED PARTIAL DERIVATIVES

If f is a function of two variables x and y, and if f_x, f_y, f_{xy}, and f_{yx} are all continuous, then $f_{xy} = f_{yx}$.

The next example illustrates how information gained from the second-order partial derivatives, as well as from the first-order partials, can be used to analyze a typical situation in economics.

Example 2.8

Suppose the daily output Q of a factory depends on the amount K of capital invested in the plant and in equipment, and also on the size L of the labor force, measured in man-hours of work. In economics, the partial derivatives $\partial Q/\partial K$ and $\partial Q/\partial L$ are known as the *marginal products* of K and L, respectively.

(a) Give economic interpretations of the two marginal products.

(b) Give an economic interpretation of the sign of the second-order partial derivative $\partial^2 Q/\partial L^2$.

Solution

(a) The marginal product of L, $\partial Q/\partial L$, is the rate at which output Q changes with respect to labor L for a fixed level K of capital investment. Hence $\partial Q/\partial L$ is (approximately) the change in output that would result if capital investment were held fixed and labor were increased by 1 man-hour. The marginal product of K has an analogous interpretation.

(b) If $\partial^2 Q/\partial L^2$ is negative, the marginal product of L *decreases* as L increases. This implies that for a fixed level of capital investment the effect on the daily output due to the addition of 1 man-hour of labor is greater when the work force is small than when the work force is large. Similarly, if $\partial^2 Q/\partial L^2$ is positive, the marginal product of L *increases* as L increases. □

We conclude this section by giving a simple geometric interpretation of the partial derivatives of a function of two variables. This geometric point of view will be useful in Section 5 when we study relative extrema of such functions.

For a fixed number y_0, the points (x, y_0, z) form a vertical plane whose equation is $y = y_0$. Suppose $z = f(x, y)$ and let y be held fixed at $y = y_0$. The corresponding points $(x, y_0, f(x, y_0))$ form a curve in three-dimensional space that is the intersection of the surface $z = f(x, y)$ with the plane $y = y_0$ (Figure 5). At each point $(x, y_0, f(x, y_0))$, the partial derivative $\partial z/\partial x$ is simply the slope of the tangent to this curve in the plane $y = y_0$.

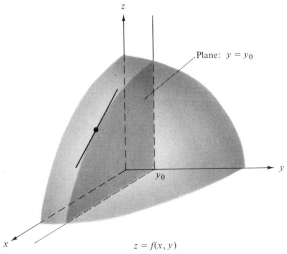

Figure 5

Similarly, if x is held fixed at $x = x_0$, the corresponding points $(x_0, y, f(x_0, y))$ form a curve that is the intersection of our surface with the vertical plane $x = x_0$ (Figure 6). Hence, at each point $(x_0, y, f(x_0, y))$, the partial derivative $\partial z/\partial y$ is the slope of the tangent to this curve in the plane $x = x_0$.

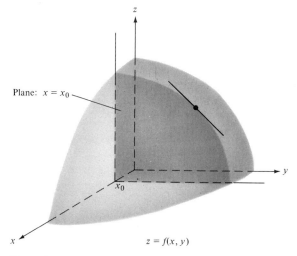

Figure 6

Problems

1. For each of the following functions, compute all the (first-order) partial derivatives.

   (a) $f(x, y) = 2xy^5 + 3x^2y + x^2 + 5$
   (b) $z = x\sqrt{y} + 3x^{-2/3}y^5 + 6y$
   (c) $f(x, y, z) = x^2yz + 3xy^2 + xz^5$
   (d) $z = (3x + 2y)^5$
   (e) $f(u, v) = (u + v)/(u - v)$
   (f) $z = (xy + 1)/xy^2$
   (g) $w = \sqrt{2xy^2 + z^3}$
   (h) $f(w, x, y, z) = \sqrt{xy + 1 - 2xz^2 + w/y}$
   (i) $f(x, y) = xe^{xy} - y^2 \log(2x + 1)$
   (j) $z = (2x^3y - 7)^5 e^{2x+y}$

2. For each of the following functions, compute the second-order partial derivatives f_{xx}, f_{yy}, f_{xy}, and f_{yx}.

   (a) $f(x, y) = 5x^4y^3 - 3x^2y^5 + 2xy + 5$ (b) $f(x, y) = (x + 1)/(y - 1)$
   (c) $f(x, y) = e^{2x^2y}$ (d) $f(x, y) = \log(x^2 + y^2)$
   (e) $f(x, y) = \sqrt{x^2 + y^2}$ (f) $f(x, y) = x^2ye^x$

3. A manufacturer has determined that the number of toasters he will sell each month is given by a function $f(x, y)$, where x is the amount of money (measured in thousands of dollars) spent on advertising during the month and y is the selling price (in dollars) of the toasters. Give economic interpretations of the partial derivatives f_x and f_y.

4. At a certain factory, the daily output of a certain commodity is approximately $2K^{1/2}L^{1/3}$ units, where K denotes the capital investment measured in thousands of dollars and L the size of the labor force measured in man-hours. Suppose the current capital investment is $900,000 and that 1,000 man-hours of labor are used each day. Estimate the effect that an additional capital investment of $1,000 would have on the daily output if the size of the labor force is not changed.

5. Suppose the daily output Q of a factory depends on the amount K of capital investment measured in thousands of dollars and on the size L of the labor force measured in man-hours of work.
    (a) Give an economic interpretation of the sign of the second-order partial derivative $\partial^2 Q/\partial K^2$.
    (b) Give an economic interpretation of the sign of the mixed partial derivative $\partial^2 Q/\partial K\, \partial L$.

6. Suppose the daily output Q of a factory depends on the amount K of capital investment and on the size L of the labor force measured in man-hours of work. A *law of diminishing returns* states that in certain circumstances, there is a value L_0 such that the marginal product of labor will be increasing for $L < L_0$ and decreasing for $L > L_0$.
    (a) Discuss the economic factors that might account for this phenomenon.
    (b) Translate this law of diminishing returns into a statement about the sign of a certain second-order partial derivative.

7. A publisher has found that in a certain city each of his salesmen will sell approximately

$$\frac{r^2}{2{,}000p} + \frac{s^2}{100} - s \quad \text{sets of encyclopedias a month}$$

where s denotes the total number of salesmen he has working in the city, p the price of a set of books, and r the amount of money spent each month on local advertising. Suppose the publisher currently employs 10 salesmen, spends $6,000 a month on local advertising, and sells the encyclopedias for $800 a set. Suppose further that the cost of producing the encyclopedias is $80 a set and that each salesman earns $600 a month. Use an appropriate partial derivative to estimate the change in the publisher's total monthly *profit* that would result if one more local salesman were hired.

8. Two commodities are said to be *substitutes* if the demand Q_1 for the first commodity increases as the price p_2 of the second increases, and if the demand Q_2 for the second commodity increases as the price p_1 of the first increases.
   (a) Give an example of a pair of substitute commodities.
   (b) If two commodities are substitutes, what must be true of the partial derivatives $\partial Q_1/\partial p_2$ and $\partial Q_2/\partial p_1$?
   (c) Suppose the demand functions for two commodities are
   $$Q_1 = 3{,}000 + \frac{400}{p_1 + 3} + 500e^{0.8p_2}$$
   and
   $$Q_2 = 2{,}000 - 100p_2 + \frac{500}{p_1 + 4}$$
   Are the commodities substitutes?

9. A bicycle dealer has found that if the price of 10-speed bicycles is x dollars and the price of gasoline is $10y$ cents per gallon, he will sell approximately $200 - 10\sqrt{x} + 4(y + 2)^{3/2}$ bicycles a month. Currently the bicycles sell for $121 apiece and the price of gasoline is 70 cents a gallon. Use an appropriate partial derivative to estimate the effect on the monthly sale of bicycles if the price of gasoline is increased to 80 cents a gallon.

10. Let f be a function of two variables x and y. Using the formal definition of f_x on page 332 as a model, define f_y as the limit of a certain quotient.

3. THE CHAIN RULE

In Example 1.4, we examined a situation in which the demand Q for a certain commodity depended on both its own price x and the price y of a competing commodity. The prices x and y were expected to increase with time t, and therefore we could regard Q as a function of t. Suppose we wanted to determine the rate of change of Q with respect to t. One approach would be to write Q explicitly as a function of t and simply compute the derivative dQ/dt directly. An alternative approach (which is often more efficient) is to compute dQ/dt using a generalized form of the chain rule. Let us try to discover what a reasonable chain rule might be in this context.

A change in t generates a change in x and a change in y, both of which induce changes in Q. Suppose we hold y fixed. Using the chain rule for functions of a single variable, we conclude that

$$\text{Rate of change of } Q \text{ with respect to } t \text{ for fixed } y = \frac{\partial Q}{\partial x}\frac{dx}{dt}$$

Similarly,

Rate of change of Q with respect to t for fixed $x = \dfrac{\partial Q}{\partial y}\dfrac{dy}{dt}$

You might suspect that the total rate of change of Q with respect to t is the *sum* of these two "partial" rates. That is, you might conjecture that the chain rule we are seeking can be expressed by the equation

$$\frac{dQ}{dt} = \frac{\partial Q}{\partial x}\frac{dx}{dt} + \frac{\partial Q}{\partial y}\frac{dy}{dt}$$

Let us test this conjecture using a specific example.

Example 3.1

Suppose $Q(x, y) = 300 - 20x^2 + 30y$, where $x = 2 + 0.05t$ and $y = 2 + 0.1\sqrt{t}$.

(a) Think of Q as a function of t and use our provisional statement of the chain rule to compute dQ/dt.

(b) Check the answer to part (a) by first writing Q explicitly as a function of t and then computing dQ/dt directly.

Solution

(a) According to our conjecture,

$$\frac{dQ}{dt} = \frac{\partial Q}{\partial x}\frac{dx}{dt} + \frac{\partial Q}{\partial y}\frac{dy}{dt}$$
$$= -40x(0.05) + 30(0.05t^{-1/2})$$

But $x = 2 + 0.05t$, so

$$\frac{dQ}{dt} = -40(2 + 0.05t)(0.05) + 30(0.05t^{-1/2})$$
$$= -4 - 0.1t + 1.5t^{-1/2}$$

(b) On the other hand,

$$Q = 300 - 20(2 + 0.05t)^2 + 30(2 + 0.1\sqrt{t})$$
$$= 280 - 4t - 0.05t^2 + 3\sqrt{t}$$

Hence,

$$\frac{dQ}{dt} = -4 - 0.1t + 1.5t^{-1/2}$$

which is precisely the answer we obtained in part (a). □

It turns out that our conjecture about the chain rule is correct. Here is a more general statement of the rule.

THE CHAIN RULE

Suppose z is a function of n variables $x_1, x_2, \ldots, x_n$, each of which is a function of a variable t. Then z can be regarded as a function of the single variable t, and

$$\frac{dz}{dt} = \frac{\partial z}{\partial x_1}\frac{dx_1}{dt} + \frac{\partial z}{\partial x_2}\frac{dx_2}{dt} + \cdots + \frac{\partial z}{\partial x_n}\frac{dx_n}{dt}$$

The chain rule is actually a statement about the derivative of a composite function. For practice, identify this composite function and rewrite the chain rule using functional notation. (See Problem 4.)

Example 3.2

Find dR/dt, if $R = 3xy^2 + 2xz + 5y$, $x = t^2 + 1$, $y = 2t$, and $z = 1 - t^2$.

Solution

We first apply the chain rule to get

$$\frac{dR}{dt} = \frac{\partial R}{\partial x}\frac{dx}{dt} + \frac{\partial R}{\partial y}\frac{dy}{dt} + \frac{\partial R}{\partial z}\frac{dz}{dt}$$

$$= (3y^2 + 2z)2t + (6xy + 5)(2) + 2x(-2t)$$

We then substitute the appropriate functions of t for x, y, and z and conclude that

$$\frac{dR}{dt} = (12t^2 + 2 - 2t^2)2t + (12t^3 + 12t + 5)(2) + (2t^2 + 2)(-2t)$$

$$= 40t^3 + 24t + 10 \qquad \square$$

The chain rule that we have just introduced and illustrated is actually a special case of the following general result.

THE GENERAL CHAIN RULE

Suppose z is a function of n variables $x_1, x_2, \ldots, x_n$, each of which is a function of m variables $t_1, t_2, \ldots, t_m$. Then z can be regarded as a function of the m variables $t_1, t_2, \ldots, t_m$, and for each $j = 1, 2, \ldots, m$,

$$\frac{\partial z}{\partial t_j} = \frac{\partial z}{\partial x_1}\frac{\partial x_1}{\partial t_j} + \frac{\partial z}{\partial x_2}\frac{\partial x_2}{\partial t_j} + \cdots + \frac{\partial z}{\partial x_n}\frac{\partial x_n}{\partial t_j}$$

Notice that in the formula for $\partial z/\partial t_j$ there is one term for each of the n original variables of the function z. The use of this general form of the chain rule is illustrated in the following examples.

Example 3.3

Let $w = xe^y + z^2$, where $x = 2r + s$, $y = s^2$, and $z = r^2s$. Think of w as a function of the two variables r and s and compute $\partial w/\partial r$.

Solution

By the chain rule,

$$\frac{\partial w}{\partial r} = \frac{\partial w}{\partial x}\frac{\partial x}{\partial r} + \frac{\partial w}{\partial y}\frac{\partial y}{\partial r} + \frac{\partial w}{\partial z}\frac{\partial z}{\partial r}$$

$$= e^y(2) + xe^y(0) + 2z(2rs)$$

$$= 2e^y + 4zrs$$

Substituting $y = s^2$ and $z = r^2s$, we conclude that

$$\frac{\partial w}{\partial r} = 2e^{s^2} + 4r^3s^2 \qquad \square$$

Example 3.4

Suppose that when apples sell for c cents a pound and bakers earn b dollars an hour the price of an apple pie at a certain chain of supermarkets is $0.25c^{1/3}b^{1/2}$ dollars. The supermarkets will be able to sell approximately $867/(p + 0.2)$ apple pies a month when the price is p dollars a pie. Currently, the price of apples is stable at 27 cents a pound. Bakers have been earning \$4 an hour, but a new contract, negotiated on the eve of a threatened nationwide strike, grants them an immediate raise of \$1 an hour. Use calculus to estimate the effect of this raise on the sale of apple pies by the supermarkets.

Solution

Let Q denote the number of pies sold by the supermarkets during a month. Then $Q = 867/(p + 0.2)$, where $p = 0.25c^{1/3}b^{1/2}$. Hence Q can be regarded as a function of the two variables c and b. To estimate the change in Q due to the unit increase in b, we simply evaluate $\partial Q/\partial b$ when $c = 27$ and $b = 4$.

Since Q was originally a function of the single variable p, the chain rule reduces to the simple formula

$$\frac{\partial Q}{\partial b} = \frac{dQ}{dp}\frac{\partial p}{\partial b}$$

Thus,

$$\frac{\partial Q}{\partial b} = \frac{-867}{(p + 0.2)^2} 0.125c^{1/3}b^{-1/2}$$

When $c = 27$ and $b = 4$,

$$p = 0.25(3)(2) = 1.5$$

Hence,

$$\frac{\partial Q}{\partial b} = \frac{-867}{(1.7)^2}(0.125)(3)(0.5) = -56.25$$

That is, the supermarkets will sell approximately 56 fewer pies a month after the bakers' pay raise goes into effect. □

Problems

1. In each of the following cases, use the chain rule to find dz/dt. Check your answer by writing z explicitly as a function of t and then differentiating directly with respect to t.

   (a) $z = 3x^2 + y^4$, $x = 5t + 1$, $y = t^2$
   (b) $z = 2x_1^3 + x_1 x_2 + 5x_3$, $x_1 = -t^2$, $x_2 = 1 - 3t$, $x_3 = t^2 - 4t + 1$
   (c) $z = (r+s)/(r-s)$, $r = t^3 + 1$, $s = 1 - t^3$
   (d) $z = e^{2xy} + x \log y$, $x = 3t^2$, $y = 1 - t$
   (e) $z = 3xy + xy^2 + y/x$, $x = e^t$, $y = \log t$

2. Use the chain rule to find the indicated derivatives.

   (a) $z = 2x^4 + 3y^2 + xy$, $x = 2u + v$, $y = v^2 - u$; $\partial z/\partial u$, $\partial z/\partial v$
   (b) $z = wxy$, $w = t^2$, $x = 2t$, $y = 1 - t$; dz/dt
   (c) $f(t) = t^2 + 2t + 1$, $t = x - 2xy^3$; f_x, f_y
   (d) $f(x, y) = x/y^2$, $x = e^{2r+s}$, $y = e^{2r-s}$; f_r, f_s
   (e) $z = (3t^2 + 2t + 5)^6$, $t = 2xy^3$; $\partial z/\partial x$, $\partial z/\partial y$
   (f) $z = e^{xy}$, $x = 2t + 1$, $y = 1 - s^2$; $\partial z/\partial t$, $\partial z/\partial s$

*3. Suppose $z = f(x, y)$, $x = at$, and $y = bt$, where a and b are constants. Think of z as a function of t and find an expression for the second derivative d^2z/dt^2 in terms of the constants a and b and the second-order partial derivatives f_{xx}, f_{yy}, and f_{xy}.

4. Consider the composite function

$$h(t) = f(g_1(t), g_2(t), \ldots, g_n(t))$$

formed from the functions $f(x_1, x_2, \ldots, x_n)$, $g_1(t), g_2(t), \ldots$, and $g_n(t)$. Use the chain rule to express the derivative dh/dt in terms of the derivatives of the functions $g_1, g_2, \ldots, g_n$ and the partial derivatives of the function f.

5. A liquor store carries two brands of inexpensive white table wine, one from California and the other from New York. If the California wine sells for x dollars a bottle and the New York wine for y dollars a bottle, the monthly demand for the California wine is approximately $300 - 20x^2 + 30y$ bottles. It is expected that t months from now the price of a bottle of the California wine will be $2 + 0.05t$ dollars and the price of a bottle of the New York wine will be $2 + 0.1\sqrt{t}$ dollars.

   (a) Use the chain rule to find an expression for the rate at which the demand for the California wine will be changing with respect to time t months from now.

   (b) At what rate will the demand for the California wine be changing 9 months from now?

6. A bicycle dealer estimates that if at the beginning of a month the price of 10-speed bicycles is x dollars and the price of gasoline is $10y$ cents per gallon he will sell approximately $200 - 10\sqrt{x} + 4(y + 1)^{3/2}$ bicycles that month. It is expected that t months after March 1, the bicycles will be selling for $106 + 5t$ dollars apiece and the price of gasoline will be $10(5 + \sqrt{3t})$ cents a gallon. Use the chain rule to estimate the difference between the number of bicycles that will be sold in July and the number that will be sold in June.

4. LEVEL CURVES

There are some practical problems in which one is interested in the possible combinations of variables x and y for which a function $f(x, y)$ will be equal to a certain constant. For example, a manufacturer whose output depends on the numbers of skilled and unskilled workers employed at his plant may want to determine the possible combinations of skilled and unskilled workers that will result in a certain desired level of output.

The combinations of x and y for which $f(x, y)$ is equal to a fixed number can often be represented geometrically as the points on a curve in the xy plane. Such a curve is said to be a *level curve* of f. The level curve $f(x, y) = C$ is the projection onto the xy plane of the curve formed by intersecting the surface $z = f(x, y)$ with the horizontal plane $z = C$. The relationship between a surface $z = f(x, y)$ and a corresponding level curve $f(x, y) = C$ is shown in Figure 7.

LEVEL CURVES

Suppose $z = f(x, y)$ and let C be a constant. The points (x, y) for which $f(x, y) = C$ form a curve in the xy plane that is said to be a *level curve* of f.

Consider the following examples.

344 FUNCTIONS OF SEVERAL VARIABLES

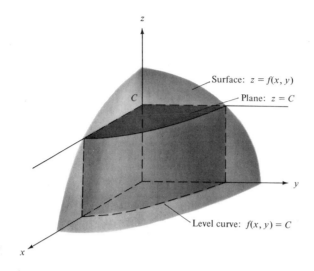

Figure 7

Example 4.1

Sketch the level curve $f(x, y) = 4$ if $f(x, y) = x^2 - y$.

Solution

We rewrite the equation $f(x, y) = 4$ as

$$x^2 - y = 4$$

or

$$y = x^2 - 4 = (x - 2)(x + 2)$$

The graph of this polynomial in the xy plane (Figure 8) is the level curve we are seeking. This curve consists of all the points (x, y) for which $f(x, y) = 4$.

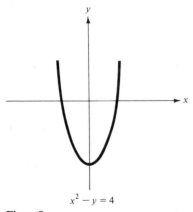

Figure 8

Example 4.2

Find the level curve of the function $f(x, y) = xy$ that passes through the point $(2, 3)$.

Solution

The equations of the level curves of f are of the form $xy = C$ or $y = C/x$, where C is a constant. To find the equation of the particular level curve that passes through $(2, 3)$, we must find the corresponding value of C. Substituting $x = 2$ and $y = 3$ into the equation of the level curve we find that $C = 6$. Hence the desired level curve (Figure 9) is the graph of the function $y = 6/x$.

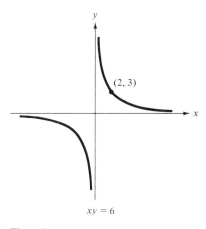

Figure 9

Frequently, the level curve of a function of two variables x and y is the graph of a certain function of the single variable x. For instance, in Example 4.1 the level curve $f(x, y) = 4$ was the graph of the function $y = x^2 - 4$, and in Example 4.2 the level curve $f(x, y) = 6$ was the graph of the function $y = 6/x$. Usually, the functions of x that arise in this way can be differentiated. The resulting derivative dy/dx gives the slope of the tangent to the level curve or, equivalently, the rate of change of y with respect to x on the level curve. In certain practical problems, this derivative gives us useful information. Here is a typical example from economics.

Example 4.3

Using $10x$ man-hours of skilled labor and $10y$ man-hours of unskilled labor, a manufacturer can produce $f(x, y) = x^2 y$ units of a certain commodity each day. Currently the manufacturer uses 160 man-hours of skilled labor and 320 man-hours of unskilled labor a day but wants to increase by 10 the number of man-hours of skilled labor used each day. Estimate the corresponding change that the manufacturer should make in the level of unskilled labor so that the total daily output will remain the same.

346 FUNCTIONS OF SEVERAL VARIABLES

Solution

Currently, $x = 16$ and $y = 32$, so the present level of production is $f(16, 32) = 8{,}192$ units per day. The corresponding level curve $f(x, y) = 8{,}192$ is the graph of the function $y = 8{,}192/x^2$ which is sketched in Figure 10 for positive values of x.

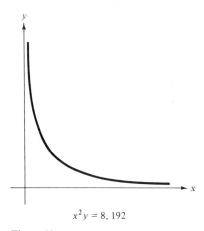

$x^2 y = 8{,}192$

Figure 10

This curve shows the relationship between the input x of skilled labor and the input y of unskilled labor that must hold if production is to be maintained at its current level. Our goal is to estimate the change in y along this curve corresponding to a unit increase in x from $x = 16$ to $x = 17$. Since the derivative

$$\frac{dy}{dx} = -\frac{16{,}384}{x^3}$$

is the rate of change of y with respect to x, the value of this derivative when $x = 16$ should be an approximation to the desired change in y. That is,

$$\text{Change in } y \approx -\frac{16{,}384}{(16)^3} = -4$$

Hence to compensate for the increase of 10 man-hours of skilled labor, the manufacturer should decrease the level of unskilled labor by approximately 40 man-hours a day. □

In the preceding example, we obtained the slope dy/dx of a level curve by first solving the equation $f(x, y) = C$ explicitly for y in terms of the variable x, and then differentiating the resulting expression for y with respect to x. Unfortunately, it is sometimes difficult or even impossible to solve the equation $f(x, y) = C$ explicitly for y. In such cases, the corresponding derivative dy/dx may be found either by implicit differentiation (as in Chapter 2, Section 5) or by the application of the following formula involving the partial derivatives of the original function f.

THE SLOPE OF A LEVEL CURVE

If the level curve $f(x, y) = C$ is the graph of a differentiable function of x, the corresponding derivative dy/dx is given by the formula

$$\frac{dy}{dx} = -\frac{f_x}{f_y}$$

Before giving a mathematical proof of this formula, let us illustrate how it is used. In our first example, we consider a level curve whose equation can be solved explicitly for y and we compute dy/dx by all three methods.

Example 4.4

Let $f(x, y) = xy$ and find dy/dx for the level curve $f(x, y) = C$

(a) By solving for y in terms of x and differentiating the resulting expression

(b) By differentiating the equation $f(x, y) = C$ implicitly

(c) By applying the formula.

Solution

(a) The level curve in question is the graph of the equation $y = C/x$. Hence $dy/dx = -C/x^2$.

(b) We differentiate both sides of the equation

$$xy = C$$

with respect to x, keeping in mind that y is really a function of x. Using the product rule we get

$$x\frac{dy}{dx} + y = 0$$

or

$$\frac{dy}{dx} = -\frac{y}{x}$$

Notice that this agrees with the answer in part (a), since $y = C/x$.

(c) Since $f_x = y$ and $f_y = x$, the formula tells us that

$$\frac{dy}{dx} = -\frac{f_x}{f_y} = -\frac{y}{x}$$

which is the same answer we obtained in part (b). □

Example 4.5

Let $f(x, y) = x^2 + 2xy + y^3$ and find dy/dx for the level curve $f(x, y) = C$.

Solution

In this case, there is no obvious way to solve for y explicitly in terms of x. Hence we must either differentiate implicitly or use the formula.
    Differentiating the equation

$$x^2 + 2xy + y^3 = C$$

implicitly with respect to x we get

$$2x + 2x\frac{dy}{dx} + 2y + 3y^2\frac{dy}{dx} = 0$$

or

$$\frac{dy}{dx} = -\frac{2x + 2y}{2x + 3y^2}$$

Using the formula we get the same answer,

$$\frac{dy}{dx} = -\frac{f_x}{f_y} = -\frac{2x + 2y}{2x + 3y^2} \qquad \square$$

Did you notice that in the preceding examples the expressions for dy/dx that were obtained by formula and by implicit differentiation did not contain the constant C that was in the original definition of the level curve? This may have disturbed you, since the numerical value of the slope dy/dx at a point on a specific level curve $f(x, y) = C$ ought to depend on C. To see how the computation of an actual slope depends on the constant C, consider the following example.

Example 4.6

Suppose $f(x, y) = xy$ and compute the derivative dy/dx when $x = 2$:

(a) For the level curve $f(x, y) = 1$.
(b) For the level curve $f(x, y) = 4$.

Solution

From Example 4.4 we know that for *any* level curve of this function, the derivative is given by the formula $dy/dx = -y/x$. To use this formula to compute dy/dx at a particular point on a particular level curve, we need the x and y coordinates of that point.

(a) When $x = 2$, we find the y coordinate of the corresponding point on the level curve $f(x, y) = 1$ by substituting $x = 2$ and $C = 1$ into the equation $xy = C$. We get $y = \frac{1}{2}$. Hence,

$$\frac{dy}{dx} = -\frac{\frac{1}{2}}{2} = -\frac{1}{4}$$

That is, when $x = 2$, the slope of the line tangent to the level curve $f(x, y) = 1$ is equal to $-\frac{1}{4}$.

(b) When $x = 2$, the y coordinate of the corresponding point on the level curve $f(x, y) = 4$ is $y = 2$, and so the slope at this point is

$$\frac{dy}{dx} = -1$$

A graph showing two level curves and corresponding tangent lines when $x = 2$ is sketched in Figure 11.

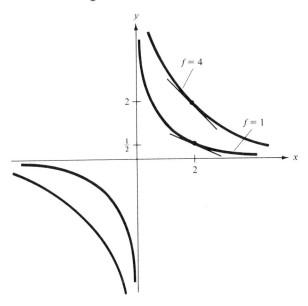

Figure 11

When the equation defining a level curve cannot be solved explicitly for y, use of the formula to find dy/dx is often slightly more efficient than the method of implicit differentiation. However, implicit differentiation is a more versatile technique that can be used in situations to which this particular formula does not apply. Hence it is worth mastering the technique of implicit differentiation, even though a prefabricated formula for dy/dx is available. The formula, on the other hand, is particularly useful in certain theoretical applications when an actual algebraic expression for the function $f(x, y)$ is not given. We shall use the formula in Section 6, when we develop a method for solving constrained optimization problems.

We conclude this section by giving a mathematical proof of the formula for dy/dx. The proof, based on the chain rule, is short but subtle.

Suppose $f(x, y)$ is a differentiable function of the two variables x and y. For any constant C the equation $f(x, y) = C$ implicitly defines y as a function of x. Hence we may regard f as a function of the single variable x. If we differentiate both sides of the equation $f = C$ with respect to x, we get

$$\frac{df}{dx} = 0$$

But, by the chain rule,

$$\frac{df}{dx} = \frac{\partial f}{\partial x}\frac{dx}{dx} + \frac{\partial f}{\partial y}\frac{dy}{dx}$$

$$= \frac{\partial f}{\partial x} + \frac{\partial f}{\partial y}\frac{dy}{dx}$$

Substituting this expression into the equation $df/dx = 0$ we conclude that

$$\frac{\partial f}{\partial x} + \frac{\partial f}{\partial y}\frac{dy}{dx} = 0$$

or

$$\frac{dy}{dx} = -\frac{\partial f/\partial x}{\partial f/\partial y} = -\frac{f_x}{f_y}$$

Problems

1. For each of the following functions, sketch the indicated level curves.
   (a) $f(x, y) = x + 2y; f = 1, f = 2, f = 3$
   (b) $f(x, y) = x^2 + y; f = 0, f = 4, f = 9$
   (c) $f(x, y) = x^2 - 4x - y; f = -4, f = 5$
   (d) $f(x, y) = x/y; f = -2, f = 2$
   (e) $f(x, y) = xy; f = 1, f = -1, f = 2, f = -2$
   (f) $f(x, y) = ye^x; f = 0, f = 1$
   (g) $f(x, y) = xe^y; f = 1, f = e$

2. For each of the following functions, find an equation of the level curve that passes through the given point.
   (a) $f(x, y) = x^2 + 2x + y; (1, 3)$
   (b) $f(x, y) = x^2 + y^2 + 1; (2, 0)$
   (c) $f(x, y) = e^y/x; (e, 0)$
   (d) $f(x, y) = x \log y; (\sqrt{2}, 1)$

3. For each of the following functions, use implicit differentiation to find an expression for dy/dx if $f(x, y) = C$ for some constant C. Check your answer by using the formula on page 347.
   (a) $f(x, y) = x^2 - 4x - y$
   (b) $f(x, y) = x/y$
   (c) $f(x, y) = xe^y$
   (d) $f(x, y) = x^2y + 2y^3 - 3x - 2y$
   (e) $f(x, y) = x + 1/y - 2xy^2$
   (f) $f(x, y) = x \log y$

4. For each of the following functions, find the slope of the indicated level curve for the specified value of x.
   (a) $f(x, y) = x^2 - y^3; f = -2, x = 5$
   (b) $f(x, y) = x^2 + 2xy + y^3; f = 8, x = 0$
   (c) $f(x, y) = x^2y - 3xy + 5; f = 9, x = 1$
   (d) $f(x, y) = (x^2 + y)^3; f = 8, x = -1$
   (e) $f(x, y) = e^y/x; f = 2, x = \frac{1}{2}$

5. Using $10x$ man-hours of skilled labor and $10y$ man-hours of unskilled labor, a manufacturer can produce $f(x, y) = 10x\sqrt{y}$ units of a certain commodity each day. Currently the manufacturer uses 300 man-hours of skilled labor and 360 man-hours of unskilled labor a day but wants to increase by 10 the number of man-hours of skilled labor used each week. Estimate the corresponding change that the manufacturer should make in the level of unskilled labor so that the total daily output will remain the same.

6. Suppose the manufacturer in Problem 5 currently uses 300 man-hours of skilled labor and 360 man-hours of unskilled labor a day but wants to increase by 10 the number of man-hours of *unskilled* labor used each week. Estimate the corresponding change that should be made in the level of skilled labor so that the total daily output will remain the same. (*Hint*: Think of x as a function of y and use the derivative dx/dy.)

7. Suppose the equation $f(x, y) = C$ defines x as a differentiable function of y. Find a formula for dx/dy in terms of the partial derivatives of f. Justify your answer.

5. RELATIVE MAXIMA AND MINIMA

For functions of several variables, partial derivatives play a role in the study of relative maxima and minima that is analogous to the role of ordinary derivatives for functions of a single variable. In our analysis of relative extrema, we shall restrict our attention to functions of *two* variables, the case in which the theory is least complicated and in which the geometric representation of functions as surfaces will help us visualize our results. We begin our discussion with the formal definitions of relative extrema.

RELATIVE EXTREMA

A function $f(x, y)$ has a *relative maximum* at (a, b) if $f(x, y) < f(a, b)$ for all points (x, y) [except $(x, y) = (a, b)$] in some neighborhood of (a, b).

A function $f(x, y)$ has a *relative minimum* at (a, b) if $f(x, y) > f(a, b)$ for all points (x, y) [except $(x, y) = (a, b)$] in some neighborhood of (a, b).

In geometric terms, a relative maximum of f is a peak; a point that is higher than any nearby point on the surface $z = f(x, y)$. A relative minimum is the bottom of a valley; a point that is lower than any nearby point on the surface. For example, when $(x, y) = (a, b)$, the function sketched in Figure 12a has a relative minimum while the function in Figure 12b has a relative maximum.

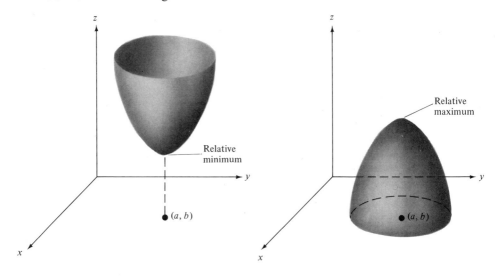

Figure 12

Using the geometric interpretation of partial derivatives that we discussed at the end of Section 2, we can derive an important relationship between the partial derivatives of a function and its relative extrema. Suppose $f(x, y)$ has a relative maximum at (a, b). Then the curve formed by intersecting the surface $z = f(x, y)$ with the vertical plane $y = b$ has a relative maximum and hence a horizontal tangent when $x = a$ (Figure 13). Since the partial derivative $f_x(a, b)$ is the slope of this tangent, it follows that $f_x(a, b) = 0$. Similarly, the curve formed by intersecting the surface with the plane $x = a$ has a relative maximum when $y = b$, and so $f_y(a, b) = 0$.

A similar argument shows that if f has a relative minimum at (a, b), then both $f_x(a, b)$ and $f_y(a, b)$ must be zero.

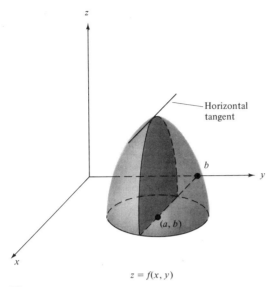

Figure 13

Points (a, b) for which both $f_x(a, b)$ and $f_y(a, b)$ are zero are said to be *critical points* of f.

CRITICAL POINTS

A point (a, b) for which both $f_x(a, b) = 0$ and $f_y(a, b) = 0$ is said to be a *critical point* of f. If the first-order partial derivatives of f are defined at all points in some region in the xy plane, then all the relative extrema of f in the region occur at critical points.

Although all the relative extrema of a function must occur at critical points, not all the critical points of a function are necessarily relative extrema. Consider, for example, the function $f(x, y) = y^2 - x^2$ whose graph (which resembles a saddle) is sketched in Figure 14. Roughly speaking, $f_x(0, 0) = 0$ in this case because the surface has a relative *maximum* (and hence a horizontal tangent) "in the x direction." On the other hand, $f_y(0, 0) = 0$ because the surface has a relative *minimum* "in the y direction." In order for a critical point to be a relative extremum, the nature of the extremum must be the same in *all* directions.

A critical point that is neither a relative maximum nor a relative minimum is called a *saddle point*. For a few very simple functions, we can use geometric arguments to decide whether a given critical point is a relative extremum or a saddle point. In most cases, however, we shall have to use a more formal technique. The following procedure for classifying critical points is the two-variable analog of the second derivative test for functions of a single variable that we developed in Chapter 2, Section 9.

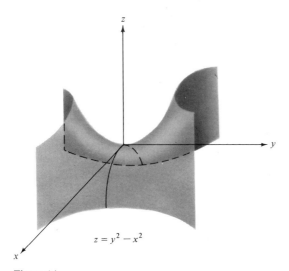

$z = y^2 - x^2$

Figure 14

THE SECOND DERIVATIVE TEST

Suppose f is a function of two variables x and y, and that all the second-order partial derivatives of f are continuous. Let

$$D = f_{xx} f_{yy} - (f_{xy})^2$$

Suppose (a, b) is a critical point of f.

(a) If $D(a, b) < 0$, then f has a saddle point at (a, b).

(b) If $D(a, b) > 0$ and $f_{xx}(a, b) < 0$, then f has a relative maximum at (a, b).

(c) If $D(a, b) > 0$ and $f_{xx}(a, b) > 0$, then f has a relative minimum at (a, b).

(d) If $D(a, b) = 0$, the test is inconclusive and f may have either a relative extremum or a saddle point at (a, b).

The sign of the quantity D that appears in the second derivative test tells us whether the function has a relative extremum at (a, b). Essentially, D tests the "consistency" of the function at (a, b). If $D(a, b)$ is positive, then the function either has a relative maximum *in all directions* or a relative minimum *in all directions*. To decide which, we may restrict our attention to *one* direction, the x direction, and use the second derivative test for functions of a single variable to conclude that f has a relative maximum at (a, b) if $f_{xx}(a, b)$ is negative and a relative minimum if $f_{xx}(a, b)$ is positive.

On the other hand, if $D(a, b)$ is negative, the behavior of the graph of f at (a, b) is not the same for all directions. In some directions, for example, the graph may have a relative maximum, while in other directions it may have a relative minimum or even an inflection point. Hence when $D(a, b)$ is negative, we can conclude immediately that f has a saddle point at (a, b) and there is nothing further to check.

We shall not give a proof of the second derivative test. (The proof is rather complicated and involves ideas that are beyond the scope of this text.) Instead, let us check that this test works when applied to the two functions whose graphs were sketched in Section 1.

Example 5.1

Classify the critical points of the function $f(x, y) = x^2 + y^2$.

Solution

Since $f_x = 2x$ and $f_y = 2y$, the only critical point of f is $(0, 0)$. To test this point, we compute the second-order partial derivatives and find that $f_{xx} = 2$, $f_{yy} = 2$, and $f_{xy} = 0$. Hence $D(x, y) = 4$ for all points (x, y) and, in particular, $D(0, 0)$ is positive. Hence f has a relative extremum at $(0, 0)$. Since $f_{xx}(0, 0) = 2$, which is positive, we conclude that the relative extremum at $(0, 0)$ is in fact a relative minimum. The graph of f is sketched in Figure 15.

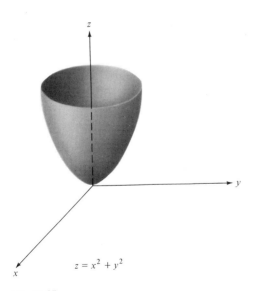

Figure 15

Example 5.2

Classify the critical points of the function $f(x, y) = y^2 - x^2$.

Solution

Since $f_x = -2x$ and $f_y = 2y$, the only critical point of f is $(0, 0)$. To test this point, we compute the second-order partial derivatives and find that $f_{xx} = -2$, $f_{yy} = 2$, and $f_{xy} = 0$. Hence $D(x, y) = -4$ for all points (x, y) and, in particular, $D(0, 0)$ is negative. Hence f has a saddle point at $(0, 0)$, as shown in the graph in Figure 16.

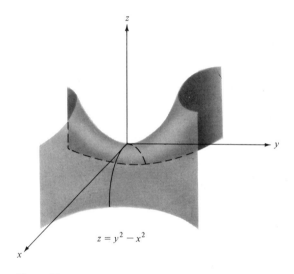

Figure 16

The ease with which we were able to locate the critical points in the two preceding examples is not typical. In the next example, the algebra leading to the identification of the critical points is slightly more involved.

Example 5.3

Classify the critical points of the function $f(x, y) = x^3 - y^3 + 6xy$.

Solution

Since $f_x = 3x^2 + 6y$ and $f_y = -3y^2 + 6x$, we find the critical points of f by solving simultaneously the two equations

$$3x^2 + 6y = 0$$

and

$$-3y^2 + 6x = 0$$

From the first equation we find that

$$y = -\frac{x^2}{2}$$

Substituting this expression for y into our second equation, we get

$$-\frac{3x^4}{4} + 6x = 0$$

which can be rewritten as

$$-x(x^3 - 8) = 0$$

The solutions of this equation are $x = 0$ and $x = 2$. These values are the x coordinates of the critical points of f. To get the corresponding y coordinates, we substitute these values of x into either one of our two original equations. We find that $y = 0$ when $x = 0$, and $y = -2$ when $x = 2$. Hence the critical points of f are $(0, 0)$ and $(2, -2)$.

The second-order partial derivatives of f are $f_{xx} = 6x$, $f_{yy} = -6y$, and $f_{xy} = 6$. Hence,

$$D = -36xy - 36 = -36(xy + 1)$$

Since $D(0, 0) = -36$, we conclude that f has a saddle point at $(0, 0)$. Since $D(2, -2) = 108$ and $f_{xx}(2, -2) = 12$, we conclude that f has a relative minimum at $(2, -2)$.

□

In the next example, we shall apply the theory of relative extrema to solve a practical optimization problem from economics. Actually, we shall be seeking the *absolute* maximum of a certain function, but it turns out that in this case the absolute and relative maxima coincide. In fact, in the vast majority of optimization problems in the social sciences involving functions of several variables, the relative and absolute extrema coincide. For this reason, we shall not pause to develop the theory of absolute extrema for functions of two variables, even though the theory is quite similar to that for functions of a single variable and is not particularly difficult.

Example 5.4

The only grocery store in a small resort community carries two brands of frozen orange juice, a local brand that it obtains at a cost of 30 cents a can, and a well-known national brand that it obtains at a cost of 40 cents a can. The owner of the store estimates that, if the local brand is sold for x cents a can and the national brand for y cents a can, he will be able to sell $70 - 5x + 4y$ cans of the local brand and $80 + 6x - 7y$ cans of the national brand each day. How should he price each brand to maximize his profit from the sale of the juice? (Assume that the absolute and the relative maxima of the profit function are the same.)

Solution

The daily profit from the sale of the orange juice is given by the function
$$f(x, y) = (x - 30)(70 - 5x + 4y) + (y - 40)(80 + 6x - 7y)$$
We compute f_x and f_y (using the product rule) and find that
$$f_x = 70 - 5x + 4y - 5x + 150 + 6y - 240$$
$$= -10x + 10y - 20$$
and
$$f_y = 4x - 120 + 80 + 6x - 7y - 7y + 280$$
$$= 10x - 14y + 240$$

Solving the equations $f_x = 0$ and $f_y = 0$ simultaneously we get $x = 53$ and $y = 55$. Hence (53, 55) is the only critical point of the profit function f.

The second-order partial derivatives are $f_{xx} = -10$, $f_{yy} = -14$, and $f_{xy} = 10$. Hence,
$$D = 140 - 100 = 40$$
which is positive. Since, in addition, f_{xx} is negative, we conclude that f has a (relative) maximum when $x = 53$ and $y = 55$. That is, the owner of the grocery store will maximize his profit if he sells the local brand of juice for 53 cents a can and the national brand for 55 cents a can. □

Problems

1. Locate the critical points of each of the following functions and classify them as relative maxima, relative minima, or saddle points.
 (a) $f(x, y) = 5 - x^2 - y^2$
 (b) $f(x, y) = 2x^2 - 3y^2$
 (c) $f(x, y) = xy$
 (d) $f(x, y) = xy + 8/x + 8/y$
 (e) $f(x, y) = 2x^3 + y^3 + 3x^2 - 3y - 12x - 4$
 (f) $f(x, y) = (x - 1)^2 + y^3 - 3y^2 - 9y + 5$
 (g) $f(x, y) = x^3 + y^2 - 6xy + 9x + 5y + 2$

*2. Let $f(x, y) = x^2 + y^2 - 4xy$. Show that f does *not* have a relative minimum at its critical point (0, 0), even though it does have a relative minimum at (0, 0) in both the x and y directions. (*Hint*: Consider the direction defined by the line $y = x$. That is, substitute x for y in the formula for f and analyze the resulting function of x.)

*3. Classify the critical points of the function $f(x, y) = (x + y)^2 + x^4$. Justify your answer.

*4. Sometimes one can classify the critical points of a function by inspecting its level curves. In each of the following cases, determine the nature of the critical point of f at $(0, 0)$.

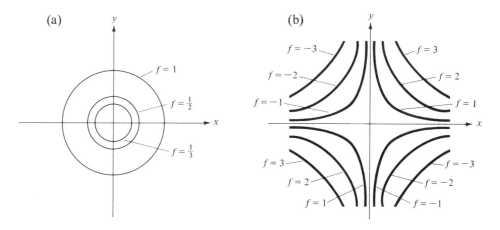

5. A small liquor store carries two competing brands of inexpensive wine, one from California and the other from New York. The owner of the store can obtain both wines at a cost of \$2 a bottle. He estimates that if he sells the California wine for x dollars a bottle and the New York wine for y dollars a bottle he will be able to sell $40 - 50x + 40y$ bottles of the California wine and $20 + 60x - 70y$ bottles of the New York wine each day. How should he price the wines to maximize his profit? (Assume that the absolute maximum and the relative maximum of the profit function are the same.)

6. The exclusive manufacturer of an efficient new type of industrial machine is planning to sell his product to both foreign and domestic firms. The price that he can expect to receive for the machines will depend on the number that he is willing to supply. (For example, if only a few of these desirable machines are made available, competitive bidding among prospective purchasers at home will tend to drive the domestic price up.) The manufacturer estimates that if he supplies x machines to the domestic market and y machines to the foreign market he will be able to sell the machines for $60 - x/5 + y/20$ thousand dollars apiece at home and for $50 - y/10 + x/20$ thousand dollars apiece abroad. If the manufacturer can produce the machines at a cost of \$10,000 apiece, how many should he supply to the domestic market and how many to the foreign market in order to maximize his total profit? (Assume that the absolute maximum and relative maximum of the profit function are the same.)

6. LAGRANGE MULTIPLIERS

An editor, constrained to stay within a fixed budget of $60,000, is faced with the question of how to divide this money between development and promotion in order to maximize the future sales of a new calculus text. A farmer, wishing to enclose 3,600 square yards of land for use as a pasture, would like to determine the dimensions of the pasture that he can enclose using the least amount of fencing. A manufacturer, whose plant can produce 500 washing machines a day, must decide what combination of standard and deluxe models will generate the largest profit.

In each of the preceding illustrations, a function of two variables is to be optimized, subject to a restriction or *constraint* on these variables. In the case of the editor, for example, suppose x denotes the amount of money allocated to development, y the amount allocated to promotion, and $f(x, y)$ the corresponding number of books that will be sold. The editor would like to maximize the function $f(x, y)$ subject to the constraint that $x + y = 60,000$.

The process of optimizing a function of two variables subject to a constraint may be visualized geometrically as follows. The function itself can be represented as a surface in three-dimensional space, while the constraint, which is an equation involving x and y, can be represented as a curve in the xy plane. In seeking the maximum and minimum values of the function subject to the given constraint, we restrict our attention to the portion of the surface that lies directly above the constraint curve. The highest point on this portion of the surface is the constrained maximum, while the lowest point is the constrained minimum. The situation is illustrated in Figure 17.

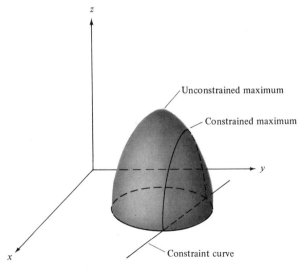

Figure 17

We have already solved some constrained optimization problems in this text. (See Chapter 2, Example 8.3, for instance.) Before we introduce an elegant alternative method for solving such problems, let us review our former technique.

Example 6.1

Minimize the function $f = x^2 + y^2$ subject to the constraint that $2x + y = -1$.

Solution

Our strategy will be to work with an equivalent minimization problem involving a function of only *one* variable. First we use the constraint equation to write y in terms of x as

$$y = -1 - 2x$$

Next we substitute this expression for y into the formula for f, getting

$$f = x^2 + (-1 - 2x)^2 = 5x^2 + 4x + 1$$

To find the value of x that minimizes f, we now set the derivative df/dx equal to zero and solve. We get

$$\frac{df}{dx} = 10x + 4 = 0$$

or

$$x = -\tfrac{2}{5}$$

The corresponding value of y can be obtained from the equation $y = -1 - 2x$. We find that $y = -\tfrac{1}{5}$ when $x = -\tfrac{2}{5}$. Hence when $2x + y = -1$, the minimum value of f occurs when $x = -\tfrac{2}{5}$ and $y = -\tfrac{1}{5}$. □

The technique for solving constrained optimization problems that was illustrated in Example 6.1 is based on the *reduction* of the original two-variable problem to an equivalent problem involving only *one* variable. A more versatile technique, called the *method of Lagrange multipliers*, involves the *expansion* of the original problem to one involving *three* variables! The chief advantage of the method of Lagrange multipliers is that its success does not depend on our ability to solve the constraint equation for one of the variables in terms of the other.

THE METHOD OF LAGRANGE MULTIPLIERS

Suppose $f(x, y)$ and $g(x, y)$ are functions whose first-order partial derivatives exist. To find the relative maximum and minimum of $f(x, y)$ subject to the constraint that $g(x, y) = k$ for some constant k, we introduce a new variable λ (called a *Lagrange multiplier*) and solve the following three equations simultaneously:

(1) $f_x(x, y) = \lambda g_x(x, y)$
(2) $f_y(x, y) = \lambda g_y(x, y)$
(3) $g(x, y) = k$

The desired relative extrema will be found among the resulting points (x, y).

The method of Lagrange multipliers is not really as mysterious as it seems. Before we discuss why this technique works, however, let us illustrate how it is used. We begin by applying the method of Lagrange multipliers to the minimization problem in Example 6.1.

Example 6.2

Use the method of Lagrange multipliers to minimize the function $f(x, y) = x^2 + y^2$ subject to the constraint that $2x + y = -1$.

Solution

If we let $g(x, y) = 2x + y$, we can rewrite the constraint as $g(x, y) = -1$. Having identified the function g, we compute the partial derivatives of f and g:

$f_x = 2x \qquad f_y = 2y \qquad g_x = 2 \qquad g_y = 1$

Using these results, we can now write down the three equations that are to be solved simultaneously:

$2x = 2\lambda$

$2y = \lambda$

and

$2x + y = -1$

From the first two equations we conclude that

$x = 2y$

Substitution of this expression for x in the third equation gives

$4y + y = -1$

or

$y = -\frac{1}{5}$

Finally, if we let $y = -\frac{1}{5}$ in the third equation, we find that $x = -\frac{2}{5}$. Hence, as in Example 6.1, we conclude that if $2x + y = -1$ the minimum value of $f(x, y)$ occurs when $(x, y) = (-\frac{2}{5}, -\frac{1}{5})$. □

You may have noticed that we cheated slightly in Example 6.2. In particular, we neglected to verify that the point $(-\frac{2}{5}, -\frac{1}{5})$ does indeed correspond to the constrained *minimum* of f. There is a rather complicated version of the second derivative test that we could have used to show that this point corresponds to a *relative* minimum of the constrained function. And, with a little more work, we could have verified that this relative minimum is actually an *absolute* minimum. We shall not develop these advanced techniques in this text. Fortunately, in almost all practical applications, the absolute and relative extrema coincide and the nature of the extremum is clear from the practical context.

Let us try one more example.

Example 6.3

Find the maximum and minimum values of the function $f(x, y) = xy$ subject to the constraint that $x^2 + y^2 = 8$.

Solution

We write the constraint as $g(x, y) = 8$, where $g(x, y) = x^2 + y^2$. From the partial derivatives

$$f_x = y \quad f_y = x \quad g_x = 2x \quad g_y = 2y$$

we are led to the three Lagrange equations

$$y = 2\lambda x$$

$$x = 2\lambda y$$

and

$$x^2 + y^2 = 8$$

We may rewrite the first equation as

$$2\lambda = \frac{y}{x}$$

provided $x \neq 0$, and the second as

$$2\lambda = \frac{x}{y}$$

provided $y \neq 0$, and conclude that

$$\frac{y}{x} = \frac{x}{y}$$

or

$$x^2 = y^2$$

provided neither x nor y is zero. It is easy to see that in this case neither x nor y can be zero if all three of the Lagrange equations are to hold. (Convince yourself of this.) Hence the equation $x^2 = y^2$ is valid, and we combine this fact with the third Lagrange equation to get

$$2x^2 = 8$$

or

$$x = \pm 2$$

If $x = 2$, we see from the equation $x^2 = y^2$ that $y = 2$ or $y = -2$. Similarly, if $x = -2$, it follows that $y = 2$ or $y = -2$. Hence, using the method of Lagrange multipliers, we have identified four points, $(2, 2)$, $(2, -2)$, $(-2, 2)$, and $(-2, -2)$, at which the constrained extrema may occur. Since

$$f(2, 2) = 4 \qquad f(-2, -2) = 4 \qquad f(2, -2) = -4 \quad \text{and} \quad f(-2, 2) = -4$$

we conclude that when $x^2 + y^2 = 8$ the maximum value of $f(x, y)$ is 4 which occurs at the points $(2, 2)$ and $(-2, -2)$, and the minimum value is -4 which occurs at $(2, -2)$ and $(-2, 2)$.

(For practice, check these answers by solving this optimization problem using our former technique.) □

Although a rigorous explanation of why the method of Lagrange multipliers works involves advanced ideas beyond the scope of this text, there is a rather simple geometric argument that you should find convincing. Suppose the constraint curve $g(x, y) = k$ and the level curves $f(x, y) = C$ are drawn in the xy plane as shown in Figure 18.

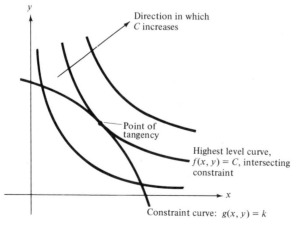

Figure 18

To maximize $f(x, y)$ subject to the constraint $g(x, y) = k$, we must find the highest level curve of f that intersects the constraint curve. As the sketch in Figure 18 suggests, this critical intersection will occur at a point where the constraint curve is tangent to a level curve; that is, where the slope of the constraint curve $g(x, y) = k$ is equal to the slope of a level curve $f(x, y) = C$. According to the formula we derived in Section 4,

$$\text{Slope of constraint curve} = -\frac{g_x}{g_y}$$

and

$$\text{Slope of level curve} = -\frac{f_x}{f_y}$$

Hence the condition that the slopes be equal can be expressed by the equation

$$-\frac{f_x}{f_y} = -\frac{g_x}{g_y}$$

or, equivalently, by

$$\frac{f_x}{g_x} = \frac{f_y}{g_y}$$

If we let λ denote this common ratio, we have

$$\lambda = \frac{f_x}{g_x} \quad \text{and} \quad \lambda = \frac{f_y}{g_y}$$

from which we get the first two Lagrange equations

$$f_x = \lambda g_x$$

and

$$f_y = \lambda g_y$$

Thus the points (x, y) that satisfy these two equations are the points at which a level curve $f(x, y) = C$ is tangent to the constraint curve $g(x, y) = k$. The third Lagrange equation

$$g(x, y) = k$$

is merely a statement of the fact that the point lies on this constraint curve.

Suppose all three of the Lagrange equations are satisfied at a certain point (a, b). Then f will reach its constrained *maximum* at (a, b) if the *highest* level curve that intersects the constraint does so at this point. On the other hand, if the *lowest* level curve that intersects the constraint does so at (a, b), then f will take on its constrained *minimum* at this point. To illustrate the situation further, let us sketch the constraint curve and optimal level curves for the problem we solved in Example 6.3. In drawing this sketch (Figure 19), we use the well-known fact from geometry that the graph of the equation $x^2 + y^2 = r^2$ is a circle of radius r centered at the origin.

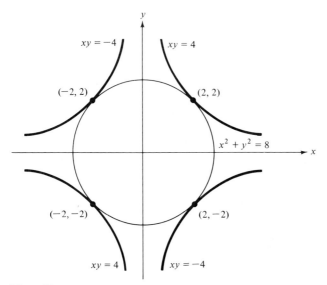

Figure 19

Notice that in this case there are four points at which the constraint curve is tangent to a level curve. Two of these points, (2, 2) and (−2, −2), maximize f subject to the given constraint, while the other two, (2, −2) and (−2, 2), minimize f.

In the next example, we apply the method of Lagrange multipliers to a typical constrained optimization problem from economics.

Example 6.4

An editor has been allotted $60,000 to spend on the development and promotion of a new calculus text. He estimates that if he spends x thousand dollars on development and y thousand dollars on promotion, approximately $20x^{3/2}y$ copies of the book will be sold. How much money should the editor allocate to development and how much to promotion in order to maximize sales?

Solution

Our goal is to maximize the function $f(x, y) = 20x^{3/2}y$ subject to the constraint $g(x, y) = 60$, where $g(x, y) = x + y$. Hence the Lagrange equations are

$30x^{1/2}y = \lambda$

$20x^{3/2} = \lambda$

and

$x + y = 60$

Equating the expressions for λ in the first two equations we get

$30x^{1/2}y = 20x^{3/2}$

Since the maximum value of f clearly does not occur when $x = 0$, we may assume that $x \neq 0$ and divide by $x^{1/2}$ to get

$x = \frac{3}{2}y$

Substituting this expression into the third equation we get

$\frac{5}{2}y = 60$

and so

$y = 24$ and $x = 36$

Hence, to maximize sales, the editor should spend $36,000 on development and $24,000 on promotion. If he does, he can expect to sell $f(36, 24) = 103,680$ copies of the book.

A graph showing the relationship between the budgetary constraint and the level curve for maximal sales is sketched in Figure 20.

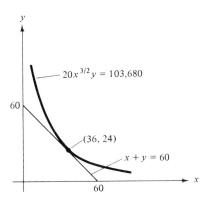

Figure 20

You have probably noticed that in using the method of Lagrange multipliers to solve constrained optimization problems we have never actually obtained a numerical value for the multiplier λ. Although one can frequently find the constrained extrema of a function without solving for λ, there are certain practical problems in which the numerical value of λ gives us useful information. This is because the Lagrange multiplier λ has the following simple interpretation.

THE SIGNIFICANCE OF THE LAGRANGE MULTIPLIER λ

Suppose M is the maximum (or minimum) value of $f(x, y)$ subject to the constraint $g(x, y) = k$. As the value k of the constraint function varies, so does the corresponding optimal value M of the function f. The Lagrange multiplier λ is the rate of change of M with respect to k. That is,

$$\lambda = \frac{dM}{dk}$$

To prove this, suppose that $M = f(a, b)$. Then the coordinates of the point (a, b) must satisfy the three Lagrange equations

$f_x(a, b) = \lambda g_x(a, b)$

$f_y(a, b) = \lambda g_y(a, b)$

and

$g(a, b) = k$

Since the third equation implies that $g(a, b) - k = 0$, we can write M as follows:

$M = f(a, b) - \lambda[g(a, b) - k]$

The coordinates a and b of the point at which the optimal value M is attained are both functions of k. Hence, by the chain rule,

$$\frac{d}{dk}[f(a, b)] = f_x(a, b)\frac{da}{dk} + f_y(a, b)\frac{db}{dk}$$

and

$$\frac{d}{dk}[g(a, b)] = g_x(a, b)\frac{da}{dk} + g_y(a, b)\frac{db}{dk}$$

Moreover, λ since is also a function of k,

$$\frac{d}{dk}(\lambda k) = \lambda + k\frac{d\lambda}{dk} = \lambda + g(a, b)\frac{d\lambda}{dk}$$

Thus

$$\frac{dM}{dk} = \frac{d}{dk}[f(a, b)] - \lambda \frac{d}{dk}[g(a, b)] - g(a, b)\frac{d\lambda}{dk} + \frac{d}{dk}(\lambda k)$$

$$= \frac{da}{dk}[f_x(a, b) - \lambda g_x(a, b)] + \frac{db}{dk}[f_y(a, b) - \lambda g_y(a, b)] + \lambda$$

But

$f_x(a, b) - \lambda g_x(a, b) = 0$

by the first Lagrange equation, and

$$f_y(a, b) - \lambda g_y(a, b) = 0$$

by the second Lagrange equation, and so

$$\frac{dM}{dk} = \lambda$$

In our final example, we return to the situation described in Example 6.4 to illustrate the use of the Lagrange multiplier λ in a practical problem.

Example 6.5

Suppose that the editor in Example 6.4 is allotted $61,000 instead of $60,000 to spend on the development and promotion of the new calculus book. Estimate how the additional $1,000 will affect the *maximum* number of copies of the book that can be sold.

Solution

In Example 6.4 we computed the maximum value of the sales function $f(x, y) = 20x^{3/2}y$ subject to the budgetary constraint $g(x, y) = k$, where $g(x, y) = x + y$ and $k = 60$. In particular, we solved the three equations

$$30x^{1/2}y = \lambda$$

$$20x^{3/2} = \lambda$$

and

$$x + y = 60$$

to find that the maximum value M of $f(x, y)$ occurred when $x = 36$ and $y = 24$. We can now use either the first or the second Lagrange equation to compute the corresponding value of λ. Using the second equation we get

$$\lambda = 20(36)^{3/2} = 4{,}320$$

Since $\lambda = dM/dk$, it follows that the unit increase in k from $k = 60$ to $k = 61$ will increase the maximal sales M of the book by approximately 4,320 copies. □

Problems

Use the method of Lagrange multipliers to solve Problems 1 through 11.

1. Find the maximum value of the function $f(x, y) = xy$ subject to the constraint that $x + y = 1$. Sketch the constraint curve and the optimal level curve.

2. Find the maximum and minimum values of the function $f(x, y) = xy$ subject to the constraint that $x^2 + y^2 = 1$. Sketch the constraint curve and the optimal level curves.

3. Find the minimum value of the function $f(x, y) = x^2 + y^2$ subject to the constraint that $xy = 1$. Sketch the constraint curve and the optimal level curve.

4. Find the minimum value of the function $f(x, y) = x^2 + 2y^2 - xy$ subject to the constraint that $2x + y = 22$.

5. Find the minimum value of the function $f(x, y) = x^2 - y^2$ subject to the constraint that $x^2 + y^2 = 4$.

6. Find the maximum and minimum values of the function $f(x, y) = 8x^2 - 24xy + y^2$ subject to the constraint that $x^2 + y^2 = 1$.

7. A farmer wishes to fence off a rectangular pasture along the bank of a river. The area of the pasture is to be 3,200 square yards, and no fencing is needed along the river bank. Find the dimensions of the pasture that will require the least amount of fencing.

8. There are 320 feet of fencing available to enclose a rectangular field. How should the fencing be used so that the enclosed area is as large as possible?

9. Prove that of all rectangles with a given area the square has the smallest perimeter.

10. The product of two positive numbers is 128. The first number is added to the square of the second. How small can this sum be?

11. A carpenter has been asked to build an open box with a square base. The sides of the box will cost $3 per square foot, and the base will cost $4 per square foot. What are the dimensions of the box of maximum volume that can be constructed for $48?

12. Use the Lagrange multiplier λ to estimate the change in the maximum volume of the box in Problem 11 that would result if the money available for construction were increased to $49.

13. A manufacturer has $8,000 to spend on the development and promotion of a new product. He estimates that if he spends x thousand dollars on development and y thousand dollars on promotion he will be able to sell approximately

$$\frac{320y}{y+2} + \frac{160x}{x+4} \quad \text{units of the product}$$

at a price of $150 per unit. Use the method of Lagrange multipliers to determine how he should allocate the $8,000 to maximize his profit if his manufacturing cost is $50 per unit.

14. Suppose the manufacturer in Problem 13 decides to spend $9,000 instead of $8,000 on the development and promotion of his product. Estimate how this change will affect his maximum possible profit.

*15. (a) If he has unlimited funds available, how much should the manufacturer in Problem 13 spend on development and promotion in order to maximize his profit?
  (b) What is the value of the Lagrange multiplier λ that corresponds to the optimal budget in part (a)?
  (c) Discuss your answer to part (b) in light of the interpretation of λ as dM/dk.

7. REVIEW PROBLEMS

1. In each of the following cases, find the indicated composite function and specify its domain.
   (a) $f(g(x, y))$, where $f(t) = (t + 1)^5$, $g(x, y) = 2x + 3y - 2$
   (b) $h(f(x, y), g(x, y))$, where $h(u, v) = u \log v$, $f(x, y) = x + y$, $g(x, y) = e^{x+y}$

2. For each of the following functions, compute all the first-order partial derivatives.
   (a) $f(x, y) = 2x^3 y + 3x\sqrt{y} + y/x$
   (b) $f(x, y, z) = (3x + 2y + z^2)^5$
   (c) $f(x_1, x_2) = 10x_1/(2 + x_1) + 20x_2/(5 + x_2) - x_1 - x_2$

3. For each of the following functions, compute the second-order partial derivatives $f_{xx}, f_{yy}, f_{xy},$ and f_{yx}.
   (a) $f(x, y) = x^2 + y^3 - 2xy$ (b) $f(x, y) = e^{x^2 + y^2}$
   (c) $f(x, y) = x \log y$

4. Use the chain rule to find the indicated derivatives.
   (a) $z = x^2 + 3xy$, $x = t^2 + 1$, $y = 1 - t$; dz/dt
   (b) $z = t^2 - 2t + 1$, $t = x - y^2$; $\partial z/\partial x$, $\partial z/\partial y$
   (c) $f(u, v) = 2u + 3v$, $u = x^2 + y^2$, $v = x^2 - y^2$; f_x, f_y

*5. Suppose $z = f(x, y)$, where $x = u + v$ and $y = u - v$. Think of z as a function of u and v and find an expression for the mixed partial derivative $\partial^2 z/\partial u \, \partial v$ in terms of the second-order partial derivatives $f_{xx}, f_{yy},$ and f_{xy}.

6. At a certain factory, the daily output of a certain commodity is approximately $4K^{1/3}L^{1/2}$ units, where K denotes the firm's capital investment measured in thousands of dollars and L the size of the labor force measured in man-hours of work. Suppose that the current capital investment is $125,000 and that 900 man-hours of labor are used each day. Use marginal analysis to estimate the effect that an additional capital investment of $1,000 would have on the daily output.

7. A publisher has found that in a certain city each of his salesmen will sell approximately

$$\frac{r^2}{2{,}000p} + \frac{s^2}{100} - s \quad \text{sets of encyclopedias a month}$$

where s denotes the total number of salesmen he has working in the city, p the price of a set of the books, and r the amount of money spent each month on local advertising. Suppose the publisher currently employs 10 salesmen, spends \$6,000 a month on local advertising, and sells the encyclopedias for \$800 a set. Suppose further that the cost of producing the encyclopedias is \$80 a set and that each salesman earns \$400 a month plus a commission of 8 percent of the price of each set he sells. Use an appropriate partial derivative to estimate the change in the publisher's total monthly profit that would result if one more local salesman were hired.

8. Two competing brands of power lawnmowers are sold in the same town. When the price of the first brand is x dollars per mower, the price of the second brand is y dollars per mower, and the average per capita income in the community is z dollars a year, the local demand for the first brand of mowers is given by the function $f(x, y, z)$.

   (a) How would you expect the demand f to be affected by an increase in x? By an increase in y? In z?
   (b) Translate your answers to part (a) into conditions on the signs of the partial derivatives of f.
   (c) Suppose $f(x, y, z) = a + bx + cy + dz$. What can you say about the signs of the coefficients b, c, and d if your conclusions in part (a) are to hold?

9. Two commodities are said to be *complementary* if the demand Q_1 for the first commodity decreases as the price p_2 of the second increases, and if the demand Q_2 for the second decreases as the price p_1 of the first increases.
   (a) Give an example of a pair of complementary commodities.
   (b) If two commodities are complementary, what must be true of the partial derivatives $\partial Q_1/\partial p_2$ and $\partial Q_2/\partial p_1$?
   (c) Suppose the demand functions for two commodities are

   $$Q_1 = 2{,}000 + \frac{400}{p_1 + 3} + 500 e^{-0.5 p_2}$$

   and

   $$Q_2 = 2{,}000 - 100 p_2 + \frac{500}{p_1 + 4}$$

   Are the commodities complementary?

10. A liquor store carries two brands of inexpensive white table wine, one from California and the other from New York. If the California wine sells for x dollars a bottle and the New York wine for y dollars a bottle, the monthly demand for the New York wine is approximately $240 - 12y^2 + 20x$ bottles. It is expected that t months from now, the price of a bottle of the California wine will be $2 + 0.05t$ dollars and the price of a bottle of the New York wine will be $2 + 0.1\sqrt{t}$ dollars. At what rate will the demand for the New York wine be changing 25 months from now?

11. For each of the following functions, sketch the indicated level curves.
(a) $f(x, y) = x^2 - y$; $f = 2, f = -2$
(b) $f(x, y) = x^2 + y^2$; $f = 1, f = 4$
(c) $f(x, y) = (1/x) \log y$; $f = 0, f = 1$

12. For each of the following functions, find an expression for dy/dx if $f(x, y) = C$ for some constant C.
(a) $f(x, y) = x^3 - 3x + y^2$ (b) $f(x, y) = ye^x$

13. For each of the following functions, find the slope of the indicated level curve for the specified value of x.
(a) $f(x, y) = x^2 - y^3$; $f = 2, x = 1$ (b) $f(x, y) = xe^y$; $f = 2, x = 2$

14. Locate the critical points of each of the following functions and classify them as relative maxima, relative minima, or saddle points.
(a) $f(x, y) = x^3 + y^3 + 3x^2 - 18y^2 + 81y + 5$
(b) $f(x, y) = x^2 + y^3 + 6xy - 7x - 6y$

15. The exclusive manufacturer of an efficient new type of industrial machine is planning to sell his product to both foreign and domestic firms. He can produce the machines for $10,000 apiece, and he estimates that if he supplies x machines to the domestic market and y machines to the foreign market he will be able to sell the machines for $150 - x/6$ thousand dollars apiece at home and for $100 - y/10$ thousand dollars apiece abroad.
  (a) How many machines should the manufacturer supply to the domestic market in order to maximize his profit at home?
  (b) How many machines should he supply to the foreign market in order to maximize his profit abroad?
  (c) How many machines should he supply to each market in order to maximize his *total* profit?
  (d) Is the relationship between the optimal level of domestic sales in part (a), the optimal level of foreign sales in part (b), and the optimal level of total sales in part (c) accidental? Explain.
  (e) Does a similar relationship hold in Problem 6 on page 359? What accounts for the difference between these two problems in this respect?

16. Use the method of Lagrange multipliers to find the maximum and minimum values of the function $f(x, y) = x^2 + 2y^2 + 2x + 3$ subject to the constraint that $x^2 + y^2 = 4$.

17. Use the method of Lagrange multipliers to prove that of all rectangles with a given perimeter the square has the largest area.

18. A manufacturer has $11,000 to spend on the development and promotion of a new product He estimates that if he spends x thousand dollars on development and y thousand dollars on promotion he will be able to sell approximately

$$\frac{250y}{y + 2} + \frac{100x}{x + 5} \quad \text{units of the product}$$

at a price of $350 per unit. How should he allocate the $11,000 to maximize his profit if his manufacturing cost is $150 per unit?

19. Suppose the manufacturer in Problem 18 decides to spend $12,000 instead of $11,000 on the development and promotion of his product. Estimate how this change will affect his maximum possible profit.

20. If he has unlimited funds available, how much should the manufacturer in Problem 18 spend on development and how much on promotion to maximize his profit?

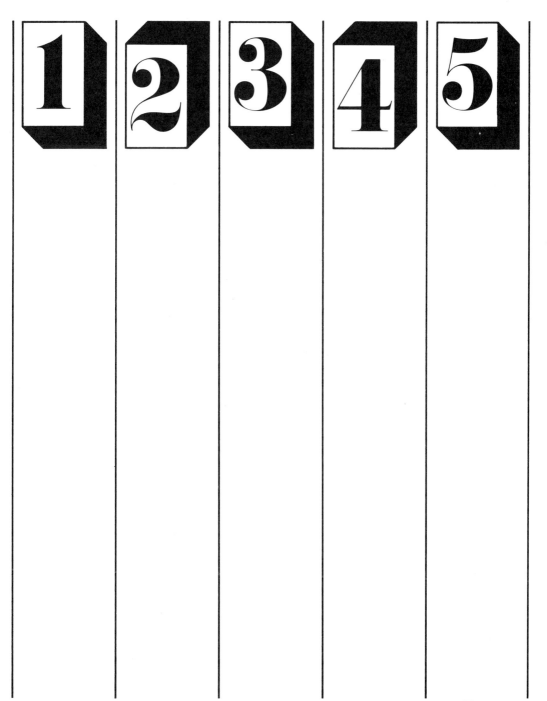

Appendix

1. EXPONENTIAL FUNCTION

x	e^x Value	e^{-x} Value	x	e^x Value	e^{-x} Value	x	e^x Value	e^{-x} Value
0.00	1.0000	1.00000	0.50	1.6487	.60653	1.00	2.7183	.36788
0.01	1.0101	0.99005	0.51	1.6653	.60050	1.20	3.3201	.30119
0.02	1.0202	.98020	0.52	1.6820	.59452	1.30	3.6693	.27253
0.03	1.0305	.97045	0.53	1.6989	.58860	1.40	4.0552	.24660
0.04	1.0408	.96079	0.54	1.7160	.58275	1.50	4.4817	.22313
0.05	1.0513	.95123	0.55	1.7333	.57695	1.60	4.9530	.20190
0.06	1.0618	.94176	0.56	1.7507	.57121	1.70	5.4739	.18268
0.07	1.0725	.93239	0.57	1.7683	.56553	1.80	6.0496	.16530
0.08	1.0833	.92312	0.58	1.7860	.55990	1.90	6.6859	.14957
0.09	1.0942	.91393	0.59	1.8040	.55433	2.00	7.3891	.13534
0.10	1.1052	.90484	0.60	1.8221	.54881	3.00	20.086	.04979
0.11	1.1163	.89583	0.61	1.8404	.54335	4.00	54.598	.01832
0.12	1.1275	.88692	0.62	1.8589	.53794	5.00	148.41	.00674
0.13	1.1388	.87809	0.63	1.8776	.53259	6.00	403.43	.00248
0.14	1.1503	.86936	0.64	1.8965	.52729	7.00	1096.6	.00091
0.15	1.1618	.86071	0.65	1.9155	.52205	8.00	2981.0	.00034
0.16	1.1735	.85214	0.66	1.9348	.51685	9.00	8103.1	.00012
0.17	1.1853	.84366	0.67	1.9542	.51171	10.00	22026.5	.00005
0.18	1.1972	.83527	0.68	1.9739	.50662			
0.19	1.2092	.82696	0.69	1.9937	.50158			
0.20	1.2214	.81873	0.70	2.0138	.49659			
0.21	1.2337	.81058	0.71	2.0340	.49164			
0.22	1.2461	.80252	0.72	2.0544	.48675			
0.23	1.2586	.79453	0.73	2.0751	.48191			
0.24	1.2712	.78663	0.74	2.0959	.47711			
0.25	1.2840	.77880	0.75	2.1170	.47237			
0.26	1.2969	.77105	0.76	2.1383	.46767			
0.27	1.3100	.76338	0.77	2.1598	.46301			
0.28	1.3231	.75578	0.78	2.1815	.45841			
0.29	1.3364	.74826	0.79	2.2034	.45384			
0.30	1.3499	.74082	0.80	2.2255	.44933			
0.31	1.3634	.73345	0.81	2.2479	.44486			
0.32	1.3771	.72615	0.82	2.2705	.44043			
0.33	1.3910	.71892	0.83	2.2933	.43605			
0.34	1.4049	.71177	0.84	2.3164	.43171			
0.35	1.4191	.70469	0.85	2.3396	.42741			
0.36	1.4333	.69768	0.86	2.3632	.42316			
0.37	1.4477	.69073	0.87	2.3869	.41895			
0.38	1.4623	.68386	0.88	2.4109	.41478			
0.39	1.4770	.67706	0.89	2.4351	.41066			
0.40	1.4918	.67032	0.90	2.4596	.40657			
0.41	1.5068	.66365	0.91	2.4843	.40252			
0.42	1.5220	.65705	0.92	2.5093	.39852			
0.43	1.5373	.65051	0.93	2.5345	.39455			
0.44	1.5527	.64404	0.94	2.5600	.39063			
0.45	1.5683	.63763	0.95	2.5857	.38674			
0.46	1.5841	.63128	0.96	2.6117	.38298			
0.47	1.6000	.62500	0.97	2.6379	.37908			
0.48	1.6161	.61878	0.98	2.6645	.37531			
0.49	1.6323	.61263	0.99	2.6912	.37158			

Excerpted from Burrington's *Handbook of Mathematical Tables and Formulas*, McGraw-Hill, New York.

2. NATURAL LOGARITHMS (BASE e)

x	$\log x$	x	$\log x$	x	$\log x$	x	$\log x$
.01	−4.60517	0.50	−0.69315	1.00	0.00000	1.5	0.4 0547
.02	−3.91202	.51	.67334	1.01	.00995	1.6	7000
.03	.50656	.52	.65393	1.02	.01980	1.7	0.5 3063
.04	.21888	.53	.63488	1.03	.02956	1.8	8779
		.54	.61619	1.04	.03922	1.9	0.6 4185
.05	−2.99573	.55	.59784	1.05	.04879	2.0	9315
.06	.81341	.56	.57982	1.06	.05827	2.1	0.7 4194
.07	.65926	.57	.56212	1.07	.06766	2.2	8846
.08	.52573	.58	.54473	1.08	.07696	2.3	0.8 3291
.09	.40795	.59	.52763	1.09	.08618	2.4	7547
0.10	−2.30259	0.60	−0.51083	1.10	.09531	2.5	0.9 1629
.11	.20727	.61	.49430	1.11	.10436	2.6	5551
.12	.12026	.62	.47804	1.12	.11333	2.7	9325
.13	.04022	.63	.46204	1.13	.12222	2.8	1.0 2962
.14	−1.96611	.64	.44629	1.14	.13103	2.9	6471
.15	.89712	.65	.43078	1.15	.13976	3.0	9861
.16	.83258	.66	.41552	1.16	.14842	4.0	1.3 8629
.17	.77196	.67	.40048	1.17	.15700	5.0	1.6 0944
.18	.71480	.68	.38566	1.18	.16551	10.0	2.3 0258
.19	.66073	.69	.37106	1.19	.17395		
0.20	−1.60944	0.70	−0.35667	1.20	.18232		
.21	.56065	.71	.34249	1.21	.19062		
.22	.51413	.72	.32850	1.22	.19885		
.23	.46968	.73	.31471	1.23	.20701		
.24	.42712	.74	.30111	1.24	.21511		
.25	.38629	.75	.28768	1.25	.22314		
.26	.34707	.76	.27444	1.26	.23111		
.27	.30933	.77	.26136	1.27	.23902		
.28	.27297	.78	.24846	1.28	.24686		
.29	.23787	.79	.23572	1.29	.25464		
0.30	−1.20397	0.80	−0.22314	1.30	.26236		
.31	.17118	.81	.21072	1.31	.27003		
.32	.13943	.82	.19845	1.32	.27763		
.33	.10866	.83	.18633	1.33	.28518		
.34	.07881	.84	.17435	1.34	.29267		
.35	−1.04982	.85	−0.16252	1.35	.30010		
.36	.02165	.86	.15032	1.36	.30748		
.37	−0.99425	.87	.13926	1.37	.31481		
.38	.96758	.88	.12783	1.38	.32208		
.39	.94161	.89	.11653	1.39	.32930		
0.40	−0.91629	0.90	−0.10536	1.40	.33647		
.41	.89160	.91	.09431	1.41	.34359		
.42	.86750	.92	.08338	1.42	.35066		
.43	.84397	.93	.07257	1.43	.35767		
.44	.82098	.94	.06188	1.44	.36464		
.45	.79851	.95	.05129	1.45	.37156		
.46	.77653	.96	.04082	1.46	.37844		
.47	.75502	.97	.03046	1.47	.38526		
.48	.73397	.98	.02020	1.48	.39204		
.49	.71335	.99	.01005	1.49	.39878		

Warning: the integer to the left of the decimal point is not shown in each line.

Tables 1 to 3 from Sherman K. Stein, "Calculus and Analytic Geometry," McGraw-Hill Book Company, New York, 1974.

3. SQUARES, CUBES, SQUARE ROOTS, AND CUBE ROOTS

n	n^2	n^3	$\sqrt{n}$	$\sqrt[3]{n}$	n	n^2	n^3	$\sqrt{n}$	$\sqrt[3]{n}$
1	1	1	1.000 000	1.000 000	50	2 500	125 000	7.071 068	3.684 031
2	4	8	1.414 214	1.259 921	51	2 601	132 651	7.141 428	3.708 430
3	9	27	1.732 051	1.442 250	52	2 704	140 608	7.211 103	3.732 511
4	16	64	2.000 000	1.587 401	53	2 809	148 877	7.280 110	3.756 286
					54	2 916	157 464	7.348 469	3.779 763
5	25	125	2.236 068	1.709 976	55	3 025	166 375	7.416 198	3.802 952
6	36	216	2.449 490	1.817 121	56	3 136	175 616	7.483 315	3.825 862
7	49	343	2.645 751	1.912 931	57	3 249	185 193	7.549 834	3.848 501
8	64	512	2.828 427	2.000 000	58	3 364	195 112	7.615 773	3.870 877
9	81	729	3.000 000	2.080 048	59	3 481	205 379	7.681 146	3.892 996
10	100	1 000	3.162 278	2.154 435	60	3 600	216 000	7.745 967	3.914 868
11	121	1 331	3.316 625	2.223 980	61	3 721	226 981	7.810 250	3.936 497
12	144	1 728	3.464 102	2.289 428	62	3 844	238 328	7.874 008	3.957 892
13	169	2 197	3.605 551	2.351 335	63	3 969	250 047	7.937 254	3.979 057
14	196	2 744	3.741 657	2.410 142	64	4 096	262 144	8.000 000	4.000 000
15	225	3 375	3.872 983	2.466 212	65	4 225	274 625	8.062 258	4.020 726
16	256	4 096	4.000 000	2.519 842	66	4 356	287 496	8.124 038	4.041 240
17	289	4 913	4.123 106	2.571 282	67	4 489	300 763	8.185 353	4.061 548
18	324	5 832	4.242 641	2.620 741	68	4 624	314 432	8.246 211	4.081 655
19	361	6 859	4.358 899	2.668 402	69	4 761	328 509	8.306 624	4.101 566
20	400	8 000	4.472 136	2.714 418	70	4 900	343 000	8.366 600	4.121 285
21	441	9 261	4.582 576	2.758 924	71	5 041	357 911	8.426 150	4.140 818
22	484	10 648	4.690 416	2.802 039	72	5 184	373 248	8.485 281	4.160 168
23	529	12 167,	4.795 832	2.843 867	73	5 329	389 017	8.544 004	4.179 339
24	576	13 824	4.898 979	2.884 499	74	5 476	405 224	8.602 325	4.198 336
25	625	15 625	5.000 000	2.924 018	75	5 625	421 875	8.660 254	4.217 163
26	676	17 576	5.099 020	2.962 496	76	5 776	438 976	8.717 798	4.235 824
27	729	19 683	5.196 152	3.000 000	77	5 929	456 533	8.774 964	4.254 321
28	784	21 952	5.291 503	3.036 589	78	6 084	474 552	8.831 761	4.272 659
29	841	24 389	5.385 165	3.072 317	79	6 241	493 039	8.888 194	4.290 840
30	900	27 000	5.477 226	3.107 233	80	4 400	512 000	8.944 272	4.308 869
31	961	29 791	5.567 764	3.141 381	81	6 561	531 441	9.000 000	4.326 749
32	1 024	32 768	5.656 854	3.174 802	82	6 724	551 368	9.055 385	4.344 481
33	1 089	35 937	5.744 563	3.207 534	83	6 889	571 787	9.110 434	4.362 071
34	1 156	39 304	5.830 952	3.239 612	84	7 056	592 704	9.165 151	4.379 519
35	1 225	42 875	5.916 080	3.271 066	85	7 225	614 125	9.219 544	4.396 830
36	1 296	46 656	6.000 000	3.301 927	86	7 396	636 056	9.273 618	4.414 005
37	1 369	50 653	6.082 763	3.332 222	87	7 569	658 503	9.327 379	4.431 048
38	1 444	54 872	6.164 414	3.361 975	88	7 744	681 472	9.380 832	4.447 960
39	1 521	59 319	6.244 998	3.391 211	89	7 921	704 969	9.433 981	4.464 745
40	1 600	64 000	6.324 555	3.419 952	90	8 100	729 000	9.486 833	4.481 405
41	1 681	68 921	6.403 124	3.448 217	91	8 281	753 571	9.539 392	4.497 941
42	1 764	74 088	6.480 741	3.476 027	92	8 464	778 688	9.591 663	4.514 357
43	1 849	79 507	6.557 439	3.503 398	93	8 649	804 357	9.643 651	4.530 655
44	1 936	85 184	6.633 250	3.530 348	94	8 836	830 584	9.695 360	4.546 836
45	2 025	91 125	6.708 204	3.556 893	95	9 025	857 375	9.746 974	4.562 903
46	2 116	97 336	6.782 330	3.583 048	96	9 216	884 736	9.797 959	4.578 857
47	2 209	103 823	6.855 655	3.608 826	97	9 409	912 673	9.848 858	4.594 701
48	2 304	110 592	6.928 203	3.634 241	98	9 604	941 192	9.899 495	4.610 436
49	2 401	117 649	7.000 000	3.659 306	99	9 801	970 299	9.949 874	4.626 065